Bridge Aeroelasticity

Sensitivity Analysis and Optimal Design

High Performance Structures and Materials

Objectives

The High Performance Structures and Materials series has been established to document the dynamic and rapid changes presently happening in the field of structural engineering. New concepts are constantly being introduced, and the series reflects the wide range of significant international research and development.

The series encompasses the following topics:

High Performance Structures
Nonlinear Structural Behaviour
Emerging Applications
Design Innovation
Smart Structures
Space Structures
Microstructures
Marine and Offshore Structures
Composite Structures
Retrofitting of Structures
Sustainability in Design
Shape and Topology Optimisation
Shock and Impact
Structural Capacity under Damage
Soil Structure Interaction
Material Technology
Dynamic Control of Materials
Smart Materials
Sensor Technology
Virtual Instrumentation
Numerical Methods
Computer Packages
Computer Modelling
Control Systems

Associate Editors

Bridge Aeroelasticity

Sensitivity Analysis and Optimal Design

J.A. Jurado, S. Hernández, F. Nieto & A. Mosquera

University of La Coruña, Spain

WIT*PRESS* Southampton, Boston

J.A. Jurado, S. Hernández, F. Nieto & A. Mosquera
University of La Coruña, Spain

Published by

WIT Press
Ashurst Lodge, Ashurst, Southampton, SO40 7AA, UK
Tel: 44 (0) 238 029 3223; Fax: 44 (0) 238 029 2853
E-Mail: witpress@witpress.com
http://www.witpress.com

For USA, Canada and Mexico

WIT Press
25 Bridge Street, Billerica, MA 01821, USA
Tel: 978 667 5841; Fax: 978 667 7582
E-Mail: infousa@witpress.com
http://www.witpress.com

British Library Cataloguing-in-Publication Data

A Catalogue record for this book is available
from the British Library

ISBN: 978-1-84564-056-9
ISSN (print): 1469-0071

Library of Congress Catalog Card Number: 2010936106

The texts of the papers in this volume were set individually by the authors or under their supervision.

Printed in Great Britain by MPG Books Group, Bodmin and King's Lynn.

This book is dedicated to...

My parents, Juan and Encarna.

Pilar, Guillermo, Carlos and Dani.

Patricia and Nora.

Juan Carlos, Maribel, Juan Carlos Jr., Isabel and Eva.

Contents

Preface

This book is dedicated to the study of an aeroelastic phenomenom of cable supported long span bridges known as flutter and proposes very innovative design methodologies such as sensitivity analysis and optimization techniques utilized successfully in automobile and aerospace industries.

The topic of long span suspension and cable stayed bridges is currently of great importance. These types of bridge pose great technical difficulty due to their slenderness and often great dimensions. Therefore, these structures tend to have problems caused by natural forces such as wind loads, some of which we have witnessed in our history. Besides their intrinsic importance, these constructions are currently of great interest and we are seeing a very high incidence of bridge construction in this historical period for overcoming geographical barriers such as bays, straits, or wide estuaries. The great majority of the obstacles are crossed by suspension or cable stayed bridges commonly called cable supported bridges. Due to their good serviceability, there are a high number of projects underway.

Therefore, it seems very convenient to write a book showing the current capability of analysis and design. Up until now, the information contained in this book could only be found partially in articles of technical journals, which are still useful for learning some specific details of scientific or technological advances, yet not enough for understanding the global panorama. Hence, we thought a text dedicated to this topic would be useful for bridge design engineers as well as researchers working in the field. This book only requires previous knowledge of structural finite element models and dynamics and it is advisable to have some previous background in bridge engineering. Nevertheless, it is very self-contained in such a way that all the information necessary to understand the theoretical developments is presented without the need of additional bibliography.

The book contains twelve chapters. The first one introduces some objectives of the book to readers and presents the most important aspects. It also states the appropriateness of this publication in this historic moment.

Chapter Two describes the collapse of the Tacoma Strait bridge in 1940, the impact caused among bridge designers and to the society at large, and the development and the evolution produced for suspension bridges since then. It narrates in detail great bridges constructed in the last decades and describes various important projects underway in Europe and Asia.

The third describes two existing methodologies for the study of flutter. One of them is based on boundary layer wind tunnel tests of a reduced model of a complete bridge, while the other is a hybrid approach consisting of two phases: the first to test a reduced deck section model in an aerodynamic wind tunnel, and the second a computational phase to sensitivity, and optimization analysis of bridges as well as their participation as advisers for important bridge projects in different continents. We tried to convey our diverse activities and abilities, from purely research to practical application, so that it will be useful in the society when needed, and we hope to have achieved it.

During these years regarding bridge aeroelasticity, the authors have discussed topics and interchanged experiences and ideas with several people. Among them, A. Davenport and P. King of the University of Western Ontario, R. Scanlan and N. Jones of the Johns Hopkins University, P. Mendes from the IST of Lisbon and the team in the Politecnico de Milano, especially G. Diana and A. Zasso.

The book also deals with optimization techniques and in that scientific field the authors have enjoyed over a long period of time the wise advice and friendship of G. Vanderplaats, CEO of Vanderplaats R&D. Also, J. Arora of the University of Iowa, R. Haftka of the University of Florida and A. Belegundu of Pennsylvania State University have been instrumental in discussing many subjects and initiatives. Therefore, the authors are grateful for the hospitality and support provided on many occasions by all of them.

The authors appreciate the interest of WIT Press and their continuous inspiration and support for the publication of the book. We would also like to thank Anita Demitroff and Ibuki Kusano for their careful translation work from the manuscript to English as well as Xian Meiras and staff at WIT Press for layout work. Finally we would be very pleased to hear any suggestions regarding the contents of the book.

The Authors, 2011

Chapter 1

Aeroelastic analysis and design optimization of cable-supported bridges

1.1 Introduction

Suspension and cable-stayed bridges are two important types of bridges frequently used to connect great distances. Even though their origins and evolution processes are different, their analysis methods are very similar at present time. Thus the term, "cable-supported bridge" is often used to refer to both types of structures and used as well in this book in many occasions.

The first steel suspension bridge was built by James Finney,[1] who built one with 21-m span over the Jacob's Creek in 1801. Figure 1.1.1 shows the bridge designed by Finney, which already included detailed structural elements used for a suspension bridge today: the curved main cable, the vertical cables connected to the deck, the truss deck and stone towers that were very common during almost all through the 19th century. It is worth mentioning that Finney realized that the lateral spans did not need to be suspended from the main cable, and as can be seen in the figure, only one of them is suspended from the main cable.

More than 200 years have passed since then, and there is no new concept in the modern design of a suspension bridge as to the scheme in the figure. However, the span length of important modern bridges is one hundred times longer than that of the first bridge, and as will be mentioned in the following chapter, there are many ambitious projects underway.

The history of cable-stayed bridges is more recent than that of suspension bridges, and it can be dated back to the beginning of the 20th century. Figures 1.1.2 and 1.1.3 show two examples of this typology, in which we can observe the complex cable configuration: the cables are not tied to the deck, but to the horizontal cable from which a group of vertical cables connect to the deck. The reason for this uncommon solution is due to the fact that if the cables connect to the deck directly, because of its inclination, it would generate large compression on the deck, which would then cause buckling problem if it is not properly designed. They avoided the problem with this proposed solution, however, in the expense of not only complicating the structure, but also the construction of the bridge, which resulted in a very difficult job. Therefore this approach was finally abandoned. After the World War II when the infrastructures of road communication needed to be reconstructed, many cable-stayed bridges were built[2] such as the Strömsund Bridge erected in 1956 shown in Figure 1.1.4. Its structural configuration is a simple cable system that goes from the towers to the deck making its calculation and construction easier.

Figure 1.1.1: Sketch of suspension bridge by J. Finney.

Figure 1.1.2: Cassagne Bridge (France).

Figure 1.1.3: Lazardrieux Bridge (France).

Today both types of bridge have increased their capacity in such a dimension that the suspension bridge over the Akashi Strait in Japan shown in Figure 1.1.5 achieved 1991 m of span length, and the cable-stayed Sutong Bridge over the Yanghse river in China in Figure 1.1.6 reached 1088-m span length.

Figure 1.1.4: Strömsund Bridge (Sweden)

For both types of bridges, even if they are small, the most important point that a designer has to consider is not the service loads produced by automobile or railroad traffic, but the seismic or wind loads, which are much more dangerous producing great stresses in the material. Therefore, this book is dedicated to the study of cable-supported bridges under wind actions, and especially to one of the most significant instability phenomena that air flow can produce.

1.2 Aeroelastic phenomena

The air flow around a bridge not only causes the bridge to move but also changes the loads that wind generates on the structure. There is an important interaction between the wind pressure and the bridge deformation as an elastic body. These phenomena are called aeroelastics, and the science that studies this interaction is called aeroelasticity[3].

Some of the most important phenomena, among all, taken place on cable-supported bridge are as follows:

- vortex shedding
- flutter
- buffeting
- cable vibration

Vortex shedding is produced when the air flow lines are modified for colliding with a bridge deck. It takes place at certain wind velocities depending on the deck geometry and produces vortices with alternating rotations, which then produce a vertical force. This force changes in direction in each vortex, thereby causing vibrations of the deck. Since such vibrations have limited amplitudes, they do not cause any catastrophes; however, they might make pedestrians uncomfortable for walking or the velocity of circulating traffic may have to be limited. They can also cause the material fatigue of the bridge.

Figure 1.1.5: Akashi Bridge (Japan).

Figure 1.1.6: Sutong Bridge (China).

Flutter is a vibration produced on a bridge by an interaction with the wind flow, and its amplitude is not only unlimited but also increases in each cycle. Once this aeroelastic instability starts, it occasionally causes destruction of bridges. In the next chapter, we will take a look at some examples of this phenomenon, which obviously could have been avoided. The most famous example is that occurred to the bridge over the Tacoma Strait in 1940 described in detail in the next chapter that triggered the development of aeroelasticity in bridge engineering.

Buffeting is a vibration phenomenon produced by a turbulent wind flow. Its effects on a bridge are wind gusts of different average velocity. Although this phenomenon does not cause instability, it should be limited just as in the case of vortex shedding so that the service life of a bridge should be adequate or as long as possible.

Cable vibration is a term for a group of fluid–structure interaction phenomena produced on the cables of cable-stayed bridges. They may occur in different velocity ranges and influenced by rain. The deformation of cables can present very diverse trajectories depending on those factors, their geometry, stiffness and mass. There has not been any case of cable rupture by this phenomenon as of today; however, there have been events of great amplitude and frequency vibrations so that corrector devices had to be installed. At present time, there is no clear classification of a group of phenomena that causes this type of vibrations, and it is still an undeveloped research area.

It should be clear that in spite of the importance of all the aeroelastic phenomena described here, flutter is the most important since it may occasionally causes the total destruction of a bridge. Therefore, it has been an objective of studies by different methods, and this book is dedicated exclusively to this aeroelastic instability.

1.3 Methodologies of flutter analysis

As mentioned briefly, studies on flutter of bridges started after the collapse of the Tacoma Bridge in the 1940s. At that time, because of the non-existence of digital computers, the best way to acquire knowledge was by an experimental method. As same as for aeronautical engineering, reduced bridge models were created and then tested in wind tunnels under air flow at different velocities to observe the aeroelastic phenomena. In order to obtain correct results, vibration frequencies used to the reduced model were equally scaled as the geometry scale of the reduced model to the complete bridge[4]. This method gave very fruitful results during various decades. There are very diverse ways of performing a testing, and during the same experiment, one could obtain a wide range of information on the structural responses of a reduced model. This will be described in detail in this book. Figure 1.3.1 shows a test of the bridge over the Messina Strait using one of those models.

The revolution of calculation tools produced by digital computers has been such a magnitude that today computer programs are the most common calculation tools in engineering. Therefore, it seems logical that the fluid–structure simulation in bridge engineering under the influence of aeroelastic phenomena is performed by computational fluid mechanics[5] (CFD), and wind-tunnel testing is not necessary any longer.

Aeronautic engineering had an earlier start in this regard in which CFD techniques had opened up their way and became an important part of analysis and design tool for engineers. They already have programs to resolve Navier–Stokes equations for wind flow disturbed by an airfoil in a closed area to obtain essential pressure and velocity distributions as well as a set of aerodynamic parameters. The geometry of airfoil is designed all for flight efficiency, which also helps the resolution of theproblem. Figure 1.3.2 shows the aircraft A380, while Figure 1.3.3 shows the results of pressure field of a National Advisory Committee for Aeronautics profile by this procedure.

However, the geometry of a bridge deck is affected by a wider range of factors than that of an airplane. The most important task of a long-span bridge, at which most likely aeroelastic phenomena take place, is let the vehicle traffic pass through it. Then a series

Figure 1.3.1: Test of a reduced model of the Messina Bridge.

Figure 1.3.2: A380 aircraft.

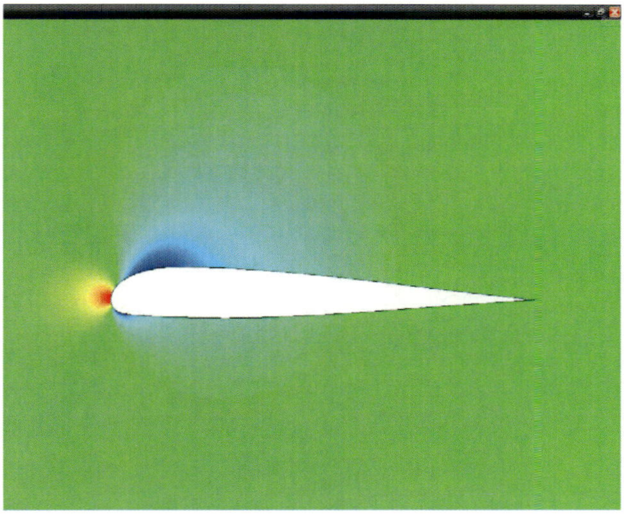

Figure 1.3.3: Static pressure field on a NACA airfoil.

of security measures associated to the task must be added such as centre dividers between two circulating directions or guide rails on the sides of the deck. In the case that pedestrians also use the bridge, there should be guide rails to separate and protect them from the circulating traffic. All these additions lead to a poorer aerodynamic quality of the deck. Figure 1.3.4 shows two bridge deck sections: one with truss section and the other with box section. Even though the latter has a smoother section, the placement of guide rails makes its aerodynamic behaviour less efficient. Therefore the fluid–structure interaction of a bridge deck is more complex than an airfoil. A set of structural elements of truss sections or guide rails change wind flow lines in large quantity and different ways causing turbulences of different sizes. Thus producing mathematical models of the phenomena that take part is rather difficult, and the formulation of turbulence associated to each type of bridge deck is still in an underdeveloped phase. This turbulence formulation is essential in order to adapt the Navier–Stroke equations appropriately. Although the number of CFD applications in bridge engineering is yet very limited to resolve aeroelastic problems, for certain types of deck sections, excluding truss sections they may be relatively useful to determine the aerodynamic coefficients or to evaluate vortex shedding. However, the obtained results are not sufficiently accurate, and consequently we need to check them experimentally, which reduces greatly the advantages of computational approach. Nevertheless, this situation will be improved little by little in the future. Figure 1.3.5 shows a CFD simulation of vortex shedding at a cross section of a bridge deck.

Nonetheless, it seems obvious that the computer calculation technique should take part in the study of general aeroelastic phenomena of cable-supported bridges, particularly in the study of flutter. This results in a methodology that combines a wind-tunnel testing with analysis by computer. This hybrid method tries to take advantage of both experimental and numerical technologies as much as possible, and it will be described in detail in later chapters. In the first experimental stage, the forces produced by wind during the vibration of a bridge are obtained, and in the second computational stage, the dynamic equilibrium of the bridge under such forces is studied. Finally a non-linear

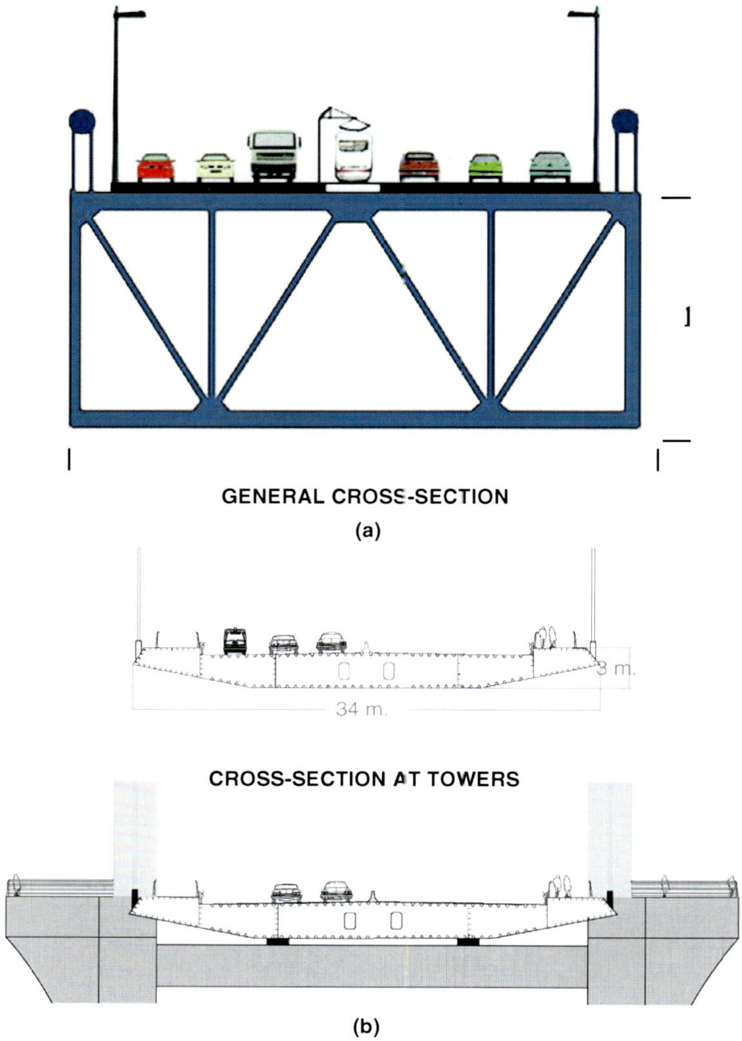

Figure 1.3.4: Types of bridge deck. (a) Truss bridge deck. (b) Box bridge deck.

eigenvalue problem is solved depending on the wind velocity. Since we can obtain the flutter velocity of suspension and cable-stayed bridges precisely and efficiently in this way, this method is well accepted for its great result quality and advantages over purely experimental method. All the theories associated to this method are described in various chapters, and examples of its applications on particular bridges, either under-construction phase or already completed, are described in this book.

The advantages of this approach will be greater if, besides flutter velocity, it provides deformation data of a bridge during the phenomenon, that is, if we can observe the geometry changes during the vibration produced by wind flow. Although this method is not designed for that purpose, the computational phase provides vibration frequencies and modes of flutter velocity, which permits the use of this data combined with the actual capacities of computer graphic presentation.

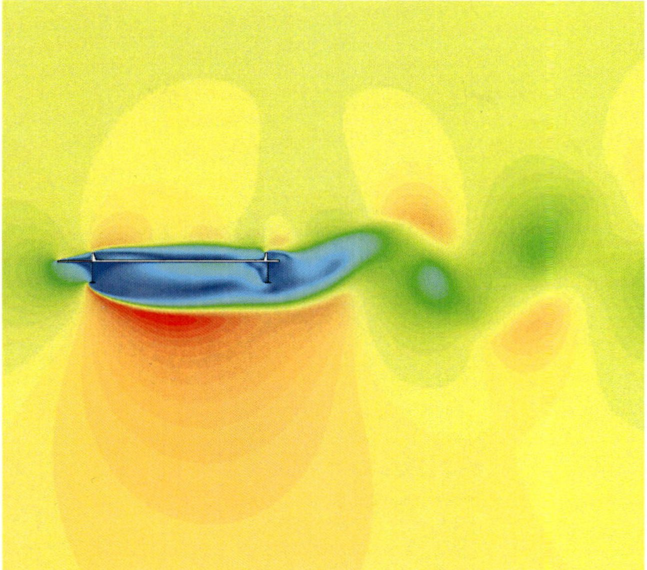

Figure 1.3.5: Velocity field of flow around a bridge deck showing vortex shedding.

We can generate a digital model of an undeformed shape of a bridge as shown in Figure 1.3.6, in which we can see the accuracy of all the bridge elements. This type of representation is naturally more detailed than a reduced model for wind-tunnel testings.
With the digital model generated along with the knowledge of vibration frequency during flutter and associated geometry of vibration modes, a computer animation can be generated to show vibration during this aeroelastic phenomenon[6].
In this way, the deformation of a bridge can be visualized clearly and graphically without performing a wind-tunnel testing of a complete bridge. Figure 1.3.7 shows an image of deformed bridge geometry during flutter by this procedure.

1.4 Sensitivity analysis: a design tool

The major part of the designing process in engineering for past decades was based on the ability and good judgment of a designer, which progressively led him to refine an initial prototype into a final design. This process was partially powered by digital computers; for defining geometry of a structure such as a bridge or a dam CAD programs might have been utilized, and finite element programs might have been employed to check the structural behaviour, which may have been supported by some testings as in the case of wind for long-span bridges.

However, in each design phase when part of prototype did not function properly and needed to be modified, the designer is the one who decided to make changes using some simple rules gained by his experiences. The designer was free to decide the changes in some aspects such as dimensions of some elements or their locations within the structure, although some of the design properties may have been predetermined. For example, if displacements of some areas of a bridge are too large, the inertia can be increased, but not the span length because it is predetermined.

Figure 1.3.6: Digital model of the original Tacoma Bridge.

Figure 1.3.7: Deformation of the digital original Tacoma Bridge model under wind flow.

In essence, a designer observes structural responses of a design in each phase and makes changes in modifiable properties hoping that they are good enough.
This is, therefore, essentially a subjective process, and its efficiency all depends on the designer's capacity and the difficulty level of the problem. Consequently, the quality of the final result may vary greatly in each case.
There have been more rational tools around for a long time to assist engineers with these tasks. They are based on so-called sensitivity analysis, which consists of obtaining the derivatives of structural response with respect to modifiable structural properties to improve the design. Let us take a look at a practical example of linear static structural analysis to understand this approach.
Static equilibrium of a structure can be written as

$$\mathbf{Ku} = \mathbf{P}$$

where $\mathbf{K}$ is the stiffness matrix of a structure, $\mathbf{u}$ the displacement vector and $\mathbf{P}$ the load vector. All of them are functions of structural variables, for example, some mechanical parameters

$$\mathbf{K}(\mathbf{x})\mathbf{u}(\mathbf{x}) = \mathbf{P}(\mathbf{x})$$

where the vector $\mathbf{x}$ contains a group of modifiable structural properties to improve the design, commonly called design variables.
By differentiating this expression,

$$d\left(\mathbf{Ku} = \mathbf{P}\right)$$

or

$$\sum_{i=1}^{n} \frac{\partial \mathbf{K}}{\partial x_i} \mathbf{u} dx_i + \mathbf{K} \sum_{i=1}^{n} \frac{\partial \mathbf{u}}{\partial x_i} dx_i = \sum_{i=1}^{n} \frac{\partial \mathbf{P}}{\partial x_i} dx_i$$

For each variable, we have

$$\frac{\partial \mathbf{K}}{\partial x_i} \mathbf{u} + \mathbf{K} \frac{\partial \mathbf{u}}{\partial x_i} = \frac{\partial \mathbf{P}}{\partial x_i}$$

or

$$\mathbf{K} \frac{\partial \mathbf{u}}{\partial x_i} = \frac{\partial \mathbf{P}}{\partial x_i} - \frac{\partial \mathbf{K}}{\partial x_i} \mathbf{u}$$

This expression is a system of equations that gives the derivatives of a displacement vector with respect to a design variable. By resolving it for each of them, we will get a matrix as

$$\frac{\partial \mathbf{u}}{\partial \mathbf{x}} = \left[\frac{\partial \mathbf{u}}{\partial \mathrm{x}_1}, \dots, \frac{\partial \mathbf{u}}{\partial \mathrm{x}_n} \right]$$

which also can be written as

$$\frac{\partial \mathbf{u}}{\partial \mathbf{x}} = \begin{vmatrix} \frac{\partial u_1}{\partial \mathbf{x}} \\ \\ \frac{\partial u_n}{\partial \mathbf{x}} \end{vmatrix}$$

Each element of this matrix contains two types of information. One is qualitative associated with the sign of the derivative; if it is positive, in order to decrease the value of the structural response, in this case, displacements of the structure, a designer has to decrease the design variable value, while if the derivative is negative, he has to do the opposite. Another is the quantitative value of the derivative, which gives an idea of how much changes one can get from modifying the variable. If the absolute value of the derivative is large, one can get significant changes by modifying that variable, while the modification causes only minor changes if the derivative is small.

The first book on sensitivity analysis of structures was written by Haug *et al.*[7] Each sensitivity analysis requires differentiation of the sate equation as in the case of the static equilibrium equation described here. In this way, different types of structural problems were studied, among which frequencies and natural vibration mode problems are found. However, the sensitivity analysis associated to aeroelastic problems of bridge engineering is only recent and it is a pioneer work of Jurado, Mosquera and Hernández[8, 9]. We will discuss the approach as well as various examples of its application to actual bridges of great importance in several chapters of this book. This approach not only allows a bridge designer to progress in the designing process faster and more rational, but also by observing carefully the results of sensitivity analysis, he can reflect on how much he could have anticipated the results by his intuition and experience. This makes an engineer mature and to improve his design skills.

1.5 Optimum design in engineering: application to bridge aeroelasticity

The techniques of design optimization in engineering have been around for the last decades. The first work of L. Schmit was published in 1960[10], in which he utilized skilfully a formulation of this method. It consisted of combining the power of finite element structural analysis already developed at that time with optimization algorithms, some of which already existed. The objective of optimization techniques was to perform design improvement process automatically, which was up till then performed subjectively by designers. An iterative process was used just as in the conventional process, but rather in a systematic manner. The process started with an initial design whose quality was checked to see if all the requirements were satisfied. When some of the requirements were not satisfied, instead of modifying the design arbitrarily, the process kept giving data of the actual design to an optimization algorithm that produced a new design, which was improved in quality from the previous one. By reiterating the process, the solution finally converged to a design that satisfied all the required constraints in the best way. Figure 1.5.1 shows schematically this method that will be described in detail in the corresponding chapter.

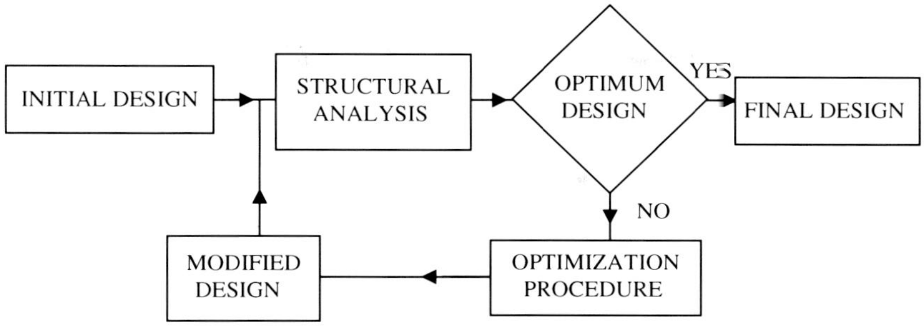

Figure 1.5.1: Flowchart of design optimization technique.

After the slow acceptance described magnificently by the author[11], little by little the method was gaining enthusiastic supports and acceptations in the academic world as well as in industries. Since then there have been numerous texts that describe the fundamental of these techniques and some of them describe practical applications in wide range of fields[12–16]. Today, there are commercial codes with optimization modules utilized continuously by automobile or aerospace companies, while in other sectors of industries, this technology is not employed very much.

The use of this design method not only provides better quality of the final design that would give greater competitiveness to the company but also helps a designer with his task. When the designing process is finished, a designer should think about the differences between the initial design and that finally produced by the optimization method. From this reflection, he should observe the changes produced and think which ones he was able to anticipate and which ones not. This will improve his skills as a designer, and we can see the optimization technique as a design course that a designer can take while he is performing his job.

In summary, we are not trying to substitute a designer with these optimization techniques, which would be impossible because of the complexity of real problems, but rather intending to help a designer not to fall into false steps that can be very probable for a design with great complexity.

Although it is appropriate to apply this technique to aeroelastic problems in bridge engineering, it is only recent to utilize them for optimization of cable-supported bridges including constraints associated with fluid–structure interaction. Specifically, Nieto, Hernández and Jurado were pioneers in dealing with optimum design of suspension bridges considering flutter phenomenon in the problem. The last two chapters of this book describe the formulation of this problem as well as its application to a real bridge.

Because of the complexity of obtaining flutter velocity, which requires successive iterations for being a non-linear problem as well as the iterative characteristics of the optimization itself, the application of this technique to aeroelastic problems demands great calculation capacity. Nevertheless, this issue will surely be less and less important with the fast evolution of computers since calculations can be carried out by sufficiently powerful machines or computer clusters that can run in parallel for distributing calculation loads.

1.6 References

[1] Brown D. J. [1993] *Bridges*, Macmillan, New York.

[2] Gimsing N. J. [1997] *Cable Supported Bridge, Concept and Design*, John Wiley & Sons, New York.

[3] Simiu E., Scanlan R. H. [1996] *Wind Effects on Structures*, John Wiley & Sons, New York.

[4] Dyrbye C., Hansen S. O. [1997] *Wind Loads on Structures*, John Wiley & Sons, New York.

[5] Wendt J. F. (ed.) [2009] *Computational Fluid Dynamics*, Springer, Berlin.

[6] Nieto F., Hernández S., Jurado J. A. [2009] *Virtual wind tunnel. An alternative approach for the analysis of bridge behaviour under wind loads.* International Journal of Advances in Engineering Software, Vol. 40, No. 3, pp 229–235.

[7] Haug E. J., Choi K. K., Komkov V. [1986] Design sensitivity analysis of structural systems, *Mathematics in Science and Engineering*, Vol. 177, Academic Press, New York.

[8] Jurado J. A., Hernández S. [2004] *Sensitivity analysis of bridge flutter with respect to mechanical parameters of the deck.* Structural and Multidisciplinary Optimization, Vol. 27, No. 4, pp. 272–283.

[9] Mosquera A., Hernández S, Jurado J. A. [2003] *Analytical sensitivity analysis of aeroelastic performance of suspension bridges under construction.* 11th International Conference on Wind Engineering. Lubbock, TX, USA.

[10] Schmit L. A. [1960] S*tructural design by systematic synthesis.* Second Conference on Electronic Computation. ASCE, Pittsburg, PA, pp. 105–132, 1960.

[11] Schmit, L. A. [1981] S*tructural synthesis. Its genesis and developement.* AIAA Journal, Vol. 19, No. 10, pp. 1249–1263.

[12] Vanderplaats, G. N. [2001] *Numerical Optimization Techniques for Engineering Design.* VR&D, Colorado Springs.

[13] Haftka R.T., Gurdal Z. [1992] *Elements of Structural Optimization.* Kluwer, Dordrecht.

[14] Arora J. S. [2005] *Introduction to Optimum Design.* Elsevier, The Netherlands.

[15] Belegundu A. D., Chandrupatla T. R [1999] *Optimization Concepts and Applications in Engineering.* Prentice-Hall, New York.

[16] Hernández S., Fontan A. N. [2002] *Practical Applications of Design Optimization.* WIT Press, Southampton.

Chapter 2

Cable-supported bridges since 1940: The Tacoma effect

2.1 Collapse of the Tacoma Narrows Bridge

On 7 November 1940, an event occurred that was to have a resounding impact on bridge engineering: the collapse of the Tacoma Narrows Bridge, on the West Coast of the United States.

The original Tacoma Narrows Bridge was situated in the state of Washington, on the West Coast of the United States of America, to the east of the city of Tahoma, as seen in Figure 2.1.2.

This suspension bridge, which came into use from 1 July 1940, had a central span 853 m long. Its deck consisted of two longitudinal steel girders with transversal ties. It had an exceptionally slender appearance because the ratio between the edge depth and span length was only 1/350. Figure 2.1.3 provides both a picture of the whole bridge and the deck cross section.

The mastermind behind this project was the great bridge engineer of the time, Leon Moisseif (1872–1943). Other brilliant examples of Moisseif's work include the Manhattan Bridge in New York (1909) and the Bay Bridge between San Francisco and Oakland (1936). Moreover, he served as consultant for the Golden Gate Bridge in San Francisco (1937). The design for the Tacoma Narrows Bridge was based on the second-order theory or – more commonly – the *deflection theory*, developed in 1888 by Josef Melan (1854–1941) in Vienna.

Working from a two-dimensional model, this calculation method stated that the deck's rigidity would work as a limitation so that the cables adopted the funicular shape associated with the acting load. In principle, only vertical actions were taken into account and analysis was limited to how the cables would lengthen and change shape because, as already mention, the second-order theory was applied. These premises gave rise to deck designs that were increasingly flexible and graceful. Indeed, a deck completely free of rigidity was considered possible provided that the cables had adequate rigidity: an idea referenced even recently by Buonapane and Billington[1] as well as Gimsing[2]. Trusses were placed to take on the horizontal loads because there was no analysis of the geometry that the cables would assume while withstanding these actions. Moisseif was not alone in applying this theory during the early 20th century. It was also utilized by other notable figures such as Steinman and Watson[3] (1887–1960) and Ammann[4] (1879–1965), the names behind the splendid St. John Bridge in Portland (1931) and the George Washington Bridge in New York (1931).

At the start of the 1930s, "*deflection theory*" was already widespread in its three-dimensional version. It proposed that the cables' geometry could be studied when both

Figure 2.1.1: Oscillation and subsequent collapse of the Tacoma Narrows Bridge.

the horizontal and vertical loads – including the static pressures of the wind – were considered. With this line of reasoning, the geometric variation of the cables would continue to be limited by the deck's rigidity, in this case, through lateral bending. Thus, it would seem that this mechanical component of the deck would also be reduced. This last premise meant that the theory's application would subsequently have dire consequences for the design of the original Tacoma Narrows Bridge. As is currently well known, the theory was suitable for analysing these structures when they are withstanding static loads, whether gravitational or wind-induced. However, it was not suitable for designs with dynamic loads because it overlooked the aeroelastic features modifying the wind force acting on the movement of the deck itself.

Ever since its opening on 1 July 1940, the Tacoma Bridge manifested, when exposed to winds of moderate speeds, vertical and symmetrical oscillations that died down once an amplitude of 1.5 m was reached (see Gimsing[2]). During its final phases of construction several workers noticed this tendency, reporting nausea and fits of vomiting, as described in Scott[5]. The bridge's strange behaviour drew the attention of not only the engineering community but also the area's inhabitants. It thus became quite an experience to cross "*Galloping Gertie*", the nickname given to the bridge during that period. However, events reached a climax on 7 November 1940. Before 10:00 in the morning, the vertical oscillations already reached their normal rate. However, after that hour and possibly after a few cable bands along the bridge's centre had broken by a wind reaching about 70 km/h, the way the bridge oscillated abruptly changed. Its movements seemed like those caused by

Figure 2.1.2: Geographical location of the original Tacoma Narrows Bridge.

asymmetrical torsion with a node located halfway along the central span, as can be seen in Figure 2.1.4.

According to Fung[9] the initial oscillation frequency was 37 cycles/min, about 3.9 rad/s, but that suddenly changed to 14 cycles/min, that is, 1.5 rad/s. At this frequency, the torsional motions of the deck increased radically and the bridge collapsed

Figure 2.1.3: Original Tacoma Narrows Bridge. General view and deck cross section.

half an hour later. In Figures 2.1.5–2.1.7, there are several photographs taken after the bridge's collapse.

The bridge's collapse caused an immediate commotion and widespread unrest. In the words of Scott[5], "*how can an apparently well-designed bridge using the latest technology to withstand winds of 161 km/h and the static horizontal pressure of winds 146 kg/m² succumb to winds at half the speed and pressures around one-sixth of the design value?*"

Figure 2.1.4: Torsion-produced oscillations of the Tacoma Bridge.

Figure 2.1.5: View of the side span following its collapse.

Immediately afterwards, the Federal Works Agency (FWA) set up a commision to analyse what happened. This included Ammann *et al.*[4],(1881–1963). The last on the list was perhaps one of the most respected scientists in the field of fluid mechanics. Among the reports about what had occurred on that fateful day of 7 November 1940, the most noteworthy was by Faquharson[7].

Figure 2.1.6: View of the deck collapsing along its central span.

Figure 2.1.7: Deck of the Tacoma Bridge after its collapse.

Of all the feasible causes for the bridge's collapse investigated by the commission, there already existed the possibility that the vibrations prior to this failure were induced by the movement of the very deck as the wind acted upon it.

Even today the collapse of the Tacoma Narrows Bridge continues to excite interest as a topic; for this reason, there are numerous studies that re-examine what happened that day. There is general concensus that the main cause for the Tacoma Bridge's failure was that the wind's forces and the deck's movements interacted so that those forces on the deck were induced by the structure itself and were therfore aeroelastic. Aeroelasticity had already been well recognized and used in other branches of engineering, such as aeronautics, but this situation led to it being applied for the first time to the field of civil engineering.

In 1971, the man who had served many years as Chair at John Hopkins University (Baltimore, USA), Scanlan and Tomko[8] supported the idea that when those aeroelastic forces generate negative damping in the structure, the deck movements continue to grow, eventually leading to collapse. This phenomenon is known as flutter. In the case of this bridge, flutter would have been produced at moderated wind speeds, leading to the first mode of asymmetrical torsion, given that the torsional rigidity of its twin-beamed deck was extremely low.

However, on many occasions over the years, the causes of the bridge's collapse had been wrongly explained. One example of this can be found in an article by Billah and Scanlan[9] for the American Association of Physics Teachers in 1990. The text is peppered with the errors – widespread among university textbooks – made while interpreting this phenomenon. The collapse is considered a typical case of resonance induced by the frequent appearance of whirlwinds or Von Kármán votices as the air flow passes through the deck section.

Recent studies, like the ones by Larsen[10] in 2000, take another look at the phenomenon and reject the idea that Von Kármán vortices were responsible for the collapse. Billah and Scanlan[9] had already dismissed the idea of vortices as the cause for failure, and studies cited by Larsen reinforced this theory. In turn, there was an innovation in the way results were obtained: studies were carried out using a computational program called DVMFLOW, resolving the phenomenon of fluid/structure interaction in two-dimensional problems. Moreover, the program's graphical interface was pleasing, as can be appreciated in Figure 2.1.8. This provides a clear visualization of the model's behaviour in the deck cross section.

On the other hand, Jurado[11] also studied this bridge using a three-dimensional finite element through the second-order theory and multimode analysis. This advanced approach produced flutter speed results similar to those described in Larsen. Moreover, through this approach, it was possible to know the role played by each vibration mode in the bridge's collapse.

With these results Hernández[12] used a very lifelike digital model of the Tacoma Narrows Bridge to achieve a three-dimensional visualization of its dynamic and aeroelastic behaviour. This enabled him to create a virtual reproduction of the process leading to the bridge's flutter-induced instability. Figure 2.1.9 offers an image from the computer animation generated with this methodology.

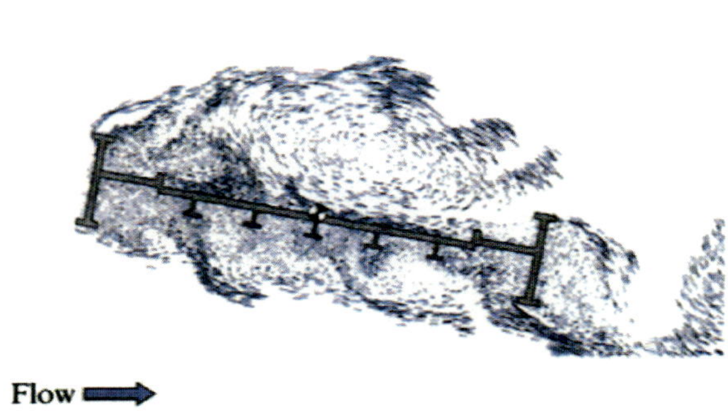

Figure 2.1.8: Image provided by the DVMFLOW code.

Figure 2.1.9: Image of the Tacoma Narrows Bridge's aeroelastic deformation.

One of the most up-to-date treatments in finding the cause of the Tacoma Narrows Bridge's collapse is found in the magnificent book *In the Wake of Tacoma*, published in 2001 by Scott[5] for the American Society of Civil Engineers (ASCE).

2.2 The "Tacoma effect"

There were, of course, numerous consequences of the Tacoma Bridge's collapse. In chronological order, these can be grouped into the following categories:

- general caution exercised by project managers;
- beginnings of aeroelasticity as applied to civil engineering;
- appearance of bridge decks with aerodynamic box girders.

The more cautious approach led to an absence of innovation and there was a reliance on tried and tested designs. Span lengths were chosen because they presented little risk and had already been put into practice. In general, loads were increased excessively due to the new uncertainty about the wind's impact, a factor that still bewildered engineers in the early days after the event.

Previous designs breathtaking in their audacity include the bridges raised in San Francisco, such as the Bay and Golden Gate bridges. The latter stood out in its time – the 1940s – for the world record – setting length of its central span.

The Bay Bridge – designed by Moisseif and inaugurated in 1936, constituted a magnificent example of a continuous double suspension bridge, with main spans of 704 m. Moreover, the truss design for the deck is the epitome of classic lines and rigidity, as can be appreciated in Figure 2.2.1. At the same time, the Golden Gate also presented a new truss design. However, it was more daring compared with that of the Bay Bridge because it had a lower ratio between the edge depth and span length, as shown in Figure 2.2.2. Designed by J. B. Strauss and opened in 1937, it had a central span of 1280 m.

Figure 2.2.1: Bay Bridge between San Francisco and Oakland.

Figure 2.2.2: Golden Gate Bridge.

What remained in the minds of the project engineers of the period, although they were unable to quantify it, was that the original Tacoma Narrows Bridge had been too slender and flexible. Therefore, the twin-beamed deck was abandoned to return to a more classical and rigid alternative deemed sufficiently trustworthy in practical terms: truss decks.

Ten years after the original one had collapsed, another bridge was constructed at the same site over the straits of Tacoma. This came to be known as the New Tacoma Bridge. As can be seen in Figure 2.2.3, the changes were significant. In the same figure one can spot the dredged up remains of one of the original bridge's towers in the lower right of the image. The twin-beamed design with a 2.4 m depth gave way to a truss with a 10 m depth, as shown in Figure 2.2.4. The bridge's width went from 11.9 to 18.3 m. Obviously, the new bridge, or rather, its deck, had excessive dimensions, but that was understandable given the earlier one's spectacular failure.

Figure 2.2.3: New Tacoma Bridge.

Figure 2.2.4: Truss deck of the new Tacoma Bridge.

The aftermath of the bridge's collapse was not only felt in new constructions; those that already had been built were subject to reinforcement measures. Two of these were the Bronx-Whitestone Bridge and the Golden Gate, already mentioned.

The Bronx-Whitestone Bridge connects the neighbourhoods of the Bronx and Queens in New York City. Opened in 1939 and designed by Ammann, it has a central span that measures 754 m. As shown in Figure 2.2.5, it first had a twin-beamed deck. Vertical vibrations were observed in the bridge, although these started on a limited scale and always decreased with time, according to Gimsing[2]. Nevertheless, the decision was made to reinforce the structure in 1946 by adding vertical trusses on each side of the deck so that its rigidity was increased. The bridge's appearance remains unchanged to this day, as can be appreciated in Figure 2.2.6.

However, these vertical trusses have been replaced, as described in Landers[13]. Instead, aerodynamic lateral barriers have been slotted along both sides of the deck, thus improving

Figure 2.2.5: Bronx-Whitestone Bridge, as it originally looked (1939).

Figure 2.2.6: Bronx-Whitestone Bridge (2002).

how the bridge withstands the wind. Figure 2.2.7 provides an illustration of these mechanisms as well as an artist's impression of what the bridge looks like since 2004.

As for the Golden Gate, it had already proved itself prone to an oscillation in windy conditions, but under strong north-westerly winds that blew for 4 h in December 1951, the vertical oscillations reached 3.3 m and bridge traffic was stopped for 3 h (see Gimsing[2]). These events led to the decision to carry out structural reinforcement. Figure 2.2.8 gives the end result. The bridge's original design, with three trusses forming an open shape, was replaced with a closed shape resulting from the addition of a new trussed surface underneath. Thus the deck enjoyed greater rigidity, especially when subjected to torsion.

Figure 2.2.9 is of the Golden Gate, offering a clear view of the section with a truss closed during the reinforcement work: a photograph taken in 1984. In that year, work

Figure 2.2.7: New aerodynamic barriers for the Bronx-Whitestone Bridge.

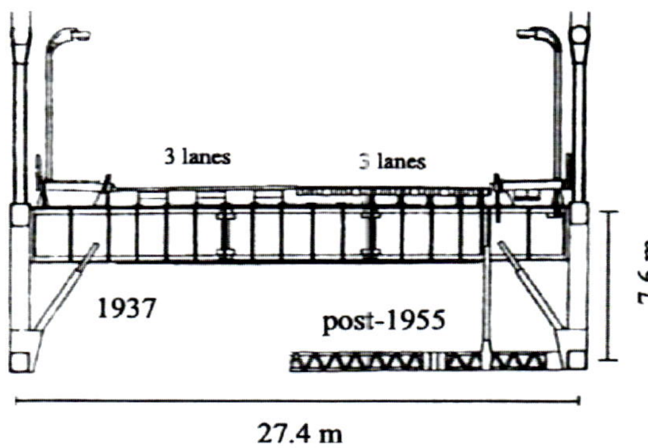

Figure 2.2.8: Deck section of the Golden Gate Bridge.

Figure 2.2.9: Golden Gate Bridge.

was carried out on the bridge. The upper part of the deck beneath the road surface was lightened to lower the deck's weight. In the last figure mentioned, one can notice how the bridge has a deformed appearance due to the replacement job that was then being carried out halfway along the main span. The gravitational loads were uncompensated, leading to the vertical asymmetrical deformation that is so apparent.

The first great suspension bridge raised after the Second World War was the Mackinac Bridge, as seen in Figure 2.2.10. The work of Steinman, it was opened in 1957 in Michigan, USA. Its central span of 1158 m does not surpass that of the Golden Gate. The deck is made up of a truss with an 11.6 m edge, making its ratio of edge depth/main span length 1/100. Therefore, it has a less slender appearance than the Golden Gate.

At first sight the bridge gives off an impression of robustness. So much care was given to its aerodynamic stability that, according to Gimsing[2], the critical wind speed value was 995 km/h, impossible to reach under natural conditions. The truss section of the deck shares the same sense of excessive dimensions observed in the New Tacoma Bridge.

At the beginning of the 1960s, great suspension bridges were first being built in Europe. One example is the Firth of Forth Bridge in Scotland. It was opened to traffic in 1964 and spanned 1006 m between its towers. Another is the 25 de Abril Bridge in Lisbon, inaugurated in 1966. Its central span is 1014 m. Figures 2.2.11 and 2.2.12 demonstrate how both have lengths that could be considered normal and trustworthy designs, relying on deck truss sections of great rigidity.

The record for main span length held by the Golden Gate since 1937 would not be challenged until 1964, when the Verrazano Strait Bridge was opened in New York. This had a central span of 1298 m, although only a modest 18 m greater than that of the Golden Gate. Ammann's design had a truss deck of 7.3 m, as seen in Figure 2.2.13. Its depth/span length ratio was thus 1/178, making it just slightly more slender than the previous record holder.

The Verrazano Bridge only meant a small step towards greater span length and deck slenderness compared with the Golden Gate. What is striking is how, after almost three decades, the typologies are nearly identical.

Figure 2.2.10: Mackinac Bridge.

Figure 2.2.11: Firth of Forth Bridge (United Kingdom).

Figure 2.2.12: 25 de Abril Bridge in Lisbon.

Figure 2.2.13: Verrazano Strait Bridge.

A possible conclusion is that in the years following the Tacoma Narrows disaster there was a period of fear and doubt, with many questions and few answers. In the words of Scott[5], "thanks to some kind of miracle human lives weren't lost in the collapse of the Tacoma Narrows Bridge, ... but the worst consequence was the loss of a whole generation of engineering practice."

This prolonged period where there was no innovation in terms of the design and construction of long-span suspension bridges saw the early development of aeroelasticity as applied to civil engineering at universities and research centres One of the main outcomes of this trend was a new deck typology for this kind of bridge: those using aerodynamic box girders, revealing an increased knowledge about how the wind interacted with the structure itself.

2.3 Recent history (1966–1988)

This period of time is characterized by the first aerodynamic box sections for suspension bridges and by the development of cable-stayed bridges.

2.3.1 Decks with aerodynamic sections

Until 1966 aerodynamic stability in long-span suspension bridges was achieved, as seen in the last section, by building trusses with deep edges to produce the deck. In that way, it was possible to reach high levels of rigidty under bending and, more importantly, under torsion compared with the outmoded and forgotten double-beamed decks.

Thanks to the advances made in the field of aeroelasticity since the collapse of the Tacoma Narrows Bridge, the Severn Bridge between Wales and England was opened to traffic in 1966. This bridge had a deck design that was unprecedented in a suspension bridge. With the main span reaching 987.5 m, the classic truss was substituted with an aerodynamic box girder 3 m deep, as shown in Figure 2.3.1. The bridge's slenderness ratio was 1/324, very close to the value 1/350 found in the original Tacoma Bridge.

This new type of deck, considered revolutionary for its time, eliminated the two main causes for the wind-induced failure experienced by the Tacoma Bridge: the absence of aerodymic features and limited rigidity under torsion of the double-beamed deck.

This bridge also relied on a system of zigzagging cables connecting the deck and the main suspension cables. While the box girder deck's great rigidity avoided any possible torsion-induced oscillations, there were fears that the vertical oscillations would be too high.

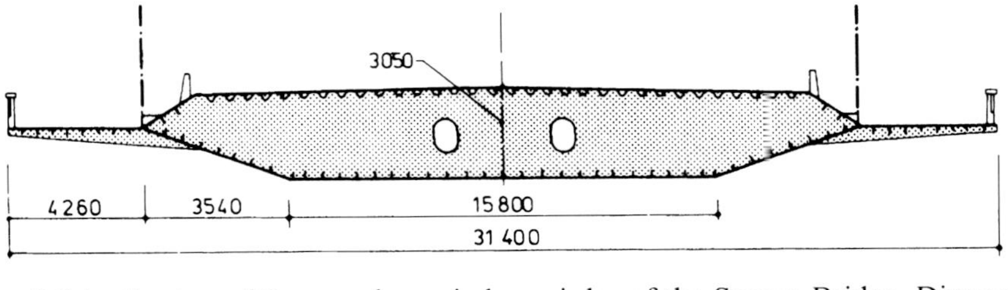

Figure 2.3.1: Section of the aerodynamic box girder of the Severn Bridge. Dimensions in millimetre.

Again in the United Kingdom, in 1981, construction of the Humber Bridge was completed. It was a suspension bridge with a main span of 1410 m, with a world record unbeaten until 1998 for central span length, surpassing the Verrazano Bridge by over 100 m. However, there was one striking difference between them. For the Verrazano Bridge, the deep-edged truss design was adopted, but the Humber Bridge inherited the aerodynamic box girder deck used on the Severn Bridge (see Figure 2.3.2). Indeed, the consultancy firm Freeman, Fox and Partners had been responsible for both projects.

Nearly three decades had to pass before the slenderness enjoyed by the original Tacoma Bridge would once again be achieved and before aerodynamic phenomenon would be understood to a reasonable degree.

Figure 2.3.2: Severn Bridge.

Figure 2.3.3: Humber Bridge.

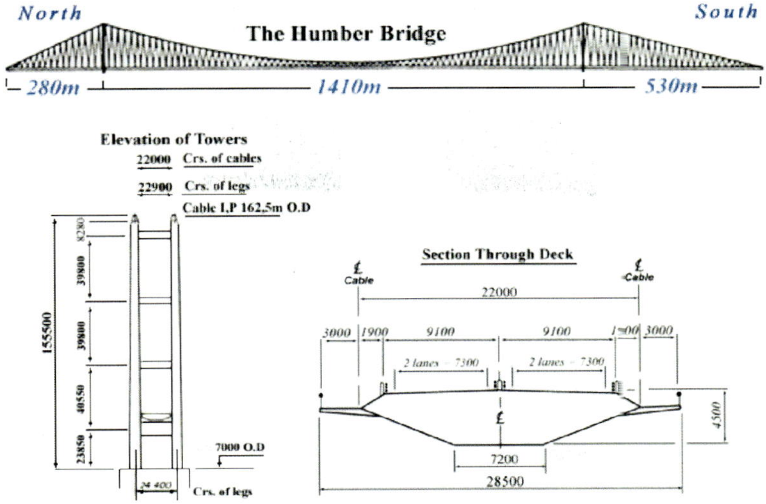

Figure 2.3.4: Humber Bridge. Elevation and section.

As observed in Figures 2.3.3 and 2.3.4 the bridge basically consisted of the already mentioned aerodynamic box girder deck joined to the main cables by means of an A-frame tower. A striking feature is the difference in lengths between the lateral spans: one is 280 m long and the other 530 m long. This was a unique feature among long-span suspension bridges. Its towers – 155.5 m in height – were made of concrete, whereas those of the Severn were steel. The ratio between edge depth and main span length was 1/314, much lower than that of the Verrazano Bridge.

Nowadays, it may be said that this bridge is at the fourth place for suspension bridges, if the criteria are based on main span length. This is an important point given that this length was only surpassed once the bridge over the Great Belt in Denmark was opened 17 years later in 1998.

1973 saw the completion of the first bridge over the Bosphorus in Istanbul, Turkey, followed by the second bridge (Fatih Sultan Mehmet) in 1988. Once again both were the projects of Freeman, Fox and Partners, and were practically identical as a result. The lengths of their central spans are 1074 and 1090 m, respectively. One of them can be seen in Figure 2.3.5. The two bridges are very similar to their counterparts over the Humber or Severn. They have an aerodynamic box girder deck and portal-like towers with horizontal inner beams, following the style of what was to become known as the "European School".

2.3.2 Cable-stayed bridges

Many cable-stayed bridge projects were carried out during this period due to the great versatility this typology offers when dealing with span lengths between 200 and 400 m. Spain boasts some magnificent creations of this kind, among which the Rande and Barrios de Luna bridges stand out.

Figure 2.3.6 presents the Rande Bridge, opened over the Vigo Estuary in 1978. Designed by Fabrizio de Miranda, it was among the first bridges of its kind with a span of 400 m. It has a steel deck and concrete towers.

Figure 2.3.5: Second Bridge over the Bosphorus.

Figure 2.3.6: Rande Bridge.

Another fine Spanish design is the Barrios de Luna Bridge by the company Carlos Fernández Casado S.L. It was opened in 1984. As shown in Figure 2.3.7, both its tower and deck are made of concrete. The 440 m length of its main span held the world record for several years.

Although these are splendid examples of cable-stayed bridge engineering from the 1980s, the length of their central spans, along with the rigidity of their cable stays, provides their decks with a high level of stability in the face of acting winds. It would not be until the 1990s when this type of bridge would include spans of up to a kilometre and the effect of the wind would simply be among the inherent features of the bridge's design.

Figure 2.3.7: Barrios de Luna Bridge.

2.4 Recent history (1989–1999)

In the 1990s the engineering of cable-supported bridges with long spans would enjoy a tremendous spurt of development; this would be true for both cable-stayed and suspension bridges. However, it must be pointed out that, apart from their original function of overcoming geographical obstacles like rivers, bridges have also been used in more versatile ways to make route connections. Rather than just one isolated structure, there are complex networks of different bridges with varied typologies.

2.4.1 Bridges of the Honshu–Shikoku route in Japan

In the late 1980s and early 1990s, the country with the greatest number and most splendid examples of cable-supported bridge engineering projects was Japan. This was mainly due to an effort that began in the 1970s to build three different links between the islands of Honshu and Shikoku, separated by the Seto Sea. As seen in Figures 2.4.1 and 2.4.2, the eastern link is between Kobe and Naruto; the central one between Kojima and Sakaide; and the one in the west between Onomichi and Imabari. Saeki *et al.*[14] has published an interesting technical overview of the bridges built between these two islands.

Along the western route between Onomichi and Imabari, as shown in Figure 2.4.3, and the north–south axis, there are three sets of suspension bridges among other kinds of structures: the Innoshima and Ohshima bridges as well as the ones over the Straits of Kurushima.

The Innoshima Bridge, completed in 1983, is a suspension bridge in the traditional American style: a deep deck with a lattice truss and a central span of 770 m (see Figure 2.4.4).

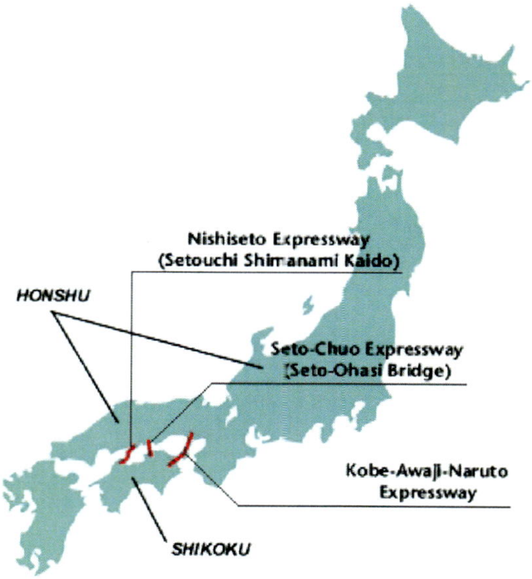

Figure 2.4.1: Routes connecting Honshu and Shikoku.

Figure 2.4.2: Routes connecting Honshu and Shikoku.

Later, in 1998, the Oshima Bridge was finished (see Figure 2.4.5). It has an aerodynamic box girder deck that is 2.2 m deep, more in keeping with the European tradition. The central span is 560 m, making this bridge one of the most recognized examples of how a civil engineering project can blend into the landscape.

To complete this route along the south, there are the Kurushima Bridges, finished in 1999. Of outstanding design, they are three consecutive suspension bridges with an aerodynamic box girder deck – found in Figure 2.4.6, portal-like steel towers with horizontal inner beams, and central spans of 600, 1020 and 1030 m.

Among the cable-stayed bridges along this route, worthy of comment are the Ikuchi and Tatara bridges. The former, inaugurated in 1991, can be seen in Figure 2.4.7. Its main span is made of steel and reaches 490 m. The cross section is an aerodynamic box girder deck.

The most important cable bridge along the network between Honshu and Shikoku is the Tatara, opened in 1998. It holds the second place for bridges of its class for its central

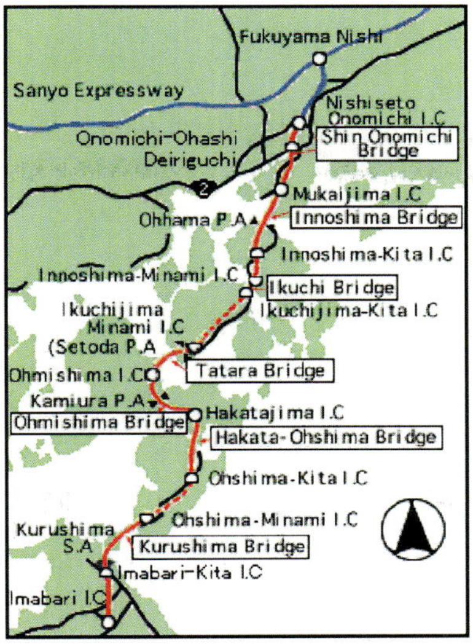

Figure 2.4.3: Western connecting route between Honshu and Shikoku islands.

Figure 2.4.4: Innoshima Bridge.

span of 890 m. Because it can hold its own alongside suspension bridges with kilometre-long spans, it is unique in the extremes to which its designers had gone while working within this typology. Figure 2.4.8 shows it is basically a steel construction, with a deck whose section is an aerodynamic box girder and rhomboidal towers. Owing to its length, the deck's aeroelastic behaviour was scrupulously examined, with both sectional tests and reduced scale models of the whole bridge, which included, as faithful as possible, a simulation of the orographical features of the surrounding terrain, as provided in Figure 2.4.9.

The central route between Kojima and Sakaide is the only one along the network that is open to both train lines and road traffic. As seen in Figure 2.4.10, the route is made up of six bridges, three of which are suspension: the Shimotsui-Seto, the Kita Bisan-Seto and the Minami Bisan-Seto. Separated by the section of a central anchorage, the last two in the list are practically identical, with one leading onto the other (see Figure 2.4.12). Originally, the bridges along this route were designed to house four railway lines in an area below their decks. Two of these lines would be for normal use and two for high-speed services, although, in the end, the faster lines were never installed.

The Shimotsui-Seto Bridge was completed in 1988. As one can observe in Figure 2.4.11, it is a suspension bridge in which only the deck of its central span (940 m long) hangs from the main cables. The deck consists of a truss with an impressive depth. It has an

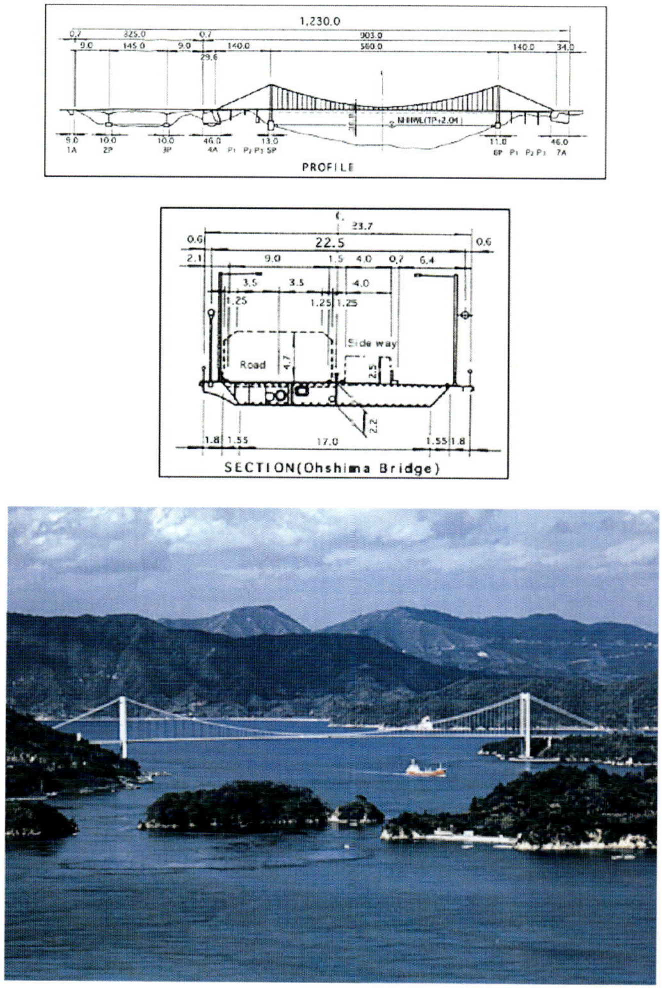

Figure 2.4.5: Oshima Bridge.

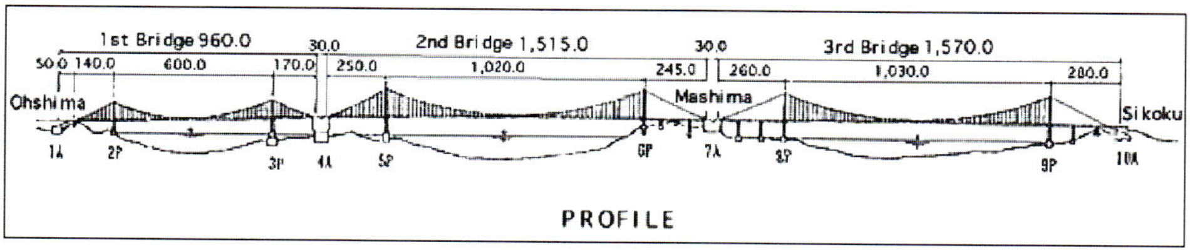

Figure 2.4.6: Kurushima Bridges.

Figure 2.4.6: *Continued.*

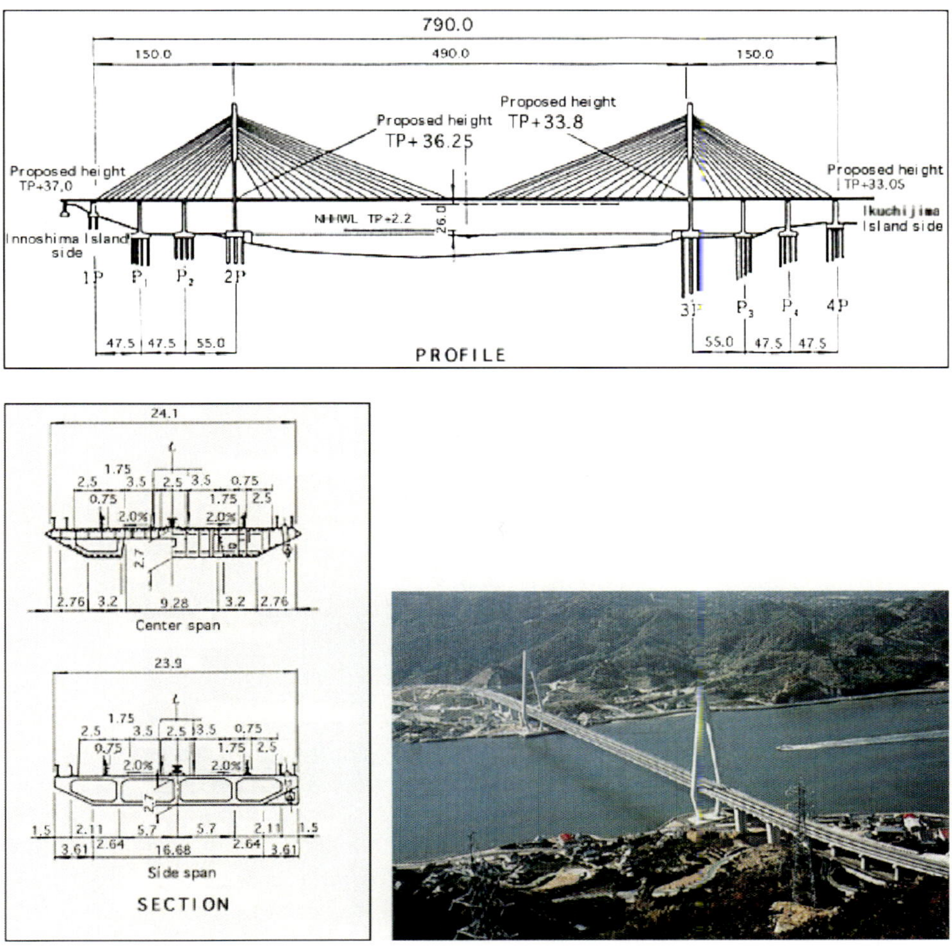

Figure 2.4.7: Ikuchi Bridge.

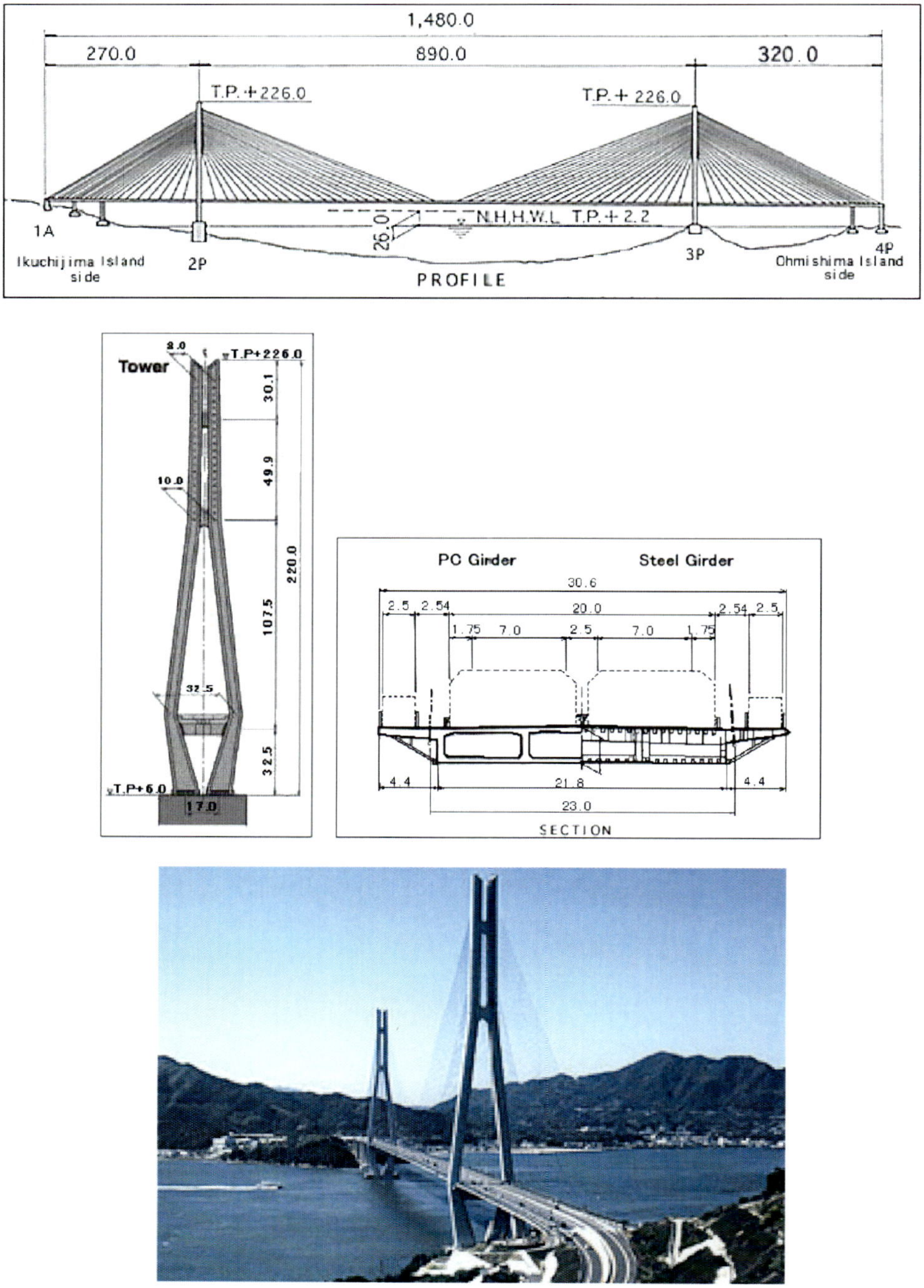

Figure 2.4.8: Tatara Bridge.

over-robust appearance because its section is exactly the same as that of the other suspension bridges along the route, although their spans are longer. Nevertheless, at the time it was built it seemed to break the mould alongside other Japanese suspension bridges. Its towers are portal-like with horizontal inner beams.

Figure 2.4.9: Tatara Bridge. Reduced scale model testing of the whole bridge.

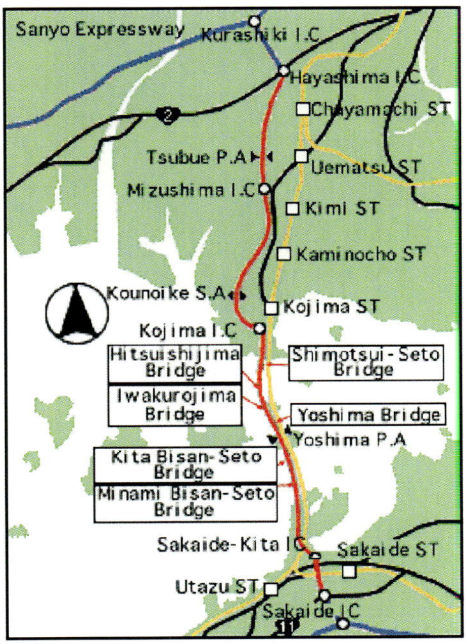

Figure 2.4.10: Central route linking Honshu and Shikoku.

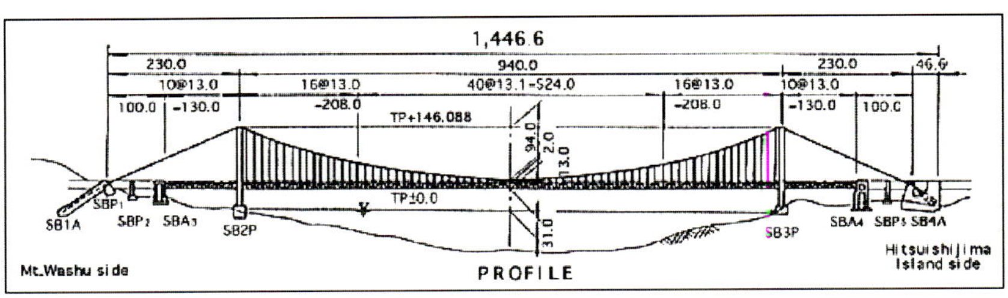

Figure 2.4.11: Shimotsui-Seto Bridge.

Figure 2.4.11: *Continued.*

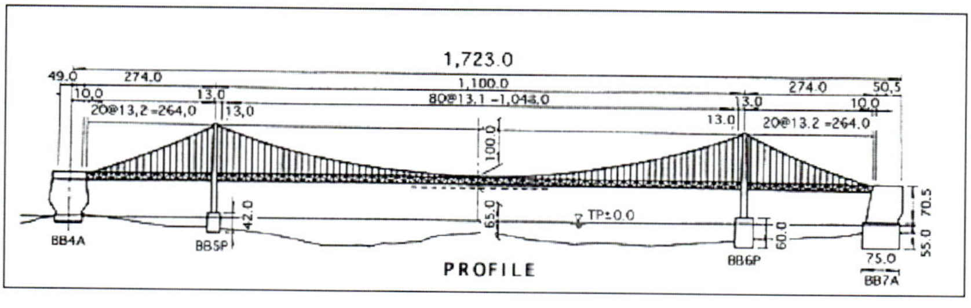

Figure 2.4.12: Kita Bisan-Seto and Minami Bisan-Seto bridges.

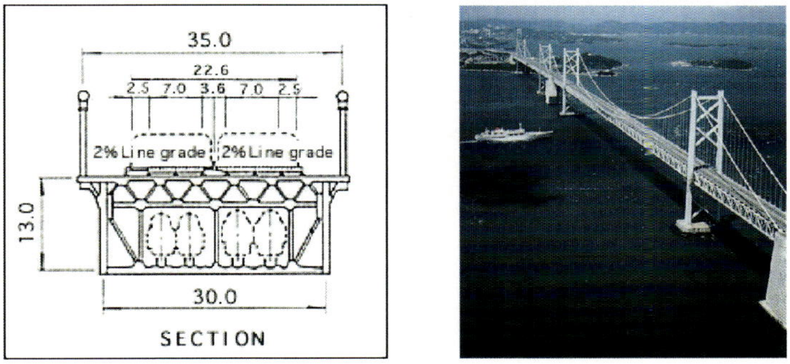

Figure 2.4.12: *Continued.*

The length of the two together is 3170 m, surpassing the 2824 m span of the Bay Bridge between San Francisco and Oakland in the USA. It should be added that the diameter of their main cables is 1.04 m, breaking world records in 1988, when they were built.

As for cable-stayed bridges along this central route, there are the Hitsuishijima and Iwakurojima bridges. They are found in Figure 2.4.13: twin bridges each with a central span of 420 m, but reaching a length of 1580 m in total. Structures entirely made of steel, they have a truss section very much in line with the others preceding them to hold the train line mentioned earlier.

There is one last route along the Honshu and Sikoku network, to the east, between Kobe and Naruto. In Figure 2.4.14 as shown, there are two suspension bridges: the Ohnaruto, and the one which today holds the world record or “gold medal” for bridges of its class, the one over the Akashi Strait.

Connecting the islands of Shikoku and Awaji to the south is the Ohnaruto Bridge, shown in Figure 2.4.15. It is a suspension bridge whose central span is 876 m. It has the typically American design often favoured by the Japanese, with a truss deck, towers with cross-tied inner beams and a deck depth that seems excessive for the combined working overload it must bear, with road traffic and a moderate central span of 876 m. The bridge opened in 1985, yet, previously, other suspension bridges with greater slenderness had already been constructed along the network. The explanation for this apparent step backwards can be found in the original planning for the bridge. At first, the lower part of the deck was to hold two lines for the high-speed train, or Shinkansen. However, in the mid-1980s, the Japanese government decided that the Akashi Straits Bridge project would be limited to road traffic. Only by building a tunnel between Shikoku and Awaji Islands would it be possible to run the train line along the Ohnaruto Bridge and, to this date, no plans have been made to carry out such a project.

The final eastern link was finished in 1998. As mentioned, the Akashi Strait Bridge still holds the world record for those of its class and is the masterpiece of Japanese bridge engineering. It has a central span of 1991 m, and two lateral ones measuring 960 m. Figure 2.4.16 reveals its conventional design, with a truss deck and towers whose inner beams are cross-tied. However, these features alone would not allow for its length; one of the reasons why the deck has a closed truss design is that, with the extreme length of its lateral spans, its cable system lacks rigidity under lateral deformation. Numerous studies about the bridge have been carried out: static, dynamic and aeroelastic. The last factor on the list has been extensively examined by Katsuchi *et al.*[15] at John Hopkins University. By applying the multimodal theory, Katsuchi analysed the wind-induced

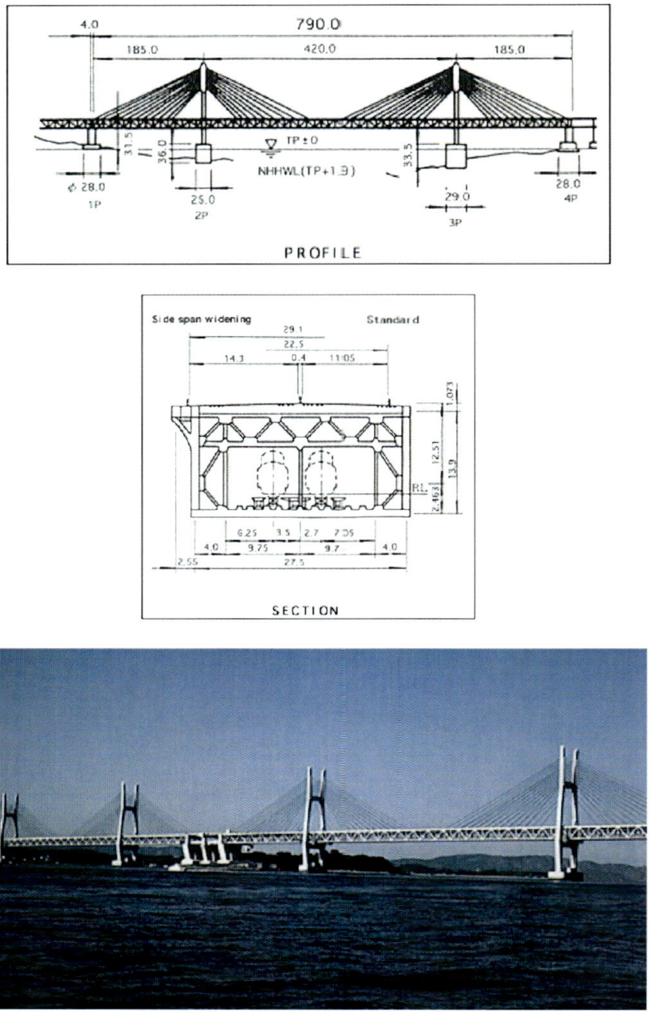

Figure 2.4.13: Hitsuishijima and Iwakurojima bridges.

aeroelastic instability. Other studies were carried out to test the bridge's behaviour, taking into account the unfavourable seismic conditions found in that region. It is curious to note that bridge construction was well underway, with the towers already raised and the main cables in place, when the Kobe earthquake occurred in 1995. The structure perfectly withstood the loads, although the towers separated by 70 cm, providing one with an idea of the impact such seismic activity can have in that zone.

2.4.2 European bridges

On the European continent, a truly elegant suspension bridge is the Hoga Küsten, completed in 1997. This Swedish bridge has a central span of 1210 m. As shown in Figure 2.4.17, it is exemplary on two fronts: its aerodynamic box girder design and of the way it harmonizes with its surroundings. Worthy of mention is the detailed analysis undertaken to verify

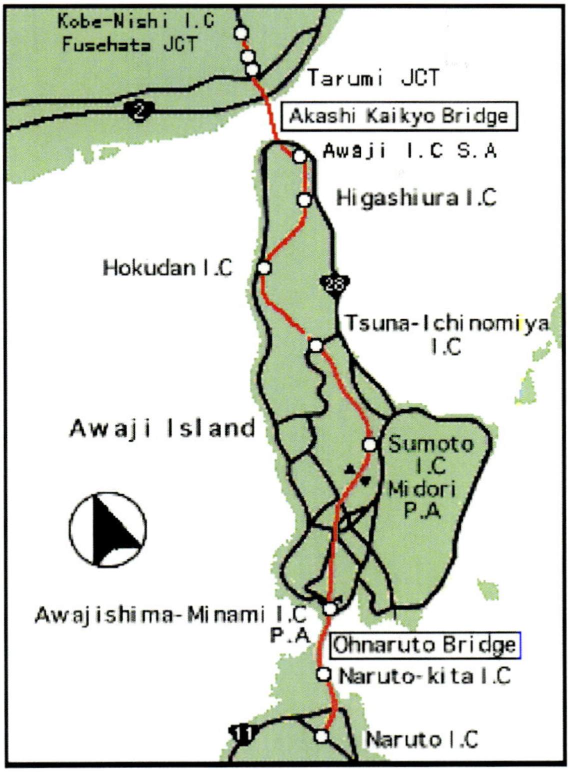

Figure 2.4.14: Eastern route connecting Honshu and Shikoku.

its aerodynamic stability during the building stage, given that it was to be situated in an area with strong winds. Tanaka[16], Cobo[17] or Ge and Tanaka[18] refer to these conditions, which influenced the bridge's construction.

However, it was in the following year, 1998, when European bridge engineering would reach new heights with the completion of what is now the "silver medal" holder for suspension bridge central span length: the Great Belt Bridge in Denmark.

Approximately 14 km long, the Great Belt Strait lies between the Funen and Zealand islands in Denmark. In the middle of this strait is the little island of Sprogø, shown in Figure 2.4.18. The solution chosen to establish this link was to use filler to make Sprogø up to four times bigger than its original size and then carry out the following work:

- Western sector: A bridge between Funen and Sprogø for both road and rail traffic.
- Eastern sector: A tunnel for the railway and a bridge for road traffic, both between Sprogø and Zealand islands.

The end product accommodates both the intense sea traffic passing through the strait and harsh weather conditions, which include severe winds and even blocks of ice that could crash against the bridge.

The most striking element within this network is the suspension bridge in the eastern sector. Graphical information about it is available in Figure 2.4.20a, b. Its central span is

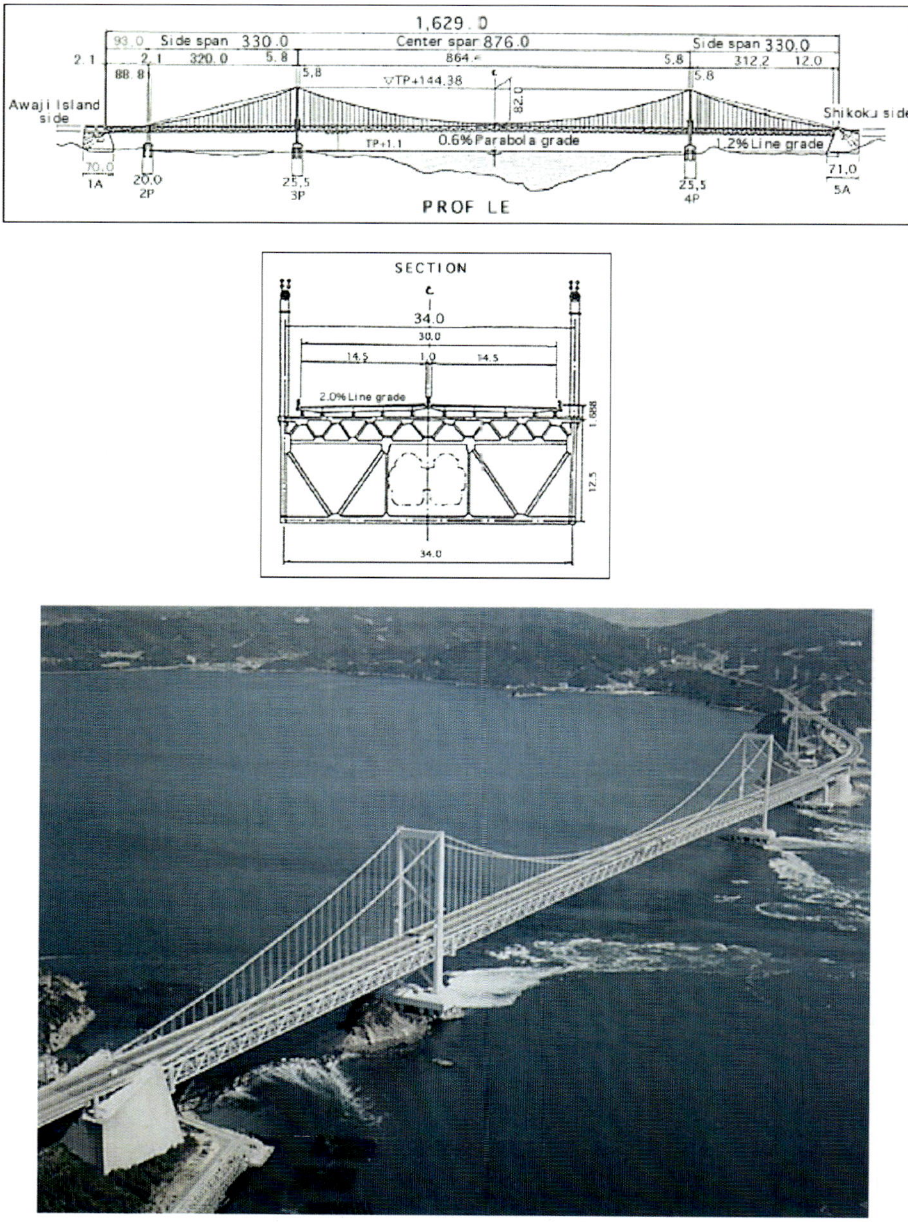

Figure 2.4.15: Ohnaruto Bridge.

1624 m: that is, 15% longer than the Humber Bridge. Moreover, with the aerodynamic box girder design for the deck, it is unique among its class. The lateral spans are 535 m, making a total length of 2700 m. Jurado[11] feels that it is outstanding in how a sense of continuity has been given to the deck between the central and lateral spans: the deck is not supported by the towers. The central part of the main cables is attached to the deck

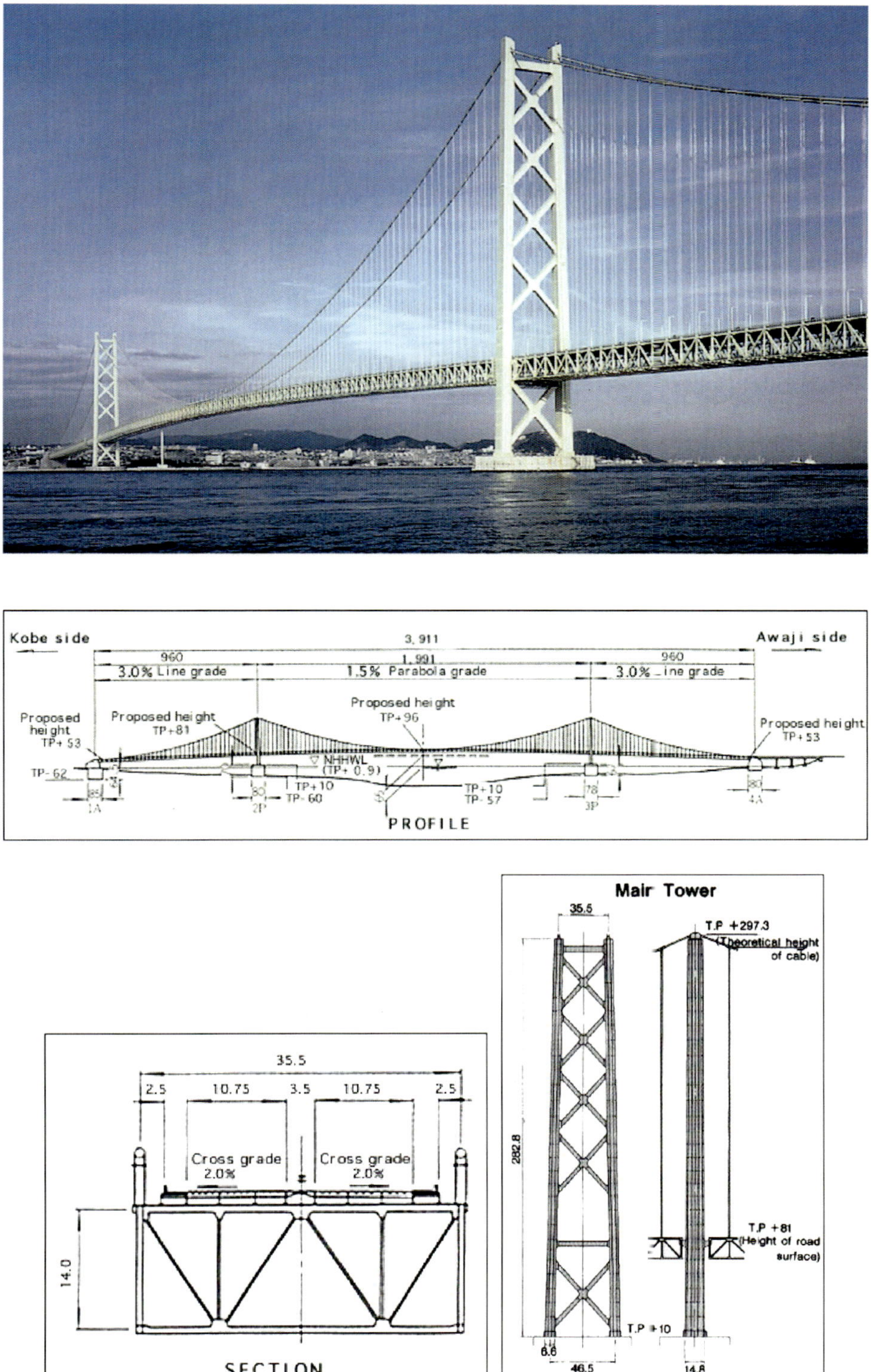

Figure 2.4.16: General view elevation and cross sections of the Akashi Bridge.

Figure 2.4.17: Location and picture of the Hoga Küsten Bridge.

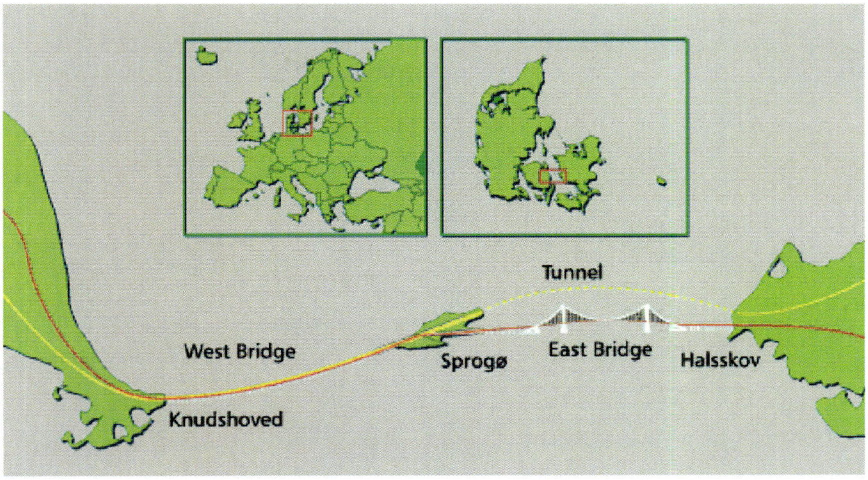

Figure 2.4.18: Location and view of the Great Belt Bridge.

by means of high rigidity elements, shown in Figure 2.4.19. Hydraulic stabilizers were placed in the supports to improve damping along the bridge. The 254-m tall towers have only two inner cross beams, both above the deck, as shown in Figure 2.4.20a. An interesting side is that it is, at present, the highest point in all of Denmark.

Another great European construction is the Øresund Bridge. Finished in 2000, shown in Figure 2.4.21, this new structure establishes a permanent link for road and rail traffic between Denmark and Sweden. The Øresund had been a physical and psychological obstacle for years, but now it is used by three million people as a connecting route.

The link consists of a submerged tunnel 4050 m long that breaks to the surface at the artificial island of Peberholm and a 7845-m long bridge. Its deck was fabricated in Spain. The central part of the cable-stayed bridge has a span of 490 m, while the side spans are 160 m, and the length of the other spans is about 140 m. The main cable-stayed bridge, built to accommodate train lines, holds the world record for main span length within cable-stayed railway bridges of this kind.

The modern European paragon of cable-stayed bridge engineering is the Normandy Bridge, completed in 1994, at the mouth of the Seine in France. This magnificent structure, whose main span is 856 m, currently holds the "bronze medal" among cable-stayed bridges, after the previously mentioned Tatara and the Sutong bridges in China. As for its design, as seen in Figure 2.4.23, the spans at each end measure 43.5 m followed by a 96 m stretch to the towers. Its deck is an aerodynamic concrete box girder that rests on the side spans and across 116 m of the main one. The remaining 624 m at the centre are made of steel in an effort to make the bridge lighter. An unusual feature is the fact that the deck is built into the towers. Thus, temperature variations produce axial and bending forces in the towers. The project engineer of this bridge, Virlogeux[19], has published interesting articles about its structural calculations.

2.4.3 Bridges in China: networks in Hong Kong

Over the past decade, Hong Kong has been one of the most dynamic areas in terms of building transportation networks that include long-span bridges supported by cables.

Figure 2.4.19: Central connection between the main cables and deck in the suspension bridge over the Great Belt Bridge.

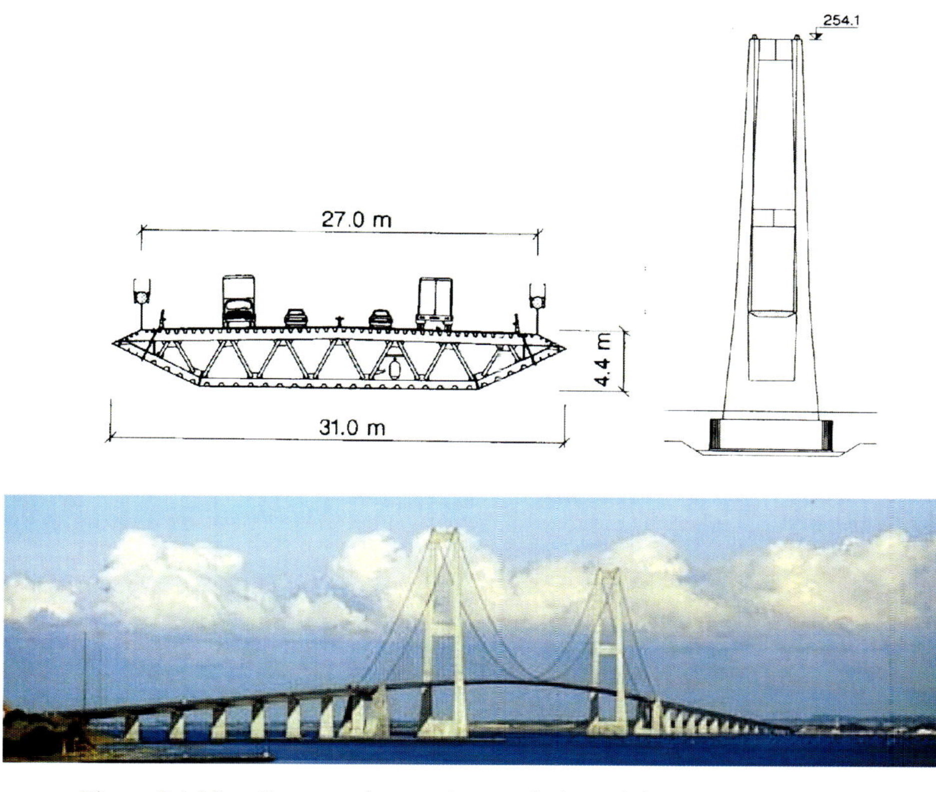

Figure 2.4.20a: Cross sections and general view of the Great Belt Bridge.

Figure 2.4.20b: Suspension bridge over Great Belt during building.

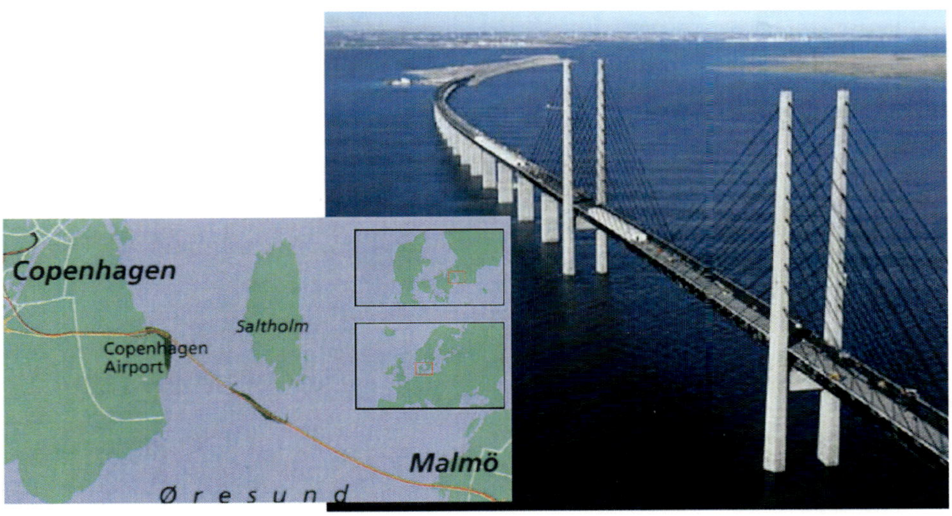

Figure 2.4.21: Øresund Bridge.

Among these ambitious routes, worthy of attention is the one that links the islands of Lantau and Tsing Yi with the mainland; the most outstanding structures are the Tsing Ma, Ting Kau and Kap Shui Mun bridges. Figure 2.4.24 indicates where these are all located.

The Ting Kau Bridge was built for road traffic and opened in 1998. As shown in Figure 2.4.25, this cable-stayed bridge is made up of three, columnar towers that are joined to the deck by means of four sets of cables. The four spans are 127, 448, 475 and 127 m, respectively, reaching a total of 1177 m. Its deck consists of steel grillages, on

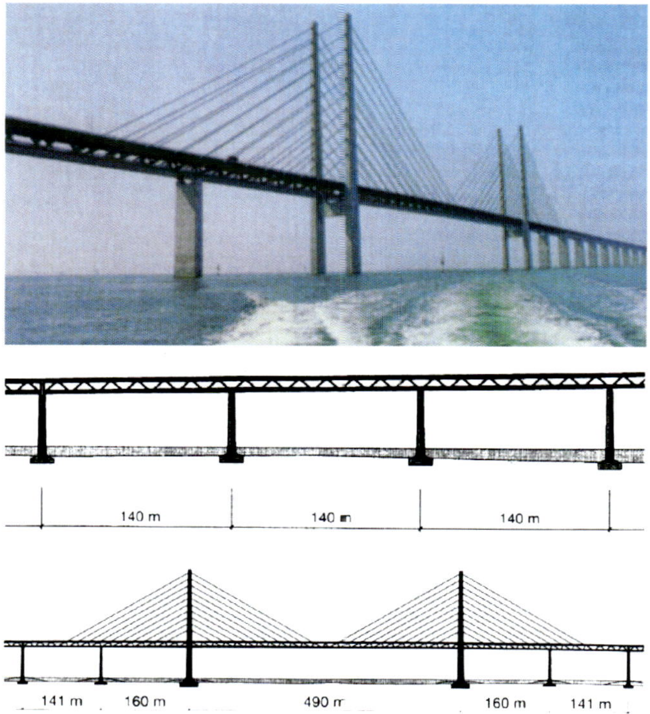

Figure 2.4.22: Main cable-stayed bridge on the Øresund.

Figure 2.4.23: Normandy Bridge.

which slabs of precast concrete rest. Each of these slabs is only 18.77 m wide and have a depth of 1.75 m. Studies on this bridge's aeroelastic behaviour were run in a wind tunnel at the University of Western Ontario in Canada, where a complete 1/180 scale model was tested.

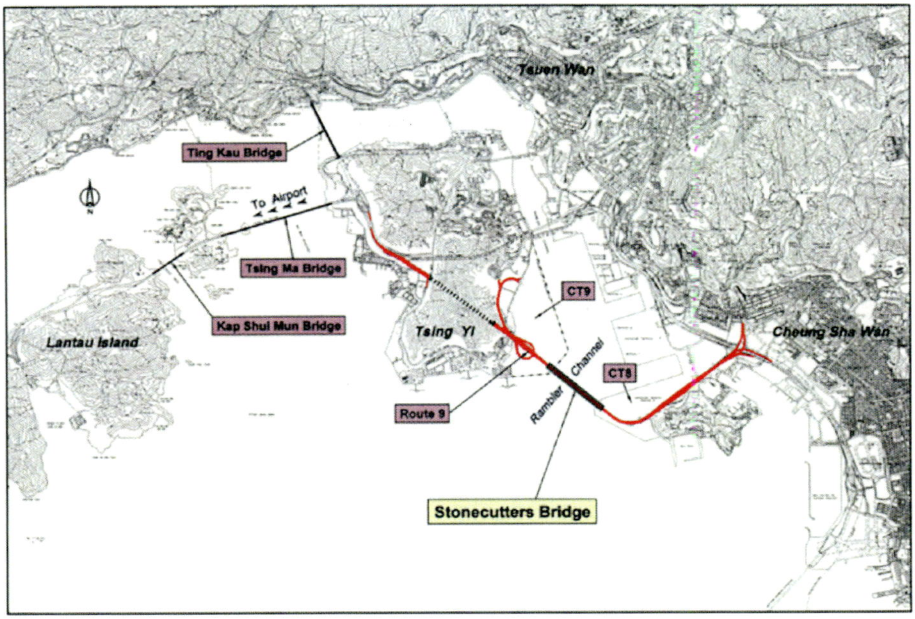

Figure 2.4.24: New communication network in Hong Kong.

Figure 2.4.25: Ting Kau Bridge.

The previous year, 1998, saw the completion of the Tsing Ma and Kap Shui Mun bridges, pictures of which have been provided in Figures 2.4.26–2.4.28. Both cater to roadways and railway lines. The Tsing Ma is a suspension bridge with a main span of 1377 m. Its style is essentially European, with an aerodynamic box girder deck, found in Figure 2.4.27, and portal-like towers that have horizontal inner beams.

Currently this bridge boasts of the world record for central span length among the bridges that support both road and rail traffic. As for the Kap Shui Mun Bridge, this is a more conventional cable-stayed bridge. Its central span is 430 m.

Figure 2.4.26: Tsing Ma Bridge.

Figure 2.4.27: Deck of Tsing Ma Bridge.

In 1999 the Jiangyin Bridge over the Yangtze River in China was completed, demonstrating how far this country had gone in terms of technological development. With a central span of 1385 m, it currently stands as the fifth longest suspension bridge after the Humber. Figure 2.4.29 shows how it has recently been given a European appearance with its aerodynamic box girder deck and concrete portal-like towers.

Figure 2.4.28: Kap Shui Mun Bridge.

Figure 2.4.29: Jiangyin Bridge.

2.5 The 21st century: achievements and projects

Over the past decade, strides in bridge engineering, particularly with design, aeroelastic studies and construction procedures, have meant that projects could be carried in places previously deemed untenable. Redefining countries and continental geography are the links, already mentioned here, that have been forged in the Japanese archipelago, between the Danish islands, or between Denmark and Sweden. Bridges whose spans nearly stretch to 2000 m have already been raised and other projects are underway. In this period of optimism, no limit seems insurmountable.

The next section will describe several ambitious projects for links that will most certainly become reality in the near future. Further details can be found in Hernández[20]:

- Stonecutters Bridge in Hong Kong
- Bridge over the Gulf of Corinth, linking Rion and Antirion
- Sutong Bridge in China
- Xihoumen Bridge in China
- Bridge over Strait of Messina
- Bridge over the Fehmarn Strait
- Links between other islands in Japan
- Bridge over Chacao Channel
- The Rias Altas link in Spain

2.5.1 Stonecutters Bridge in Hong Kong

Recently the Department of Motorways within the Government of the Special Administrative Unit of Hong Kong opened a call for bids to construct Stonecutters Bridge. This bridge – the real star among other structures – forms part of the ambitious network linking the Islands of Lantau and Tsing Yi with Kowloon, shown in Figure 2.4.24. The contract for this international project was awarded to a consortium led by the Halcrow Group Limited, which had presented a design for a bridge with a main span of 1018 m, two 290-m high columnar towers and a steel deck made up of two independent aerodynamic box girders joined by crossbeams, as shown in Figure 2.5.1.

Figure 2.5.1: Digital image of Stonecutters Bridge.

Obviously, once completed, this bridge will become the second-longest span for cable-stayed bridges. Thus, in the early years of the 21st century, this project is meeting, along with that of the Sutong Bridge, the objective of building a bridge of this class with a span superior to 1 km.

2.5.2 Bridge over the Gulf of Corinth, linking Rion and Antirion

The bridge between Rion and Antirion in Greece, shown in Figures 2.5.2 and 2.5.3 will be of great historical symbolism and, at the same time, have an impact on the economy of that region. It would form the second route connecting mainland Greece with the Peloponnesian Peninsula and foster development in both regions.

The bridge's design, as seen in Figure 2.5.4, envisages five cable-stayed spans: three side ones of 560 m each and two at each end, each measuring 42 m. With an additional nine spans over Rion and six over Antirion, the total length would be 2920 m. The cross section of the cable-stayed zone is made up of two lateral solid-web girders and cross-beams, 4 m each. The edge depth of the section is 2.5 m, while the platform width is 25 m.

The conditions for laying the foundation are particularly precarious due to the poor mechanical properties of the first 20 m of terrain and the zone's seismic risk. For this reason,

Figure 2.5.2: Geographical location.

Figure 2.5.3: Location map.

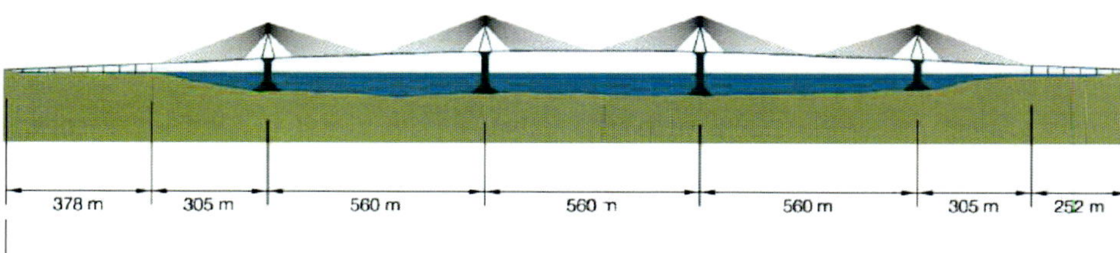

Figure 2.5.4: Side elevation.

Figure 2.5.5: Plan for foundation.

cylindrical foundations, 90 m in diameter, were needed; under each of these, about 400 piles had to be placed, which can be observed in Figure 2.5.5.

Figures 2.5.6 and 2.5.7 provide an image of the whole bridge, as well as its construction phase.

2.5.3 Sutong Bridge in China

The Sutong Bridge is nowadays the "gold medal" in cable-stayed bridges as it has surpassed the length of the main span of the Tatara Bridge with its 1088 m in 2008. The length of the side spans is 300 m while the deck is an aerodynamic cross section. The bridge spans the Changjiang River, which is a tributary of the Yangtze River, in the Chinese province of Jiangsu.

Figure 2.5.6: Image of the Rion-Antirion Bridge.

Figure 2.5.7: Construction of the Rion-Antirion Bridge.

Another figure that gives an idea of the importance of this project is the height of the bridge towers. They are 306 m high, therefore they are the tallest in the world. In Figure 2.5.8, an image of the bridge is shown

The completion of this bridge has allowed the improvement of the communications between Shanghai and Nantong, making unnecessary the former ferry route. A key characteristic of this bridge, which makes clear the level reached by the Chinese civil engineering, is the fact that the construction contractor has been a Chinese company. The role played by international companies has been constrained to planning and advising activities for the project team.

2.5.4 Xihoumen Bridge in China

The Xihoumen Bridge has been constructed in China and it holds the second place in the list of the longest suspension bridges, surpassing the Great Belt Bridge by 26 m to reach a main span of 1650. It is the main bridge in a series of five bridges located in the Zhoushan Archipielago in the Chinese Zhejiang Province. It was opened to traffic on December 2009 (Figure 2.5.9).

This suspension bridge has a central-slotted girder with a gap between boxes. The aerodynamic performance of this bridge has been carefully studied by Yang and Ge[21]. In fact, it has been found that the slot width has the most significant influence in the bridge behaviour under wind effects. Moreover, since heaving and torsional vortex-induced vibrations were found at low wind speeds, vibration control measures have been analysed by means of wind-tunnel tests.

2.5.5 Bridge project over the Strait of Messina

The Strait of Messina separates Sicily from mainland Italy. Because there is a considerable distance between the two, and the sea in the strait is particularly deep, constructing a

Figure 2.5.8: Image of the Sutong Bridge during construction.

Figure 2.5.9: Xihoumen Bridge under construction.

bridge always seemed an untenable project. However, the area's geology, with its high seismic risk, eliminated the possibility of building a tunnel as a permanent link between the island and peninsula.

Nevertheless, recent advances in bridge design and construction meant that preliminary studies could be carried out to consider a single span suspension bridge that would do away with Sicily's isolation. Examples include Brancaleoni and Diana[22] or Diana[23].

The proposal that met with approval would allow for both rail and road traffic. It would have a 3300 m span and be capable of withstanding wind speeds of up to 216 km/h and earthquakes at 7.1 on the Richter scale. Thus, it is, up to now, the most ambitious bridge project ever carried out. The following figures (Figures 2.5.10 to 2.5.16) offer various details about the proposal: its location map and links as well as the deck cross section and longitudinal elevation. For the deck, there are plans to include three independent aerodynamic box girder sections connected with a steel girder. All the deck elements will be stiffened with thick edged crossbeams placed at intervals of 30 m.

2.5.6 Fehmarn Strait link project

This is to connect two islands: Lolland in Denmark with Fehmarn in Germany. When complete, this bridge, along with those which already exist in the Øresund, would mean that the Scandinavian Peninsula would be linked with Germany over the Baltic Sea.

Figure 2.5.10: Geographical location of the proposed Messina Bridge.

Figure 2.5.11: Location map.

This link – planned to be 19 km long – will support both rail and road traffic. Any bridge proposal must leave a wide navigation channel open due to the heavy maritime traffic found in the area. The two alternatives that have been considered are provided in the next section.

Figure 2.5.12: Artist's impression of the bridge over the Messina Strait.

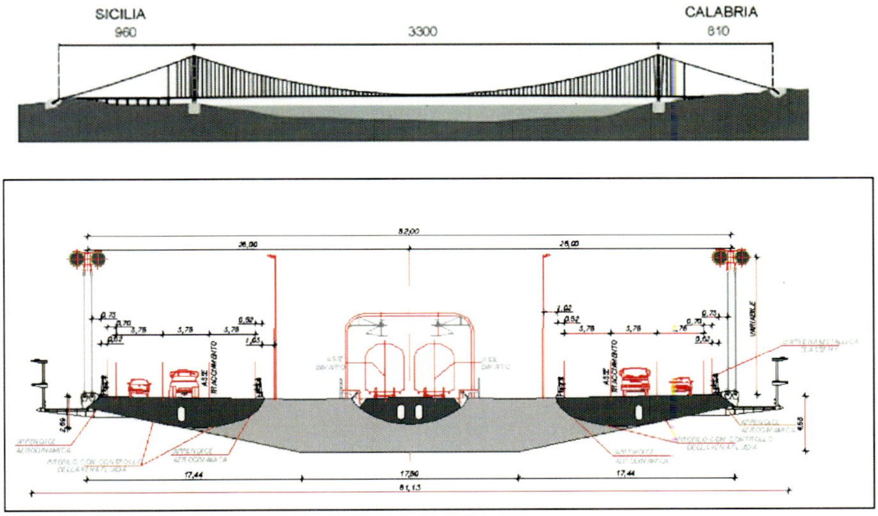

Figure 2.5.13 Longitudinal elevation and deck cross section.

Figure 2.5.14: Geographical location.

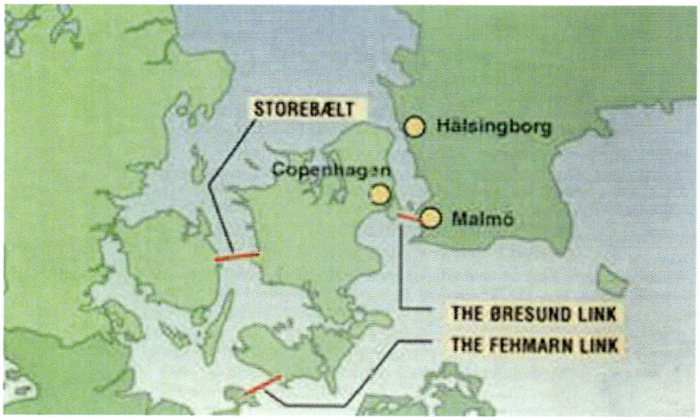

Figure 2.5.15: Detailed geographical location.

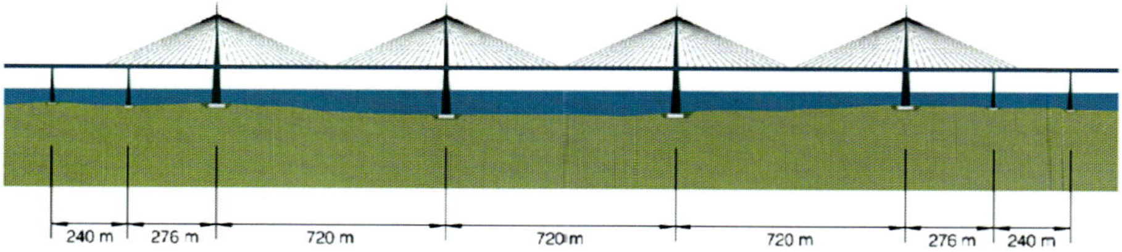

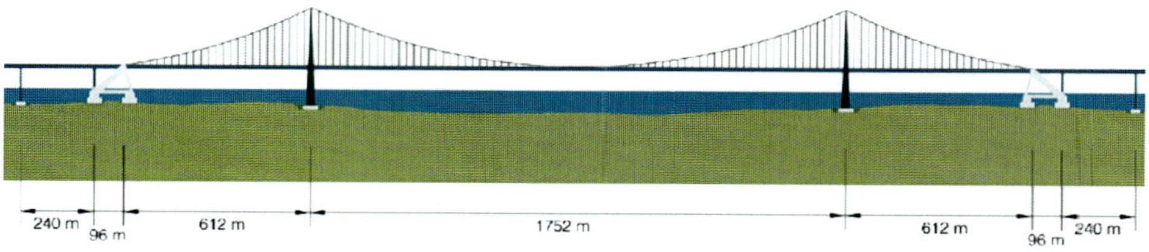

Figure 2.5.16: Longitudinal elevations of the proposals.

The proposal which is more likely to be carried out consists of a series of cable-stayed bridges. Figure 2.5.17 provides a digital image of its model.

2.5.7 Projects to link Japanese islands

The impressive structures raised along the network between Honshu and Shikoku have already been described here. Once these were completed, the Japanese government turned to a new challenge. It now plans to make various connecting routes between its main islands as it can be seen in Figure 2.5.18. See Ito[24]. These projects include

- a bridge at the entrance of Tokyo Bay,
- a bridge in Ise Bay,
- links across Kitan Bay,
- links across the Ho-Yo Strait,
- links across the Tsugaru Strait.

Figure 2.5.17: Digital image of the proposal for a multi-spanned, cable-stayed bridge.

Figure 2.5.18: Geographical location of the various links planned for Japan.

2.5.7.1 Bridge planned for the entrance of Tokyo Bay

This construction project aims to consolidate the various elements of the infrastructure underlying one of Japan's most prosperous areas, in recognition of Tokyo's importance.

Figure 2.5.19: Geographical location of the bridge project for the entrance of Tokyo Bay.

The shortest distance at the entrance of Tokyo Bay is about 7 km. However, as geological conditions and concerns about sea traffic must be taken into account, the trajectory covered by the proposed bridge is 12 km long. It goes from Yokosuka on the Miura Peninsula to Futtsu on the Bo-So Peninsula as it is shown in Figure 2.5.19.

As the seabed in the western half of the route is relatively level and shallow, the plan's main structure is a suspension bridge with three spans that stretch across an international channel for navigation of 1750 m. For this reason, the main span will be 2250 m. Figure 2.5.20 provides a glimpse of this magnificent bridge. The maximum water depth for the towers will be around 45 m, and bedrock is expected to be struck at 10 m below sea level.

2.5.7.2 Ise Bay Bridge project

Nagoya is perched at one end of the Ise Bay, as seen in Figure 2.5.21. At the bay's entrance stretches across about 20 km, between the city of Toba on the west coast and Irako Cape to the east. However, because there are two relatively large islands near Toba and a smaller one, Kami, at the centre, and, as all three are inhabited, the trajectory can be broken up with various structures.

Irako Channel, between the cape with the same name and Kami Island, has a maximum depth of 100 m, and will be used as a channel for navigation. It has been proposed that a suspension bridge with three spans – the main one being 2100 m – be built across this channel.

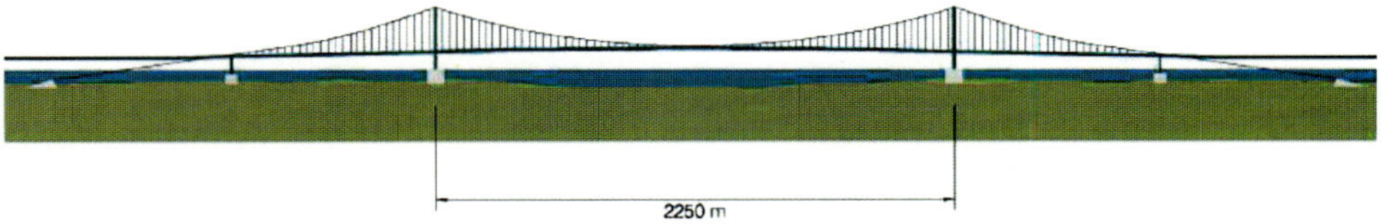

Figure 2.5.20: Longitudinal elevation of the Tokyo Bay proposal.

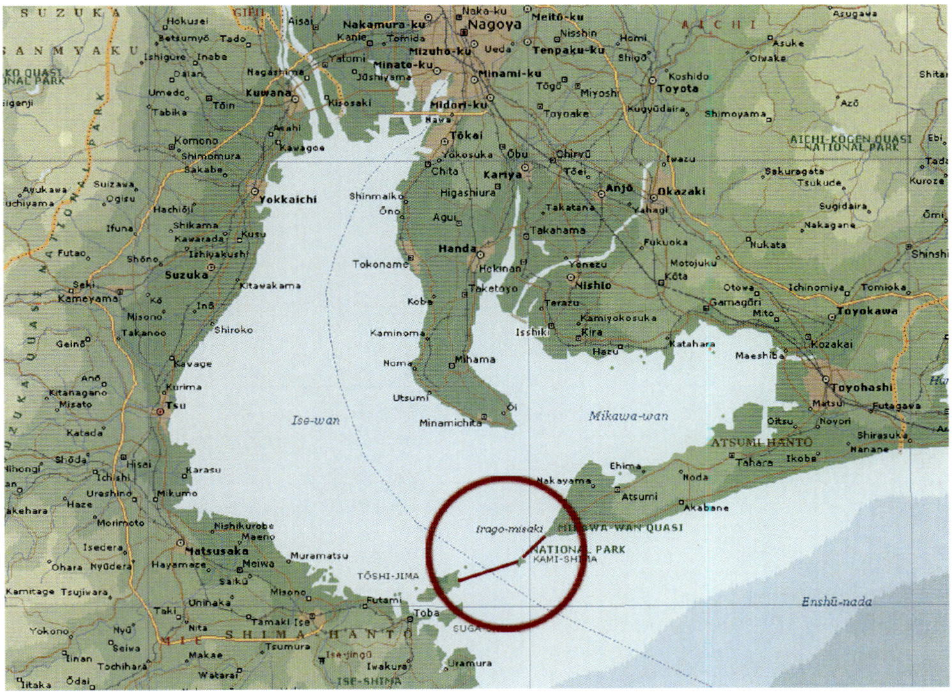

Figure 2.5.21: Geographical location of the Ise Bay Bridge project.

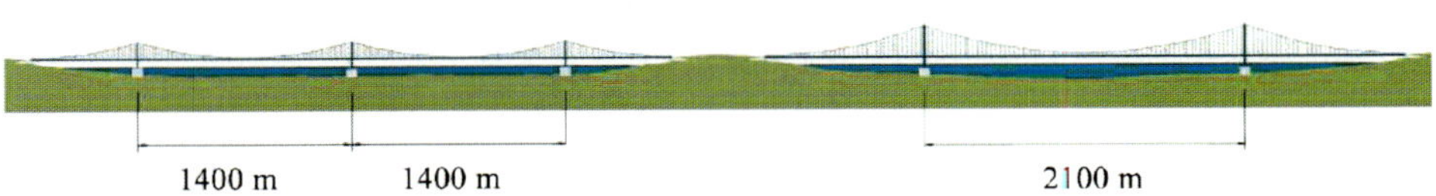

Figure 2.5.22: Longitudinal elevation of the Ise Bay project.

As a route out of Kami to the west, two alternatives were suggested: one from the north, the other from the south. The former option passes through areas with deep waters. Therefore, a suspension bridge, made up of four continuous spans was considered to reduce the number of deep water foundations. Shown in Figure 2.5.22, its two main spans measure 1400 m. The maximum depth reached by the water at foundation level is nearly 60 m.

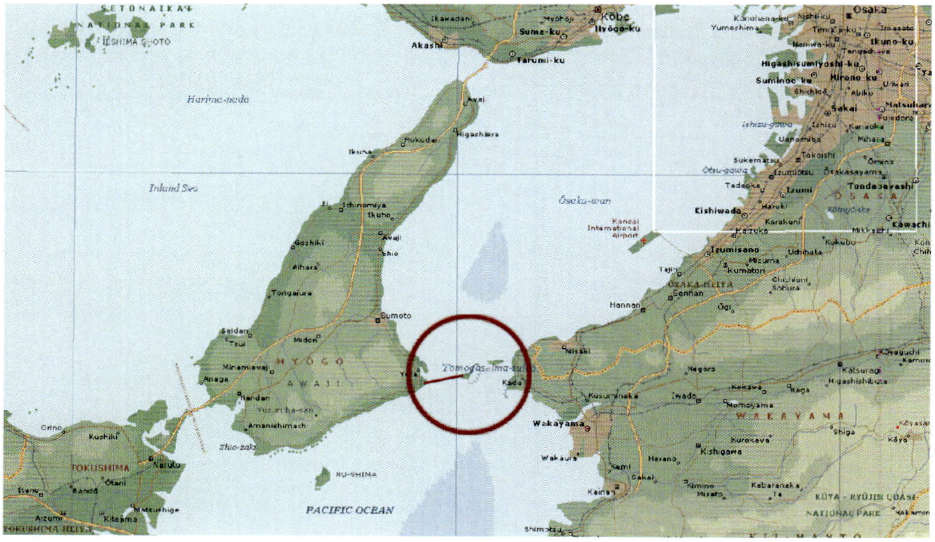

Figure 2.5.23: Geographical location of the Kitan Strait project.

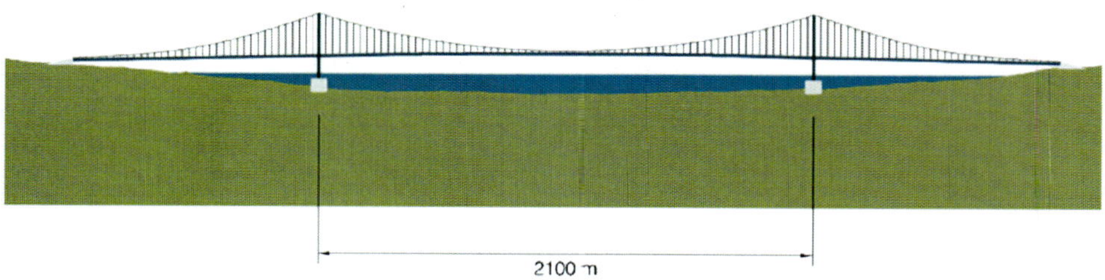

Figure 2.5.24: Longitudinal elevation of the Kitan Strait project.

2.5.7.3 Link over the Kitan Strait

The Kitan Strait separates the two biggest islands in the Japanese archipelago, Awaji and Honshu as it can be seen in Figure 2.5.23. Along with the Akashi Bridge in Osaka Bay, this proposed link would mean a new connection between these islands and Shikoku, completely closing the gap. It would also considerably improve road communication in the Kansai area, Japan's second-largest metropolis.

The distance between Awaji Island and the Kii Peninsula along the strait is around 11 km., but the presence of other islands in between means that only 6 km of open sea has to be crossed.

The widest channel is the Yura Seto, on the western shore of the strait. In this zone, there are plans to build a suspension bridge whose span will be at least 2100 m, with the foundations of the tower pillars reaching a depth between 60 and 70 m (see Figure 2.5.24). At its westernmost point, the network will include two further suspension bridges with spans between 450 and 750 m, respectively.

2.5.7.4 Project for Ho-Yo Strait link

With this project, a direct route will exist between Kyushü and Shikoku, the southernmost of the main islands in the Japanese Archipelago.

Figure 2.5.25 shows the Sada-Misaki Peninsula, in Shikoku, stretching 15 km across the Strait of Ho-Yo in Seki-Zaki, which is part of Kyushü. There is an uninhabited island in between. However, because this island is a protected natural area with a very steep western coastline, it cannot be used as an alighting point within the viaduct system. Thus, the chosen route has to be the most direct one between Sada Cape and Sekizaki.

Two alternatives were considered as elevations; graphical data for these are included in Figures 2.5.26 and 2.5.27. Both entail building the first bridge with a span of about 2000 m, crossing a distance of around 4000 m until reaching an island that serves as a foundation. One of the two proposals envisages another two suspension bridges with spans of 2000 and 2800 m, and the foundations of their towers would have a depth of up to 150 m. The last bridge would require drawing up a second proposal that adds a suspension bridge with two spans, each one measuring 3000 m. This solution would reduce the maximum foundation depth of the towers to 100 m.

2.5.7.5 Project for the Tsugaru Strait link

This project entails crossing the strait that runs between Honshu and Hokkaido islands in Japan, a distance of nearly 20 km.

The plan would be the most ambitious one in existence given that the initial design foresees a series of bridges on a grand scale as it can be appreciated in Figure 2.5.28. At the centre, there would be a suspension bridge with two main spans of 4000 m and lateral spans of 2000 m. At each extreme, two more suspension bridges would be raised; their dimensions would be on par with those of the Akashi, whose world record has been upheld until now. Once this project and the one on the Ho-Yo Strait have been completed, all four of the main islands in Japan will be interconnected by means of bridges.

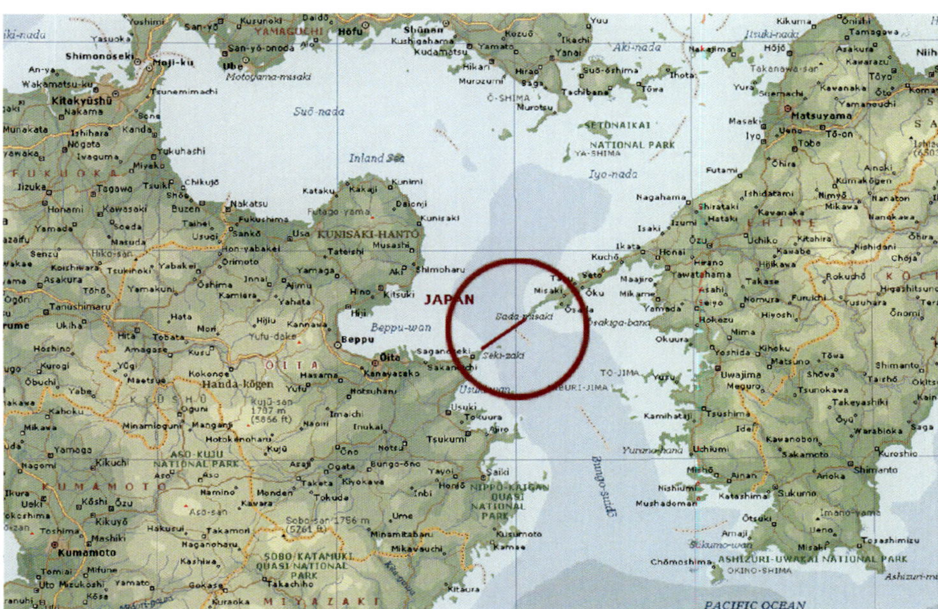

Figure 2.5.25: Geographical location of the Ho-Yo Strait link project.

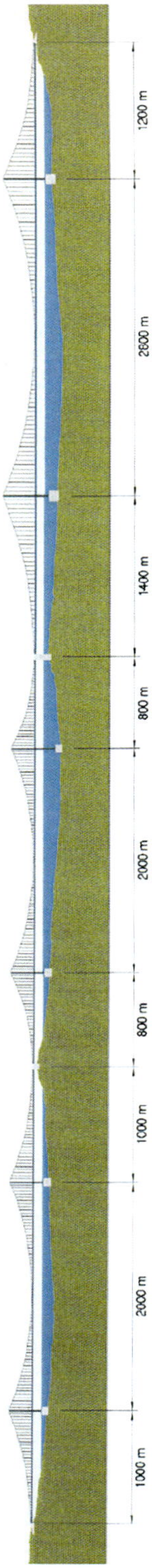

Figure 2.5.26: First proposal for the longitudinal elevation.

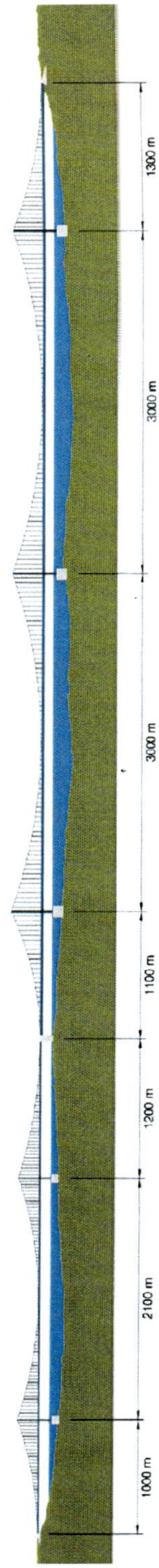

Figure 2.5.27: Second proposal for the longitudinal elevation.

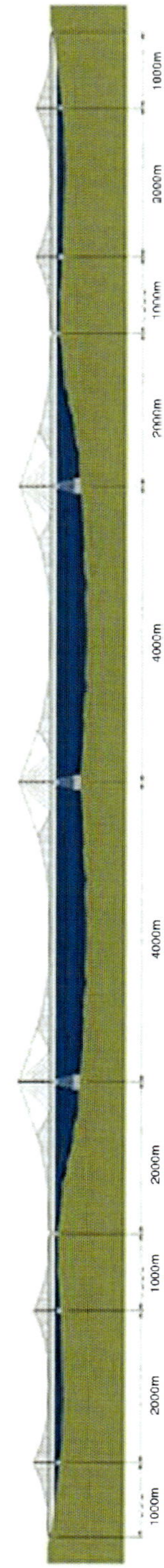

Figure 2.5.28: Longitudinal elevation of the Tsugaru Strait proposal.

Figure 2.5.29: Geographical location of the Tsugaru Strait link project.

2.5.8 Bridge project over the Chacao Channel

Chacao Channel separates Chiloé Island from mainland Chile as it can be seen in Figure 2.5.30. To cross this body of water, one must rely on a ferry line that is part of the roadway Route 5. Chiloé Island's population is around 130,000, but increased traffic along Route 5 has led to overcrowding on the ferry line.

At its narrowest point, this channel has a width of 2.5 km. Its depth is around 100 m, except in its central zone – known as *Roca Remolinos* or Whirlpool Rock – which is only 5 m deep. In this shallow area, there was once a small island that became submerged in the Valdivia earthquake of 1960. In the initial proposals, a range of options were brainstormed, from a suspension bridge with a span of 2200 m to a multi-spanned, cable-stayed bridge. The final proposal put forward by COWI – see Forsberg and Petersen[25] – consists of a multi-spanned suspension bridge with an aerodynamic box girder design. Its two main spans of 1055 and 1100 m would make the most of the shallow area within Roca Remolinos to lay foundations for a large three-dimensional tower, thus providing sufficient rigidity. It also eliminates the need to build a central anchorage section for two twin bridges, a more classical solution already found in structures like the San Francisco Bay or Kurushima Bridges, mentioned earlier. A digital picture of the proposed solution is presented in Figure 2.5.31.

2.5.9 The Rías Altas Link in Spain

Hernández[20, 26] has proposed an ambitious crossing located at the Atlantic coast of Northwestern corner of Spain spanning three straits and connecting the cities of La Coruna and

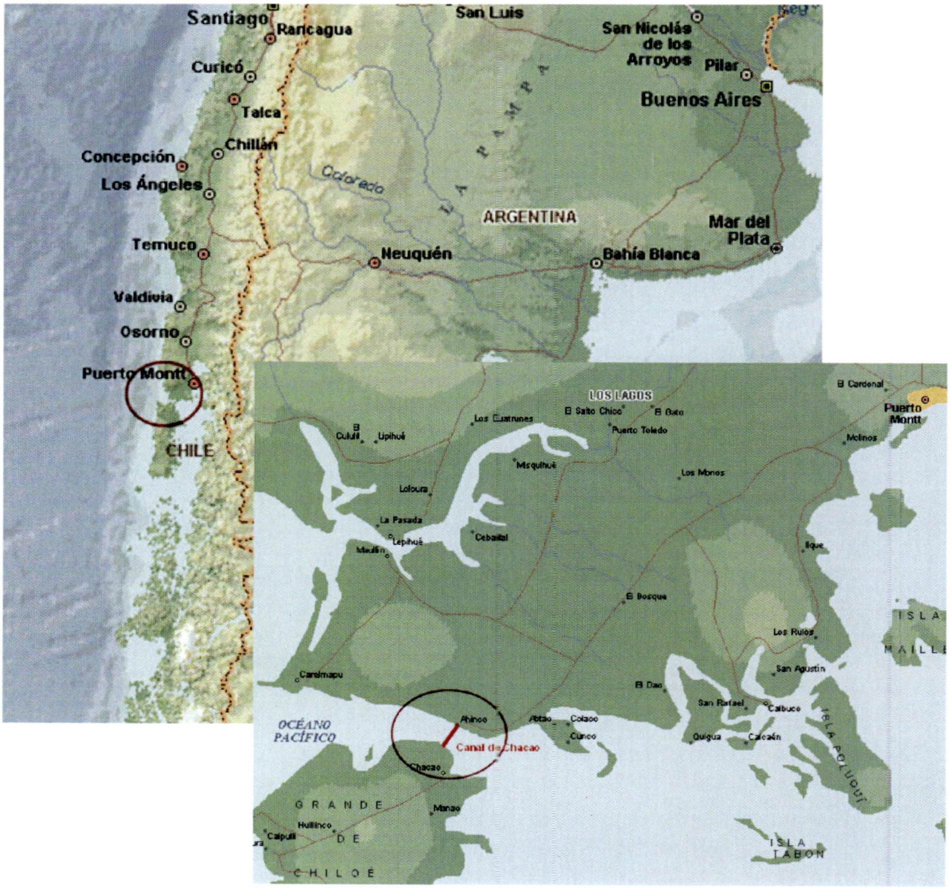

Figure 2.5.30: Geographical location of the Chacao Channel link project.

Figure 2.5.31: Digital image of the solution proposed for the Chacao Channel link.

Ferrol. They are currently communicated by a road along a distance of 62 km and the proposed crossing will decrease the distance to 13.9 km. The link will have to overcome three *rías*, a Spanish word for fjord.

Figures 2.5.32 and 2.5.33 show a map of Spain and a detail of the area where de Rias Altas crossing is projected. In Figure 2.5.34 the layout of the link is presented.

Figure 2.5.32: Map of Spain.

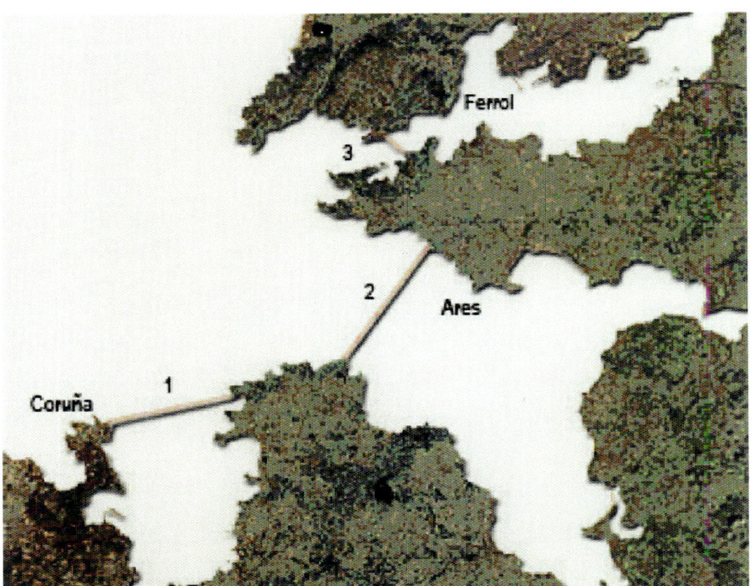

1. La Coruña Bridge
2. Ares Bridge
3. Ferrol Bridge

Figure 2.5.33: A detail of Rías Altas Region.

2.5.9.1 Suspension bridges

Two suspension bridges are included in this crossing. They are the Coruña and the Ares bridges. Both have a very long central span; specifically, main span length of 2016 m for the Coruña bridge and 2198 m for the Ares bridge have been considered.

Tower's height is related to span length and reaches 260 m for the Coruña Bridge and 296 m for the Ares Bridge. The elevations of the bridges are displayed in Figures 2.5.35 and 2.5.36 and data about the towers are shown in Figure 2.5.37.

Regarding the cross section, three different concepts are under preliminary study; namely a three-dimensional truss, a single box aerodynamic shape and a two-box aerodynamic geometry with open granting between boxes have been investigated. Conceptual designs of all of them are included in Figure 2.5.38.

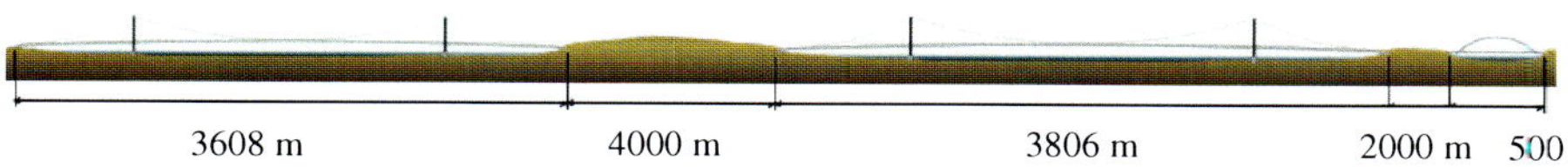

Figure 2.5.34: Rías Altas crossing.

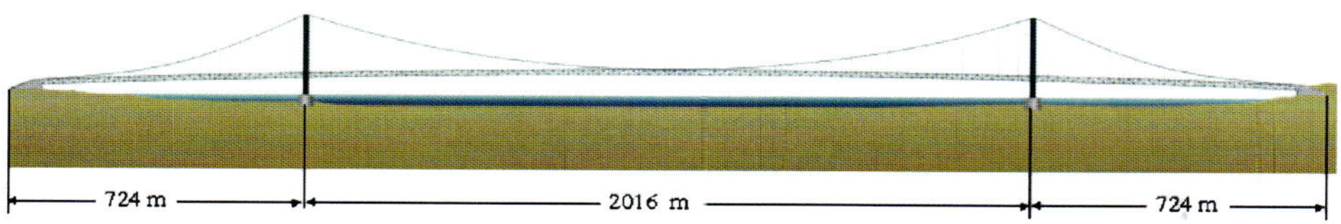

Figure 2.5.35: Elevation of Coruña Bridge.

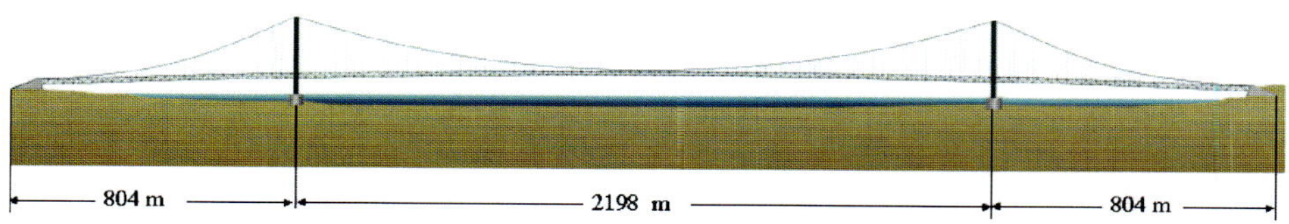

Figure 2.5.36: Elevation of Ares Bridge.

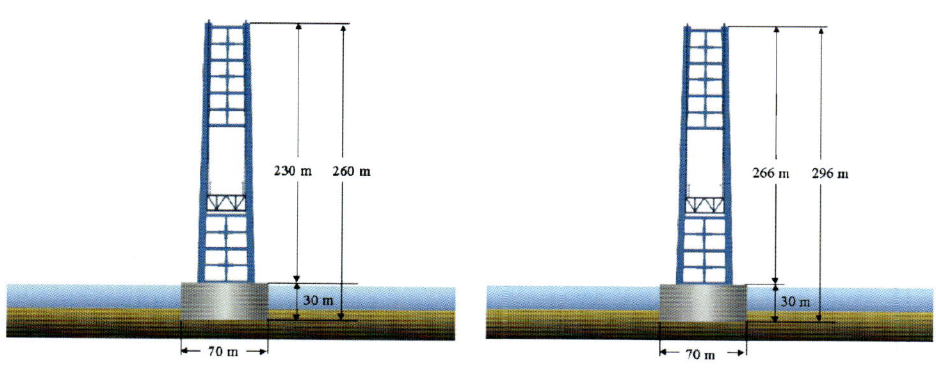

a) Coruña Bridge

b) Ares Bridge

Figure 2.5.37: Towers of suspension bridges. (a) Coruña Bridge. (b) Ares Bridge.

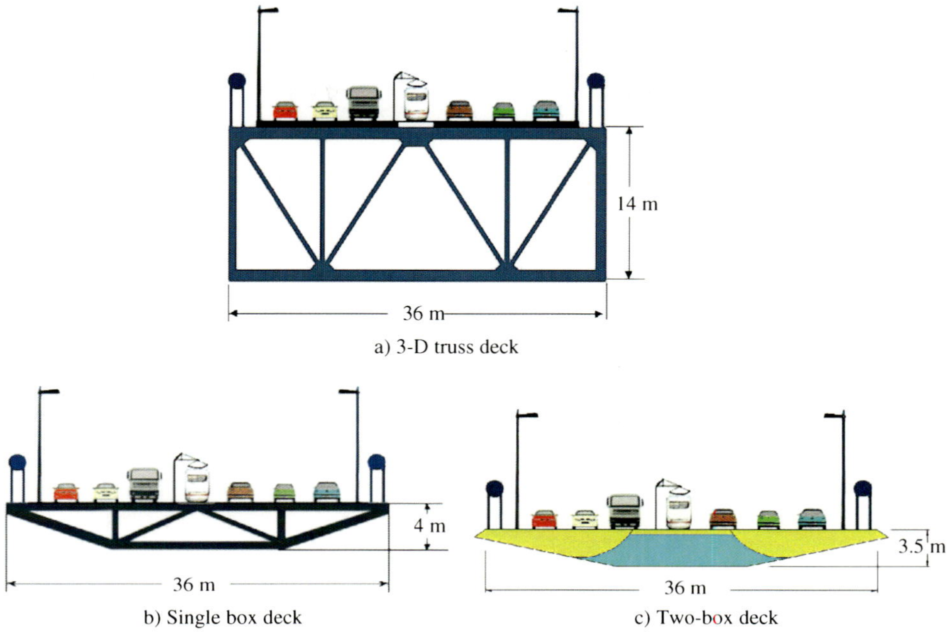

Figure 2.5.38: Cross section of suspension bridges.

Figure 2.5.39: Artist impression of Ares Bridge.

A picture of the location of one of these bridges incorporating a computer-generated image is shown in Figure 2.5.39.

2.5.9.2 Arch bridge

The Ferrol Bridge, corresponding to the northernmost position of the Rias Altas link is an arch bridge with a provisional length of 500 m. Deck is at an intermediate location of the arch and has an aerodynamic cross section aiming to favour efficiency against wind loadings.

Stiffening elements between two arches of the bridge have the same pattern than the stiffening of the towers of the suspension bridges. A computer-generated picture and some technical information of the bridge is presented in following figures (2.5.40 and 2.5.41).

Figure 2.5.40: Computer picture of Ferrol arch Bridge.

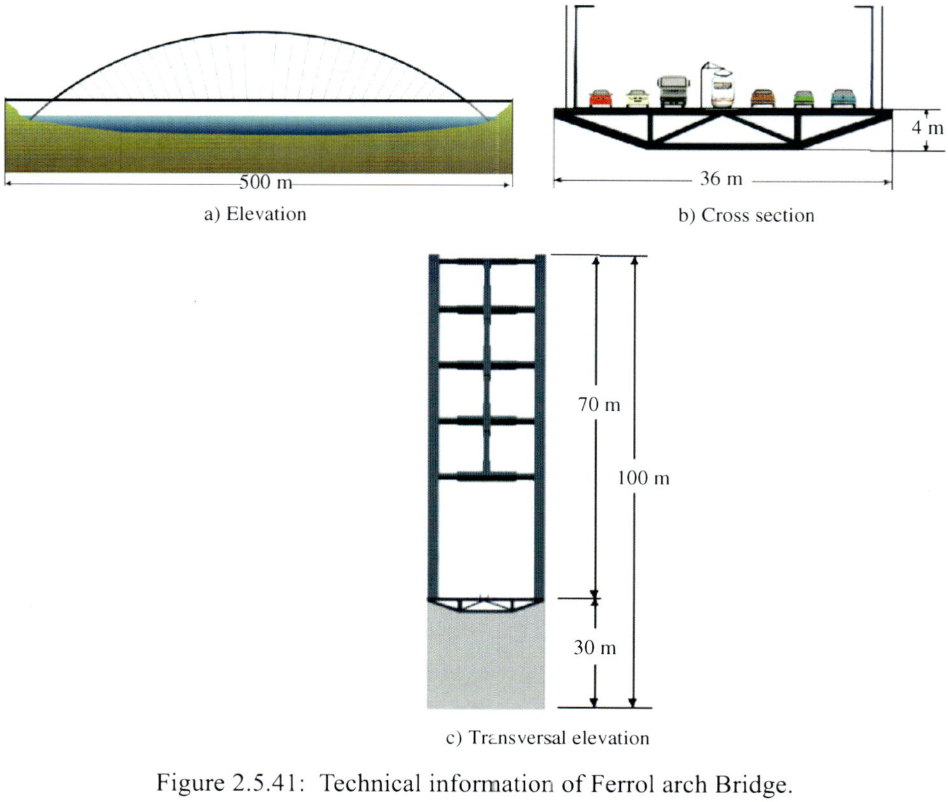

Figure 2.5.41: Technical information of Ferrol arch Bridge.

2.6 References

[1] Buonapane S. G., Billington D. P. [1993] *Theory and history of suspension bridge design from 1823 to 1940*. Journal of Structural Engineering. Vol. 119, No. 3, pp. 954–977.

[2] Gimsing N. J. [1997] *Cable Supported Bridges*. John Wiley & Sons, West Sussex, England.

[3] Steinman D. B., Watson S. R. [1957] *Puentes y sus constructores*. Turner, Spanish Society of Civil Engineers, Madrid, Spain.

[4] Ammann O. H., Von Karman T., Woodruff B. B. [1941] *The Failure of the Tacoma Narrows Bridge*. Report to the Federal Works Agency, 28 March.

[5] Scott, R. [2001] *In the Wake of Tacoma*. ASCE Press, Virginia, USA.

[6] Fung, Y. C. [1955] *An Introduction to the Theory of Aeroelasticity*. Dover, New York.

[7] Faquharson F. B. [1949] *Aerodynamic stability of suspension bridges*. Parts I-V, Bulletin No. 116, University of Washington Engineering Experiment Station, Seattle, 1949–1954.

[8] Scanlan R. H., Tomko J. J. [1971] *Airfoil and bridge deck flutter derivatives*. Journal of Engineering and Mechanical Division, ASCE, Vol. 97, No. EM6, pp. 1717–1737.

[9] Billah K. Y., Scanlan R. H. [1990] *Resonance, Tacoma Narrows bridge failure, and undergraduate physics textbooks*. American Journal of Physics. Vol. 59, No. 2, 1991.

[10] Larsen A. [2000] *Aerodynamics of the Tacoma Narrows Bridge – 60 years later*. Structural Engineering International, Vol. 10, No. 4, pp. 243–248.

[11] Jurado J. A. [2001] *Análisis Aeroelástico y de Sensibilidad del Fenómeno del Flameo en Puentes Soportados por Cables*. Ph.D. Thesis, School of Civil Engineering. University of A Coruña, Spain.

[12] Hernández S. [2001] *Virtual visualization of aeroelastic behaviour of suspension bridges*. IABSE Conference in Cable Supported Bridges. Challenging Technical Limits. Seoul, 12–14 June, 2001.

[13] Landers J. [2002] *Wind fairings will streamline and strengthen New York span*. Civil Engineering. Vol. 72, No. 1, p. 20.

[14] Saeki S., Okakawa A., Ohashi H. [1998] *Honshu-Shikoku Bridges*. Structural Engineering International. Vol. 8, No. 1, pp. 10–15.

[15] Katsuchi H., Jones N., Scanlan R. H. [1997] *Multi-mode flutter analysis of the Akashi Kaikyo Bridge*. Proc. of 8th USNCWE, John Hopkins University, Baltimore, MD, USA.

[16] Tanaka H. [1998] *Aeroelastic stability of suspension bridges during erection*. Structural Engineering International. Vol 8, No. 2, pp. 118–123.

[17] Cobo del Arco D. [1998] *An analysis of wind stability. Improvements to the response of suspension bridges*. Ph.D. Thesis, Technical University of Catalunya.

[18] Ge Y. J., Tanaka H. [2000] *Aerodynamic stability of long-span suspension bridges under erection*. Journal of Structural Engineering. Vol. 126, No.12, pp. 1404–1412.

[19] Virlogeux M. [1992] *Wind design and analysis for the Normandy Bridge*. Aerodynamics of Large Bridges, Balkema, Rotterdam.

[20] Hernández S. [2001]. Thes rías altas. A Challenging crossing. *Strait Crossing 2001*. Jon Krokeborg (ed.), Balkema, Rotterdam, pp. 407–414.

[21] Yang, Y., Ge, Y.J. [2008] *Aerodynamic investigation on a long-span suspension bridge with central-slotted box girder.* 17th IABSE Congress, Creating and

Renewing Urban Structures – Tall Buildings, Bridges and Infrastructure, Chicago, September 17–19.

[22] Brancaleoni F., Diana G. [1993] *The aerodynamic design of the Messina Strait Bridge*. Journal of Wind Engineering and Industrial Aerodynamics. Vol. 48, pp. 395–409.

[23] Diana G. [1995] *Comparisons between wind tunnel tests on a full aeroelastic model for the proposed bridge over Stretto di Messina and numerical results*. Journal of Wind Engineering and Industrial Aerodynamics. Vol. 54/55, pp. 101–113.

[24] Ito, M. [1998] 21st century super long bridges in Japan. *Bridge Aerodynamics*, Larsen & Sdahl (eds.), Balkema, Rotterdam.

[25] Forsberg T., Petersen A. [2001] The challenge of constructing a bridge over the Chacao channel. *Straits Crossings 2001*, Krokeborg (ed.). Balkema, Rotterdam.

[26] Nieto F., Hernández S., Baldomir A., Jurado J. A. [2007] *Designing challenging bridges in northwest Spain: The art of resisting extreme natural forces*. S. Hernández and C. A. Brebbia (ed.), pp 1–10, WIT Press, Southampton, UK.

Chapter 3

Methodologies of flutter analysis for cable-supported bridges

3.1 Introduction

There are two main approaches that facilitate the study of a cable-supported bridge's behaviour under wind loads. *The first methodology includes experimental aeroelasticity methods*, based on wind-tunnel testing with reduced models. On the other hand, *analytical aeroelasticity methods*, which try to solve deck movement equations, evolve around classic aeroelasticity concepts that have already been applied to aeronautical engineering.

It is important to obtain wind speeds at the onset of flutter-induced instability. However, this is just one of many calculations that must be done when designing a bridge to ensure a reasonable response to static or dynamic loads. Most of these analyses use a structural model of finite elements that simulate stiffness characteristics in the bridge, as well as its shape conditions and external loads. Several types of dynamic analyses, including aeroelastic calculation, require that the frequencies and the structure's natural vibration modes are determined beforehand. These can be obtained by either applying a first-order theory or by considering a geometric type of non-linearity, known as second-order theory.

If the flutter study is part of a more general analysis of the bridge's structure, there is a drawback in that none of the terminology used to analyse this phenomenon is similar to the ones usually employed in structural calculations.

For instance, Jones *et al.*[1] and Scanlan[2] denote some of the deck movements with the letters p and h, when these movements are normally allocated the u, v and w as they occur along the coordinates X, Y and Z. Another example of unconventional notation can be seen in Dyrbye and Hansen[3], who refer to the vertical deck movement with the Greek letter ξ. Even today there is probably no common terminology because the analytical study of bridge aeroelasticity has only been developed in recent years. Nor do the existing formulations make clear how the aeroelastic forces alter the stiffness and damping characteristics of the assembled structure. One purpose of this book is to adapt the terminology used for the aeroelastic analysis of bridges to the one used with the most common structural calculations. Another objective is to offer extensive information about dynamic problems for cable-supported bridges. Most publications fall short here. It is assumed that the modes and natural vibration frequencies are known, yet these are essential parameters for aeroelastic analysis and they require complex calculation.

This chapter first describes experimental aeroelasticity methods. It then provides a thorough explanation of the principles of analytical aeroelasticity. Furthermore, it sets out the equations determining the deck's aeroelastic behaviour and applies modal analysis to obtain the structural response. Finally, it describes how to obtain the critical parameters associated with the onset of flutter-induced instability, critical wind speed and vibration frequency.

Figure 3.2.3: Complete model of the Maysville Bridge, tested by Rowan Williams Davies & Irwin, Inc. in Ontario (Canada).

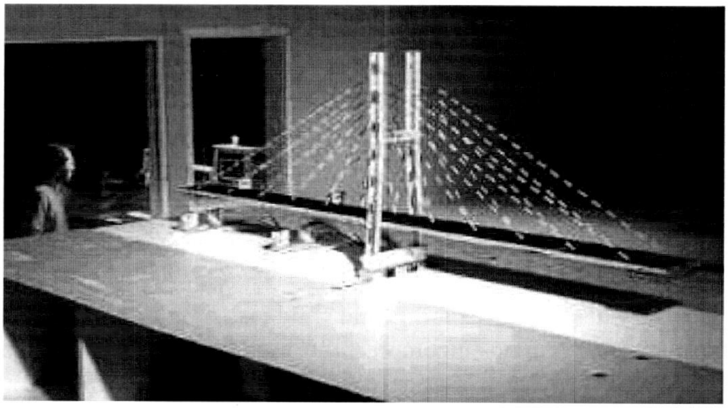

Figure 3.2.4: Model of the Maysville Bridge in the construction phase, as tested by Rowan Williams Davies & Irwin, Inc. in Ontario (Canada).

the case with cable-stayed bridges in the construction phase due to the balanced cantilever system. There could be an asymmetrical response. The towers could withstand loads superior to those experienced at the completed stage, when the side spans are connected to the anchor piers. Gamble and Irwin[7] have looked at this awkward situation, to which cable-stayed bridges are susceptible.

Nor are suspension bridges without their drawbacks during the construction phase. Decks which have been stiffened to resist torsion will not completely reach their maximum stiffness until they are connected to the tower. Thus, they are more vulnerable to instability, like the one produced by flutter. Partial models are often made to check stability in the construction phase. Figure 3.2.3 shows a complete bridge test and figure 3.2.4 shows a construction phase test.

Other types of testing are carried out with isolated structural elements, such as the cables of cable-stayed bridges, which often experience vibration caused be the dual effect of wind and rain. This test is shown in Figure 3.2.5. Occasionally some of the truss bars used in the deck also show the effects of vibrations, which should also be analysed. Test

Chapter 3

Methodologies of flutter analysis for cable-supported bridges

3.1 Introduction

There are two main approaches that facilitate the study of a cable-supported bridge's behaviour under wind loads. *The first methodology includes experimental aeroelasticity methods*, based on wind-tunnel testing with reduced models. On the other hand, *analytical aeroelasticity methods*, which try to solve deck movement equations, evolve around classic aeroelasticity concepts that have already been applied to aeronautical engineering.

It is important to obtain wind speeds at the onset of flutter-induced instability. However, this is just one of many calculations that must be done when designing a bridge to ensure a reasonable response to static or dynamic loads. Most of these analyses use a structural model of finite elements that simulate stiffness characteristics in the bridge, as well as its shape conditions and external loads. Several types of dynamic analyses, including aeroelastic calculation, require that the frequencies and the structure's natural vibration modes are determined beforehand. These can be obtained by either applying a first-order theory or by considering a geometric type of non-linearity, known as second-order theory.

If the flutter study is part of a more general analysis of the bridge's structure, there is a drawback in that none of the terminology used to analyse this phenomenon is similar to the ones usually employed in structural calculations.

For instance, Jones *et al.*[1] and Scanlan[2] denote some of the deck movements with the letters p and h, when these movements are normally allocated the u, v and w as they occur along the coordinates X, Y and Z. Another example of unconventional notation can be seen in Dyrbye and Hansen[3], who refer to the vertical deck movement with the Greek letter ξ. Even today there is probably no common terminology because the analytical study of bridge aeroelasticity has only been developed in recent years. Nor do the existing formulations make clear how the aeroelastic forces alter the stiffness and damping characteristics of the assembled structure. One purpose of this book is to adapt the terminology used for the aeroelastic analysis of bridges to the one used with the most common structural calculations. Another objective is to offer extensive information about dynamic problems for cable-supported bridges. Most publications fall short here. It is assumed that the modes and natural vibration frequencies are known, yet these are essential parameters for aeroelastic analysis and they require complex calculation.

This chapter first describes experimental aeroelasticity methods. It then provides a thorough explanation of the principles of analytical aeroelasticity. Furthermore, it sets out the equations determining the deck's aeroelastic behaviour and applies modal analysis to obtain the structural response. Finally, it describes how to obtain the critical parameters associated with the onset of flutter-induced instability, critical wind speed and vibration frequency.

3.2 Experimental aeroelasticity in long-span bridges

The field of experimental aeroelasticity studies the methods for carrying out wind-tunnel testing with reduced models of the structures. It attempts to simulate the effects of phenomena which the wind produces on those structures. This section will look at how the various tests already in place have been applied to study the aeroelastic phenomena produced in long-span bridges. It will also describe some of the wind tunnel types used to carry out these tests. A fairly detailed book on wind tunnel experimentation is published by William and Pope[4], although its main focus is on aeronautical engineering.

3.2.1 Applications of wind-tunnel testing on bridge engineering

To check the structural strength of a long-span bridge under wind loads, as well as assure its stability, it is necessary to set a wind speed limit so that the bridge's behaviour can be considered safe. The purpose of studies done by Irwin *et al.*[5] is to establish some of the criteria for flutter phenomenon and vortex-shedding excitation. Criteria will differ for the bridge once it has been completed and while it is still in the construction phase. During the latter stage, for a time much shorter than its working life, it has a distinct structure that is normally much less resistant than the final project. It is also essential to lay down the maximum oscillation amplitudes for the phenomenon of vortex-shedding formation.

Another task is to regulate the wind speed at which traffic is to be restricted or prohibited. Criteria will include turbulence intensity and the integral scale for the location. These criteria will come into play when studying aerodynamic stability and evaluating buffeting loads, caused by gusts of wind. In many cases, the geographical area determines how design criteria are interpreted. An analysis of meteorological data has to be carried out to determine the working wind speeds. For mountainous areas or where hills are close to the bridge, reduced models, on a 1:2000 or 1:5000 scale, are made, based on the topography of the terrain. These help determine the nature of the wind within a given area. Figure 3.2.1 gives an example of this kind of test.

Figure 3.2.1: Reduced model with the orography of the Feroe Islands. Danish Maritime Institute.

It is essential to guarantee that the design criteria are fulfilled for all bridges so that there is a positive response to phenomena such as flutter and vortex shedding. Reduced models in wind tunnels play an important role in obtaining that security.

The physico-mathematical theories explaining aeroelastic phenomena such as flutter shed light on them, but they do so through parameters that can only be calculated through wind-tunnel testing. One of the more widely applied theoretical breakthroughs of recent years is a hybrid theory for the aeroelastic stability of flutter. It is developed in a text by Simiu and Scanlan[6], which will be examined in depth throughout the following sections.

This theory does not attempt to provide a detailed account of the flow around the decks to obtain aerodynamic forces. Based on flutter derivatives, it instead calculates the forces occurring when the deck is moving under wind action. Flutter derivatives have to be considered through experimental tests of sectional models for the deck, as the one shown in Figure 3.2.2. Each deck shape is sufficiently distinct to make it necessary to re-obtain the derivatives because they are sufficiently sensitive to any geometrical change in their cross section. Moreover, the effect of turbulence on vortex shedding and flutter appear to make it difficult to apply theories; it is increasingly necessary to depend on experimental data.

Wind-tunnel tests provide various kinds of data. On the one hand, they enable the bridge's global stability to be studied, preventing phenomena like flutter from happening. On the other, they can also be used to evaluate the wind loads found in various load conditions.

In this last case, the purpose of wind-tunnel testing is to predict wind loads accurately by calculating aerodynamic parameters, such as the lift, drag and moment derivatives for the deck, as well as its variations in the attack angle. In addition, wind loads will be evaluated in other components of the structure, such as the towers or cables. Even vehicle traffic is modelled sometimes.

It is also necessary to study the stability and wind loads for the construction phase. The deck's cross section often remains incomplete during this phase so that there are no wind shields, asphalt surface or railway tracks. The absence of these components means that the dynamic behaviour is different than it would be at the final stage. This is especially

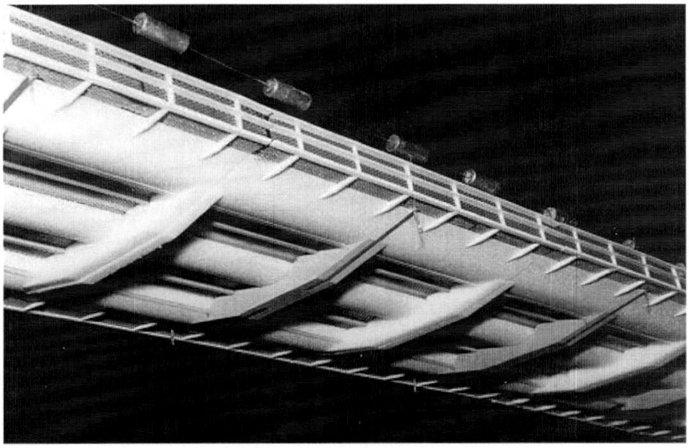

Figure 3.2.2: Cross-sectional model for the bridge planned for the Straits of Messina. Tests by the Danish Maritime Institute.

Figure 3.2.3: Complete model of the Maysville Bridge, tested by Rowan Williams Davies & Irwin, Inc. in Ontario (Canada).

Figure 3.2.4: Model of the Maysville Bridge in the construction phase, as tested by Rowan Williams Davies & Irwin, Inc. in Ontario (Canada).

the case with cable-stayed bridges in the construction phase due to the balanced cantilever system. There could be an asymmetrical response. The towers could withstand loads superior to those experienced at the completed stage, when the side spans are connected to the anchor piers. Gamble and Irwin[7] have looked at this awkward situation, to which cable-stayed bridges are susceptible.

Nor are suspension bridges without their drawbacks during the construction phase. Decks which have been stiffened to resist torsion will not completely reach their maximum stiffness until they are connected to the tower. Thus, they are more vulnerable to instability, like the one produced by flutter. Partial models are often made to check stability in the construction phase. Figure 3.2.3 shows a complete bridge test and figure 3.2.4 shows a construction phase test.

Other types of testing are carried out with isolated structural elements, such as the cables of cable-stayed bridges, which often experience vibration caused be the dual effect of wind and rain. This test is shown in Figure 3.2.5. Occasionally some of the truss bars used in the deck also show the effects of vibrations, which should also be analysed. Test

Figure 3.2.5: Testing the cables of a cable-stayed bridge over the Øresund Straits.

models have shed light on the causes of these vibrations, thus facilitating the development of systems to avoid them. Among the most rigorous studies are those by Saito *et al.*[8] and Yamaguchi and Fujino[9].

A major concern that has to be taken account of when building a bridge in a very windy zone is the effect those winds have on road traffic. For this reason the second bridge over the Severn has 3-m high barriers on either side of the roadway to lessen this impact. Hamphreys and Barker[10], and King *et al.*[11] worked towards solving this problem by making wind-tunnel models.

3.2.2 Types of wind tunnel

For the aeronautical industry, who is pioneer in these kinds of experiments, and professionals from other fields, various wind-tunnel tests are carried out with a range of configurations according to shape, size, power, and flow characteristics. Noteworthy among non-aeronautical applications is the wind-tunnel testing for structural engineering in which buildings, bridges, towers, chimneys and other constructions are tested.

Environmental engineering also makes use of this tool to study problems like the pollution related to harmful gases and soil erosion. It goes without saying that the motor vehicle industry has traditionally relied on wind-tunnel testing to make their motorbikes, cars and lorries more aerodynamic. A further application of wind-tunnel testing is done in the hope of improving the performance of top athletes, such as cyclists, skiers and skaters.

Such variety means that the size of test sections can be anywhere between 50 cm and more than 50 m in width. Turbine power also varies greatly, depending if the installation has open-return, letting in and releasing air on a continuous basis, or closed-return system, in which the air is re-circulated and not renewed. Flow speed, reaching between 5 and 50 m/s, is thus a determining factor in dimensional analysis as it affects the scale of the reduced models. The tunnels used within bridge engineering fall into two very clear classifications.

- This may be according to tunnel purpose:
  - boundary layer wind tunnels
  - aerodynamic wind tunnels
- Or according to air flow:
  - closed-return tunnels.
  - open-return tunnels.

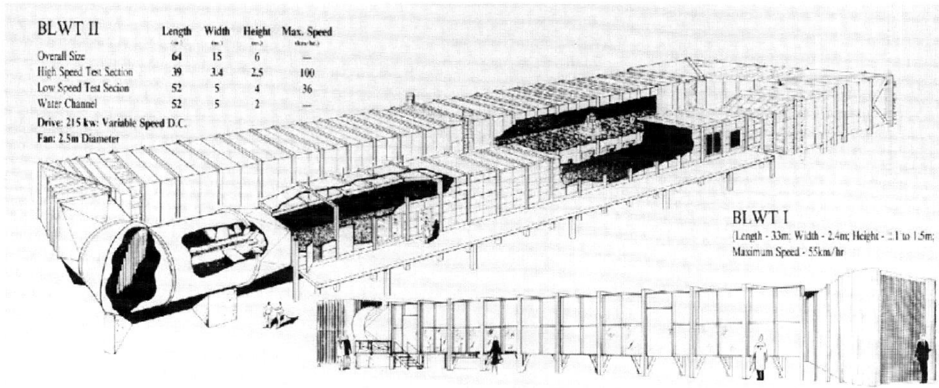

Figure 3.2.6: Wind tunnels at the University of Western Ontario (Canada).

Boundary layer wind tunnels (BLWT) have test areas whose vastness simulates an atmospheric boundary layer so that the three-dimensional effects of the flow wind on the structure can be seen. Here the complete models of bridges on a reduced scale are tested, along with a relief map of the surrounding area on a scale of up to 1/100. The University of Western Ontario in London (Canada), boasts examples of this kind of wind tunnel. The team, led by Davenport and Isyumov[12] has tested many of the most renowned bridges recently constructed across the globe. Figure 3.2.6 provides a plan for the two boundary layer tunnels found in London. The older one, set up in 1965, has an open-return flow system. Its test section has a width of 2.4 m and an adjustable height between 2.1 and 1.5 m. In it, flow speeds can go up to 15 m/s and the engine power range is 30 kW. The closed-return model has two test sections. One, the high speed model, has a cross section that is 3.4 m wide and 2.5 m high. Here a rate of 30 m/s can be reached. Its low speed counterpart is 5 m wide and 4 m high, and reaches speeds of up to 10 m/s. The power of the fan for the latter facility is 215 kW.

These are what are called extra-wide BWLTs with hanger-like test chambers. Among these is the Martin Jensen Boundary Layer Wind Tunnel of the Danish Maritime Institute. It is 15 m long, 13.6 m wide and 1.7 m high. It produces wind speeds of up to 7 m/s. This figure is lower than the rates for traditional wind tunnels because its fans must force out countless cubic meters of air. These extraordinary dimensions mean that large scale full models of bridges can be analysed. This was done with the Great Belt Bridge, whose model, reduced to a scale of 1/200, can be seen in Figure 3.2.7. Another example is shown in figure 3.2.8 for the Akashi Bridge in Japan.

Within the field of bridge engineering, the main use for the other, aerodynamic kind of tunnel is to study the behaviour of the deck's cross section under wind action. To this end, a reduced model is made so that its length is between three and ten times its width and its scale is around 1/100. The dimensions needed for the test section are therefore considerably smaller than those needed for full bridge models. It is enough to have a section of around a square meter. A good example of an aerodynamic air tunnel can be found at the Civil Engineering Department of A Coruña University, in the Centre for Technological Research in Building and Civil Engineering of the same university.

As the diagram in Figure 3.2.9 shows, the air flow is produced here by a 60 HP fan. Once it passes through various rectifier and compression elements, the air enters a $1 \times 1\ m^2$

Figure 3.2.7: Model of Great Belt Bridge in the Martin Jensen Boundary Layer Wind Tunnel at Danish Maritime Institute.

Figure 3.2.8: Model of Akashi Strait Bridge. Research and Development Centre, Nagasaki.

test section to reach flow speeds of up to 30 m/s. Figure 3.2.10 provides a photograph of this tunnel. Good accessibility means that it is easy to handle the models.

Another example can be found in Figure 3.2.11, which features the tunnel at the Instituto Superior Técnico in Lisbon. Its test SECTION is $1.5 \times 1.5\,m^2$ and in it wind speeds of

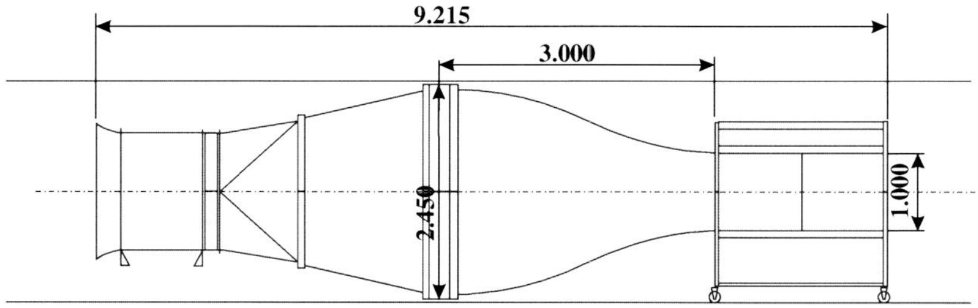

Figure 3.2.9: Plan of the wind tunnel at the School of Civil Engineering, University of Coruña, Spain.

Figure 3.2.10: Wind tunnel at the School of Civil Engineering, University of Coruña, Spain.

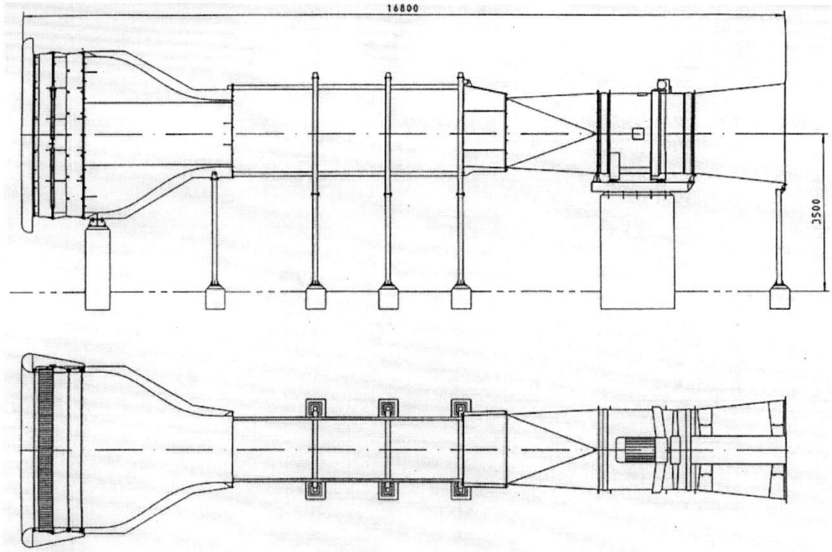

Figure 3.2.11: Elevation and ground plan of the tunnel at the Instituto Superior Técnico, Lisbon, Portugal.

up to 40 m/s are reached. Here the fan, located at the end of the circuit, draws in the air. In this way, it is surrounded on one side by the contraction element and, on the other, by the fan turbine. This layout has worse accessibility, but its turbulence levels may be as low as 0.5%. The photograph in Figure 3.2.12 was taken inside the test chamber; the sectional model is of the deck from the cable-stayed bridge in Coimbra, Northern Portugal.

3.2.3 Sectional tests of bridge decks

It has already been mentioned that this kind of experiment deals with reduced models of a deck to study its response to wind action. Two kinds of tests may be carried out.

3.2.3.1 Aerodynamic tests

A fixed model may be used: that is, there is no kind of movement. The aim is to measure the forces exerted on the model by the air flow, as represented in Figure 3.2.13. These

Figure 3.2.12: Test chamber of the tunnel at the Instituto Superior Técnico, Lisbon.

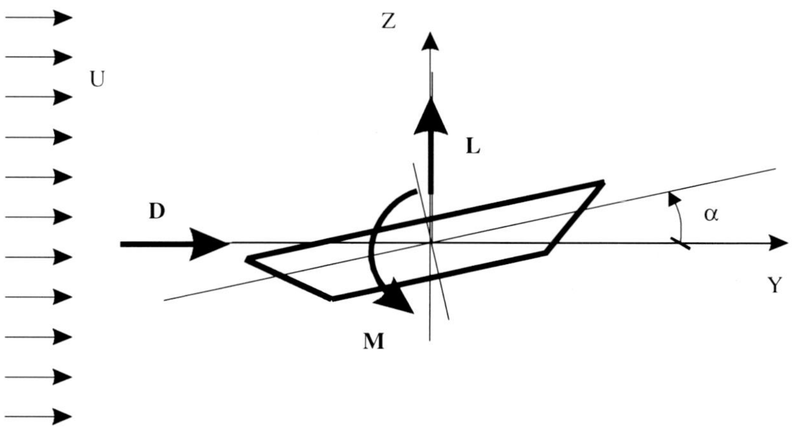

Figure 3.2.13: Aerodynamic forces on the deck of a bridge with an angle of attack α.

forces are usually non-dimensionalized and are expressed according to the following coefficients:

$$C_D = \frac{D}{\frac{1}{2}\rho U^2 B}; \quad C_L = \frac{L}{\frac{1}{2}\rho U^2 B}; \quad C_M = \frac{M}{\frac{1}{2}\rho U^2 B^2} \qquad (3.2.1)$$

where D is the drag force per unit of length along the deck; L, the lift and M, the pitching moment in the torsion-induced gyration direction, also per unit of length. These components also remain non-dimensionalized with the dynamic air flow pressure defined as $p = 1/2\rho U^2$ and the deck width B. This test is carried out for different angles of attack α so that it is also possible to obtain the derivatives for the aerodynamic coefficients in terms of this parameter.

3.2.3.2 Aeroelastic testing

In this kind of test, the model should simulate as precisely as possible the stiffness characteristics found in the original prototype. Therefore it is necessary to carry out a dimensional analysis determining the mass and mass moment of the model, as well as the position and stiffness of the elastic springs used in its support. Figure 3.2.14 offers a

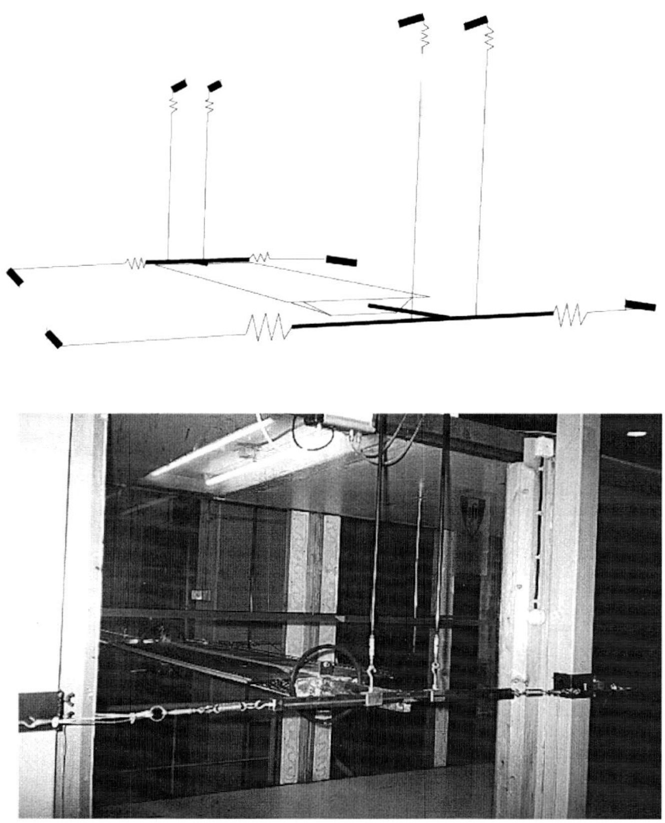

Figure 3.2.14: System used to support a sectional model for a bridge deck tested at the Instituto Superior Técnico in Lisbon.

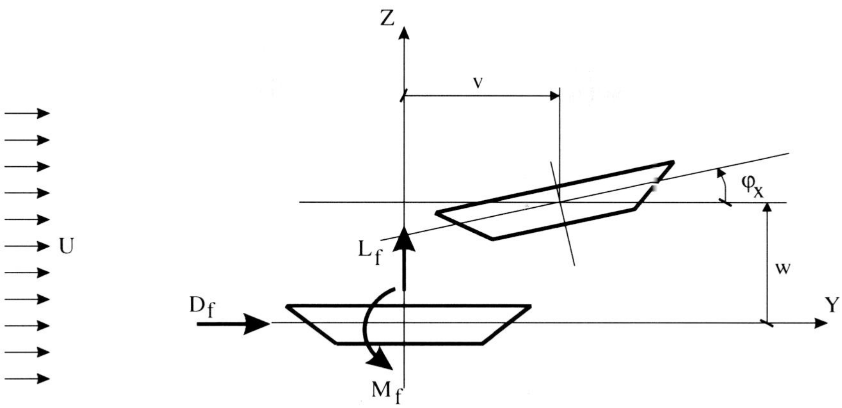

Figure 3.2.15: Forces and displacements in the aerodynamic test.

possible configuration for this kind of test. The aim of this test is to find the flutter derivatives that relate the forces occurring in the support system to possible deck displacements (u, w, φ_x). As can be seen below, this relationship takes the form

$$
\begin{aligned}
D_f &= \frac{\rho U^2 B}{2}\left(KP_1^* \frac{\dot{v}}{U} + KP_2^* \frac{B\dot{\varphi}_x}{U} + K^2 P_3^* \varphi_x + K^2 P_4^* \frac{v}{B} + KP_5^* \frac{\dot{w}}{U} + K^2 P_6^* \frac{w}{B} \right) \\
L_f &= \frac{\rho U^2 B}{2}\left(KH_1^* \frac{\dot{w}}{U} + KH_2^* \frac{B\dot{\varphi}_x}{U} + K^2 H_3^* \varphi_x + K^2 H_4^* \frac{w}{B} + KH_5^* \frac{\dot{v}}{U} + K^2 H_6^* \frac{v}{B} \right) \\
M_f &= \frac{\rho U^2 B^2}{2}\left(KA_1^* \frac{\dot{w}}{U} + KA_2^* \frac{B\dot{\varphi}_x}{U} + K^2 A_3^* \varphi_x + K^2 A_4^* \frac{w}{B} + KA_5^* \frac{\dot{v}}{U} + K^2 A_6^* \frac{v}{B} \right)
\end{aligned}
\qquad (3.2.2)
$$

D_f, L_f and M_f are the three components of the aeroelastic force while the values H_i^*, A_i^*, P_i^* $i = 1, ..., 6$ represent the 18 flutter derivatives. Figure 3.2.15 outlines the criteria for the signs considered.

A test along these lines was first proposed by Scanlan and Tomko[13] in 1971, but they confined themselves to only two degrees of freedom (w, φ_x) and had failed to obtain all of the derivatives. In 1977, Ibrahim and Mikulcik[14] have developed a numerical method for solving systems whose responses are exponentially diminishing. Also using this method Sarkar and Jones[15] developed a method for simultaneously obtaining the 18 flutter derivatives

3.3 Basic principles of analytical aeroelasticity

From the 1960s onwards, Robert H. Scanlan, lecturing at John Hopkins University (Baltimore, USA) worked alongside researchers like Tomko, Sazevari[16] Gade[17], Lin[18], Huston[19], Kumarasena[20], Jones[1] and others. They produced a series of studies on mathematical modeling for the aeroelastic instability found in long-span bridges. Scanlan can therefore be considered a pioneer in the field of aeroelasticity for long-span bridges.

Studies like these require an in-depth knowledge of the methodology for dynamic structural calculation. An analysis of flutter-induced instability may be done by considering a single kind of vibration featuring in the bridge. An alternative is to study the combination of various vibration modes so that it is possible to take into account any interaction existing among different components within each mode. The later method, called mulitmodal, has been shown to provide safer and more accurate results when it comes to dealing with bridges whose span length is between 1500 and 2000 m.

In this section, the theory of Theodorsen[21] will be revised and applied to the flutter of a flat plate. The equations for the displacement along the deck of a long-span bridge will be provided. These equations will be compared with those from the Theodorsen theory. Lastly, various theories for analysing flutter will be explained. Only two aeroelastic forces will be considered (vertical and the torsional moment) or, additionally, drag, to take into account the full range of wind action on the deck.

3.3.1 Theodorsen´s theory applied to flutter in flat plates

As was the case in the Tacoma Bridge's collapse, deck flutter causes the greatest worry in the field of long-span bridge design. Flutter can be defined in mathematical formulae with equations derived from classic aeroelasticity. This discipline, within Aeronautical Engineering, studies the dynamic behaviour of the shapes used for aircraft wings. It combines the theories of Fluid Dynamic and Deformable Solid Mechanics. The classic text on aeroelasticity by Fung[22] develops the equations governing the phenomenon of flutter instability. Theodorsen´s theory for flat plates can be considered the inspiration behind later formulations made in the field of cable-supported bridges.

This theory works with an aerodynamic shape, as in Figure 3.3.1, which undergoes displacement at a uniform horizontal speed U away from its original position. It then reaches the one after a certain amount of time t has passed. It is supposed that the shape has a chord with the length $2b$ and an infinitesimal angle of attack α.

Next to be considered is a type of movement with two degrees of freedom: a vertical displacement h that would produce bending along the bend and a pitching rotation α around a localized axis at a distance α halfway along the wing's chord. In Figure 3.3.2 these movements are shown for a flat plate.

With this theory it is also supposed that the plate movements are oscillatory in nature, so that $h = h_0 e^{i\omega t}$ and $\alpha = \alpha_0 e^{i\omega t}$, where h_0 and α_0 are complex and determine the movement's amplitude and phase angle with respect to the forces of excitation and ω the frequency. Using a non-dimensionalized time variable $\tau = Ut/b$, the movements are written as $h(\tau) = h_0 e^{ik\tau}$ and $\alpha(\tau) = \alpha_0 e^{ik\tau}$, where $k = \omega b/U$ is the reduced frequency response. With this notation, the aeroelastic forces along the flat plate are, according to Theodorsen´s theory

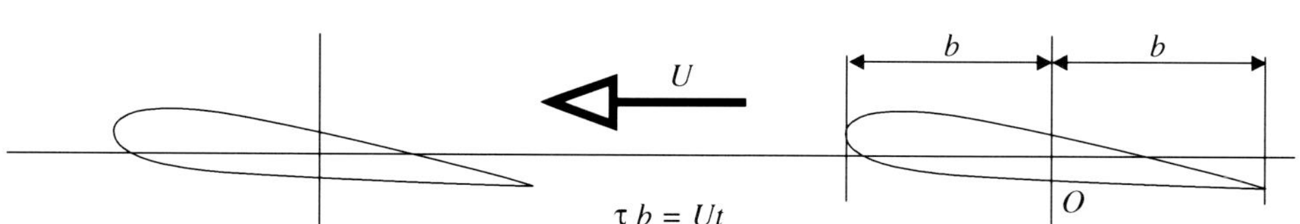

Figure 3.3.1: Displacement of an aerodynamic shape.

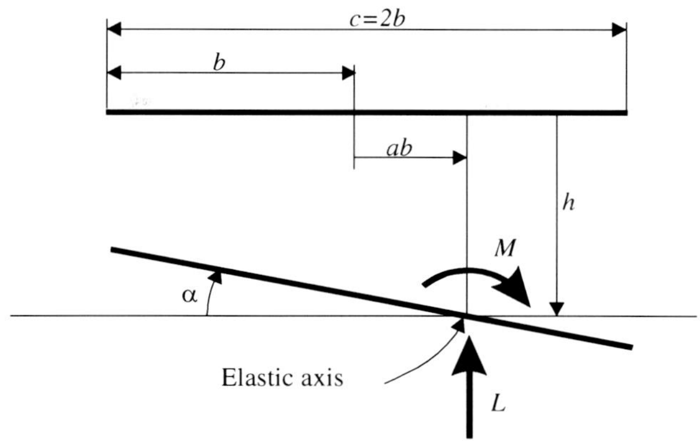

Figure 3.3.2: Movements considered for a flat plate.

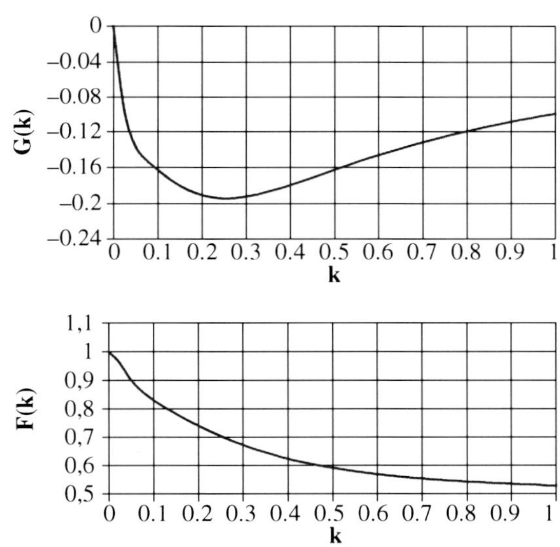

Figure 3.3.3: Real and imaginary parts of the Theodorsen function.

$$
\begin{aligned}
L &= 2\pi b\rho U^2\left\{(F+iG)\left[\alpha_0+\frac{i}{b}kh_0+\left(\tfrac{1}{2}-a\right)ik\alpha_0\right]-\frac{1}{2}k^2\left(\frac{h_0}{b}-a\alpha_0\right)+\frac{1}{2}ik\alpha_0\right\}e^{ik\tau}\\
M &= 2\pi b^2\rho U^2\left\{\left(\tfrac{1}{2}+a\right)(F+iG)\left[\alpha_0+\frac{i}{b}kh_0+\left(\tfrac{1}{2}-a\right)ik\alpha_0\right]\right.\\
&\left.-\frac{1}{2}k^2a\left(\frac{h_0}{b}-a\alpha_0\right)+\left(\tfrac{1}{2}-a\right)ik\alpha_0+\frac{k^2}{8}\alpha_0\right\}e^{ik\tau}
\end{aligned}
\tag{3.3.1}
$$

where U is the average speed of the acting wind, ρ the air density and F and G are, respectively, the real and imaginary parts of the Theodorsen function.

$$C(k) = F(k) + iG(k) = \frac{H_1^{(2)}(k)}{H_1^{(2)}(k) + iH_0^{(2)}(k)} = \frac{K_1(ik)}{K_0(ik) + K_1(ik)} \tag{3.3.2}$$

where H and K being, respectively, the modified Hünkel and Bessel functions. The real part $F(k)$ and the imaginary $G(k)$ of $C(k)$ are represented in Figure 3.3.3.

3.3.2 Linearization of aeroelastic loads through flutter derivatives

One can use the same equations governing flutter in flat plates to mathematically model the flutter-induced aeroelastic instability found on the decks of either suspension or cable-stayed bridges. However, unlike flat plates, the geometry determined by the design of bridge deck sections cannot be considered an aerodynamic shape, which makes it impossible to have a complete analytical formulation. This is why aeroelastic forces tend to be expressed in a linear function of the same two movements considered in Thedorsen's theory, multiplied by what is known as the flutter derivative. The resultant model, as shown in Figure 3.3.4, is therefore two-dimensional.

The coefficients multiplying the degrees of freedom h and α or their first derivatives cannot be obtained analytically, but it is possible to do so experimentally through wind-tunnel tests, as explained by Scanlan and Tomko[13]. The two forces are lift L and the moment M. These are defined as:

$$\begin{aligned} L &= \frac{1}{2}\rho U^2 B\left(KH_1^*\frac{\dot{h}}{U} + KH_2^*\frac{B\dot{\alpha}}{U} + K^2H_3^*\alpha + K^2H_4^*\frac{h}{B}\right) \\ M &= \frac{1}{2}\rho U^2 B^2\left(KA_1^*\frac{\dot{h}}{U} + KA_2^*\frac{B\dot{\alpha}}{U} + K^2A_3^*\alpha + K^2A_4^*\frac{h}{B}\right) \end{aligned} \tag{3.3.3}$$

where the parameters A_i^* and H_i^* are the flutter derivatives. In contrast with Theodorsen's theory, the aeroelastic analysis of long-span bridges uses the complete dimension for the section $B = 2b$. Consequently the reduced frequency is $K = \omega B/U = \omega 2b/U = 2k$. Also used is the non-dimensionalized time variable. In order to compare the aeroelastic forces according to Theodorsen's theory with the formulation proposed by Scanlan, as well as obtain the flutter derivatives for the flat plates, it is necessary to transform equations in

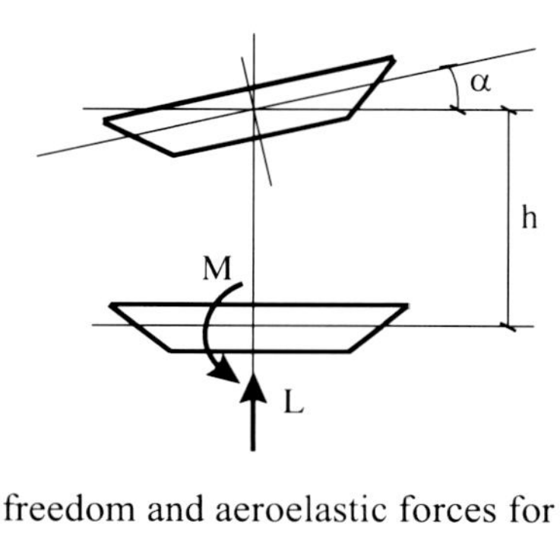

Figure 3.3.4: Degrees of freedom and aeroelastic forces for a two-dimensional model.

(3.3.1), expressing them according to the new variables. Operating with the complex numbers, the following is reached:

$$L = \frac{1}{2}\rho U^2 B \left\{ 2\pi F \frac{\dot{h}}{U} + \frac{\pi}{2}\left[1 + \frac{4G}{K} + 2\left(\frac{1}{2} - a\right) F \right] \frac{B}{U}\dot{\alpha} \right.$$

$$\left. + \pi \left[2F - \left(\frac{1}{2} - a\right) GK + \frac{K^2 a}{4} \right] \alpha - \frac{\pi}{2} K^2 \left[1 + \frac{4G}{K} \right] \frac{h}{B} \right\} \qquad (3.3.4)$$

$$M = \frac{1}{2}\rho U^2 B^2 \left\{ \pi F \left(\frac{1}{2} + a\right) \frac{\dot{h}}{U} - \frac{\pi}{2}\left[\frac{1}{2}\left(\frac{1}{2} + a\right) - \left(\frac{1}{2} + a\right)\frac{2G}{K} + F\left(a^2 - \frac{1}{4}\right) \right] \frac{B}{U}\dot{\alpha} \right.$$

$$\left. + \frac{\pi}{2}\left[\frac{K^2}{4}\left(a^2 + \frac{1}{8}\right) + 2F\left(\frac{1}{2} + a\right) + GK\left(a^2 - \frac{1}{4}\right) \right] \alpha - \frac{\pi}{2}\left[\frac{K^2 a}{2} + \left(\frac{1}{2} + a\right) 2GK \right] \frac{h}{B} \right\}$$

In these expressions the functions F and G continue to be evaluated in $k = K/2$. When $a = 0$, as is the case with decks with symmetrical sections, the flutter derivatives of the flat plate can be obtained by relating expressions (3.3.3) and (3.3.4), so that

$$\begin{aligned} KH_1^* &= -2\pi F & KA_1^* &= \frac{\pi}{2} F \\ KH_2^* &= -\frac{\pi}{2}\left[1 + \frac{4G}{K} + F \right] & KA_2^* &= -\frac{\pi}{2}\left[\frac{1}{4} - \frac{G}{K} + \frac{F}{4} \right] \\ K^2 H_3^* &= -\pi \left[2F - \frac{GK}{2} \right] & K^2 A_3^* &= \frac{\pi}{2}\left[\frac{K^2}{32} + F - \frac{KG}{4} \right] \\ K^2 H_4^* &= \frac{\pi}{2} K^2 \left[1 + \frac{4G}{K} \right] & K^2 A_4^* &= -\frac{\pi}{2} KG \end{aligned} \qquad (3.3.5)$$

It is important to notice that two different criteria have been applied: the one from Theodorsen's theory for the signs within the flat plane and the one chosen here for the bridge deck as dictated by the flutter derivative method. Figure 3.3.5 shows that both movement directions are altered. Due to the change the H_i^* signs of equation (3.3.5).

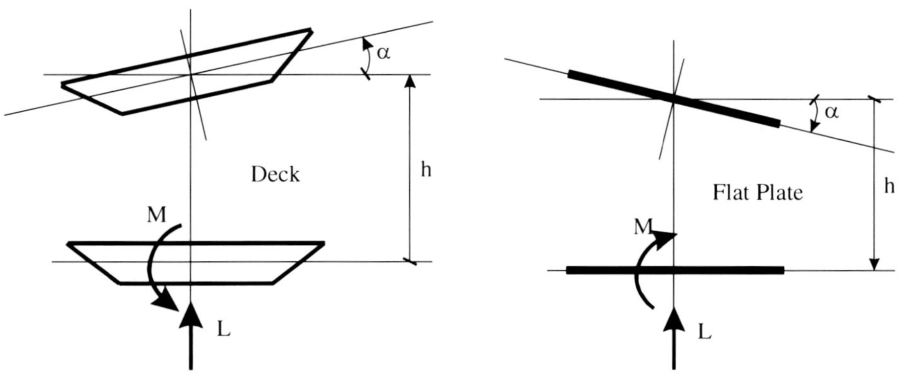

Figure 3.3.5: Different sign criteria for the two methods: flutter derivative analysis for bridge decks and Theodorsen's theory for flat plates.

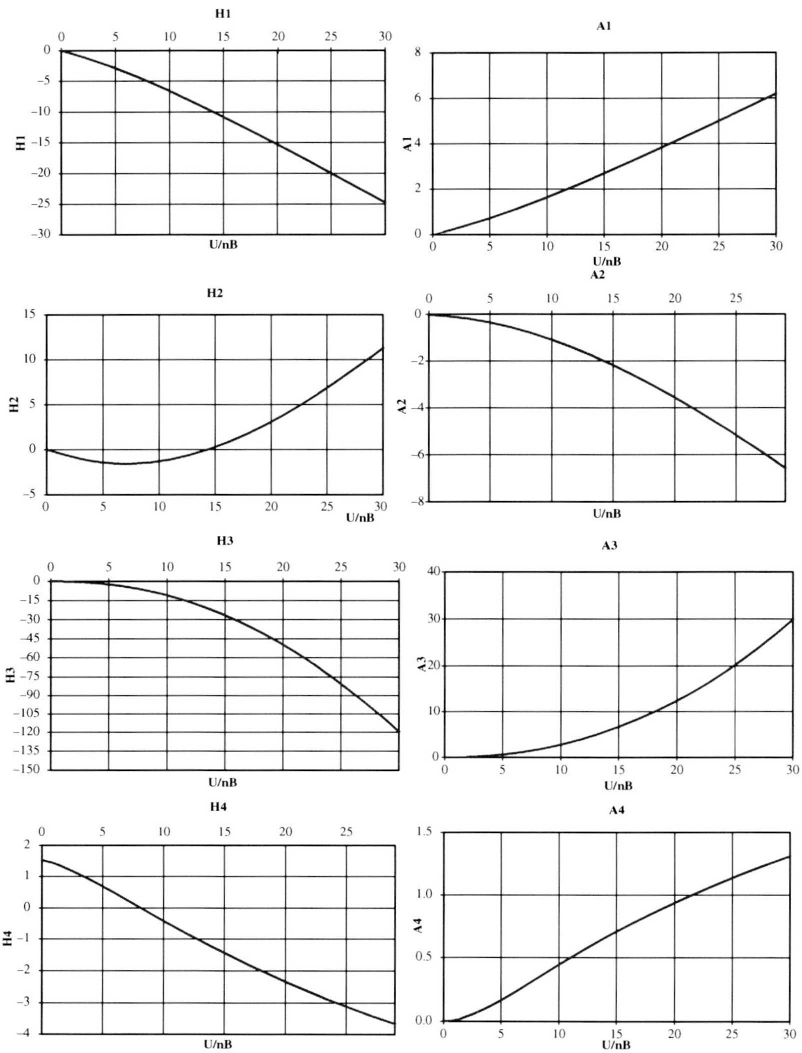

Figure 3.3.6: Flutter derivatives for a flat plate as a function of *U/nB*.

The flutter derivatives for a flat plate are analytically obtained by means of expressions (3.3.5). These tend to be represented in graphical form according to the reduced wind speed function *U/nB*, *n* being the frequency in Hz. Figure 3.3.6 provides an example of this. Moreover, the flutter derivatives for a bridge that are obtained experimentally through wind-tunnel tests are also represented in this way. The purpose for doing so is to compare the variation of each flutter derivative with the ideal values attributed to the aerodynamic shape of the flat plate.

3.3.3 Bridge flutter considering three aeroelastic forces

In 1995 Jain[23], alongside Jones and Scanlan, made further headway with this theory, by adding to the two classic movements of flutter (h and α), the degree of freedom p,

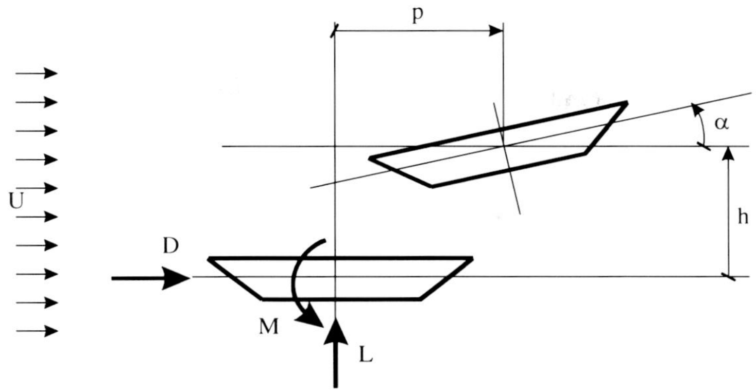

Figure 3.3.7: Model with three aeroelastic forces.

corresponding to the horizontal movement following the direction of the wind. This direction proved to play a vital role in how long-span bridges behaved, as such a long structure is normally very flexible in this direction. Furthermore, the first vibration mode of a suspension bridge is usually lateral, and is also influenced by the wind direction. A noteworthy case in point is the Akashi Strait Bridge, which has, according to Katsuchi *et al.*[24] a first natural frequency of 0.03881 Hz corresponding to a lateral vibration mode. Likewise, of the six first modes of vibration, four are lateral and all of these have frequencies that are lower than the torsional modes. In Figure 3.3.7 one finds the three existing deck movements, as well as the forces considered: lift L, drag D and the moment M.

As described in the previous sections, the aeroelastic forces causing flutter can be described as a linear combination of the deck movements. The expressions of the three forces are

$$
\begin{aligned}
L &= \frac{1}{2}\rho U^2 B\left(KH_1^*\frac{\dot h}{U} + KH_2^*\frac{B\dot\alpha}{U} + K^2H_3^*\alpha + K^2H_4^*\frac{h}{B} + KH_5^*\frac{\dot p}{U} + K^2H_6^*\frac{p}{B}\right)\\
D &= \frac{1}{2}\rho U^2 B\left(KP_1^*\frac{\dot p}{U} + KP_2^*\frac{B\dot\alpha}{U} + K^2P_3^*\alpha + K^2P_4^*\frac{p}{B} + KP_5^*\frac{\dot h}{U} + K^2P_6^*\frac{h}{B}\right)\\
M &= \frac{1}{2}\rho U^2 B^2\left(KA_1^*\frac{\dot h}{U} + KA_2^*\frac{B\dot\alpha}{U} + K^2A_3^*\alpha + K^2A_4^*\frac{h}{B} + KA_5^*\frac{\dot p}{U} + K^2A_6^*\frac{p}{B}\right)
\end{aligned}
\tag{3.3.6}
$$

where ρ is the air density; U, the speed of the acting wind; $K = B\omega/U$, the reduced frequency; and ω the circular frequency. Along with the three forces, more flutter derivatives crop up in the equations than was the case when the deck was analysed through the vertical movement and twist. There are 18 derivatives H_i^*, P_i^*, A_i^* $i = 1,\ldots,6$, six for each force, the three that multiply the displacement (h, p, α) and the other three that multiply the speeds $(\dot h, \dot p, \dot\alpha)$. All of these are functions of K. As already has been mentioned, they are obtained experimentally by testing reduced sectional models of the bridge deck in wind tunnels. The next section will look at in detail how to solve the system of equations representing the deck's dynamic equilibrium.

3.4 Movement equations for bridge decks

The aeroelastic analysis of a bridge deck is also a dynamic one in which the acting forces are determined by the deck's movements under wind flow. To solve this problem in a satisfactory manner, it is necessary to make a three-dimensional structural model of the complete bridge, like the one shown in Figure 3.4.1.

Three-dimensional bar elements subject to bending are used to model the deck. That is, the deck nodes have six degrees of freedom and the bars can undergo six forces: one axial, two shearing forces, two bending moments and one torsional moment. To ensure that the deck movements are compatible with those of the hangers and suspension cables, one must define a series of fictitious bars perpendicular to the deck. These must connect the bar elements to the ends of the hangers (or cables in the case of cable-stayed bridges). This type of structural model appears in a large number of studies; recent examples include the work of Larsen[25], Namini and Aldbrecht[26] or Semiáo and Mendes[27].

The wind subjects the deck to three aeroelastic forces; these can be broken down into three actions by length unit:

- Drag, producing lateral bending in the deck along the *XY* plane.
- Lift, which causes vertical bending in the deck in the direction that runs counter to gravitational forces.
- A moment along the *X*-axis, that produces torsional movements along the deck's longitudinal axis

Expressions (3.4.6) for these forces have been written according to the notation used by Jones and Scanlan[1]. Switching over to the notation traditionally used in structural calculations, the forces and movements acting on any node within the structure will be defined through the vectors

$$\mathbf{f} = (F_x, F_y, F_z, M_y, M_z)^T \tag{3.4.1}$$

$$\mathbf{u} = (u, v, w, \varphi_x, \varphi_y, \varphi_z)^T \tag{3.4.2}$$

The loads distributed along the unit of length are normally designated the letter "*p*". On the other hand, distributed moments are marked as "*m*". They are placed in the direction

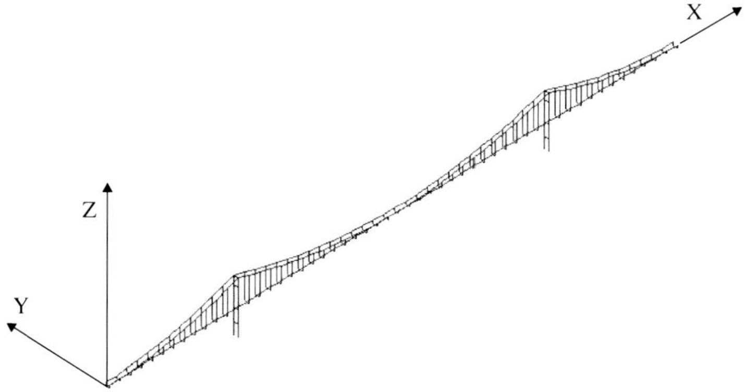

Figure 3.4.1: Three-dimensional structural model of a suspension bridge.

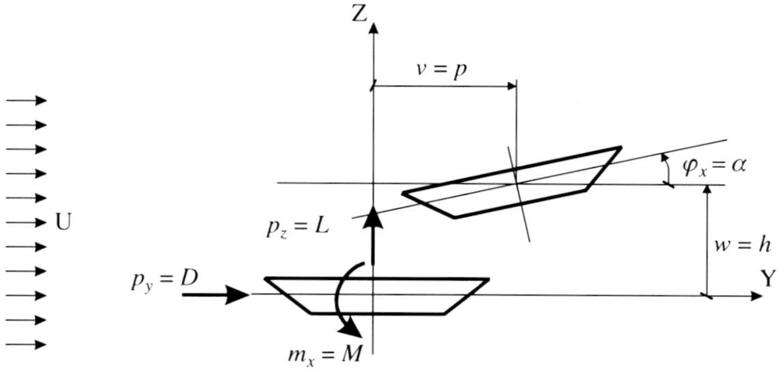

Figure 3.4.2: Changes in the notation for deck movements and forces.

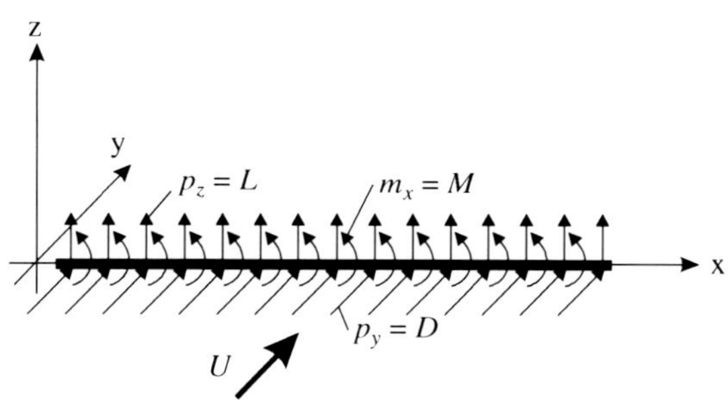

Figure 3.4.3: Aeroelastic forces distributed along the deck.

in which they act according to the order of their subscripts along the axis, as seen in the samples given in Figures 3.4.2 and 3.4.3.
Using this system of notation, expressions for (3.3.6) are

$$\begin{aligned} D = p_y &= \frac{\rho U^2 B}{2}\left(KP_1^* \frac{\dot{v}}{U} + KP_2^* \frac{B\dot{\varphi}_x}{U} + K^2 P_3^* \varphi_x + K^2 P_4^* \frac{v}{B} + KP_5^* \frac{\dot{w}}{U} + K^2 P_6^* \frac{w}{B} \right) \\ L = p_z &= \frac{\rho U^2 B}{2}\left(KH_1^* \frac{\dot{w}}{U} + KH_2^* \frac{B\dot{\varphi}_x}{U} + K^2 H_3^* \varphi_x + K^2 H_4^* \frac{w}{B} + KH_5^* \frac{\dot{v}}{U} + K^2 H_6^* \frac{v}{B} \right) \\ M = m_x &= \frac{\rho U^2 B^2}{2}\left(KA_1^* \frac{\dot{w}}{U} + KA_2^* \frac{B\dot{\varphi}_x}{U} + K^2 A_3^* \varphi_x + K^2 A_4^* \frac{w}{B} + KA_5^* \frac{\dot{v}}{U} + K^2 A_6^* \frac{v}{B} \right) \end{aligned} \quad (3.4.3)$$

The aeroelastic forces are a linear combination of the displacements and velocities. Working with a structural model of the bridge, whose deck is broken down into bar elements, it is possible to define a load vector acting on each element *e* as a matrix so that

$$\mathbf{f}_{ae} = \mathbf{K}_{a,e}\mathbf{u}_e + \mathbf{C}_{a,e}\dot{\mathbf{u}}_e \quad (3.4.4)$$

where subscript *a* refers to the aeroelastic origin of the terms. Expressing the aeroelastic forces as matrices has also been dealt with in articles by Katsuchi *et al.*[24], Astiz[28] and Mendes[29].

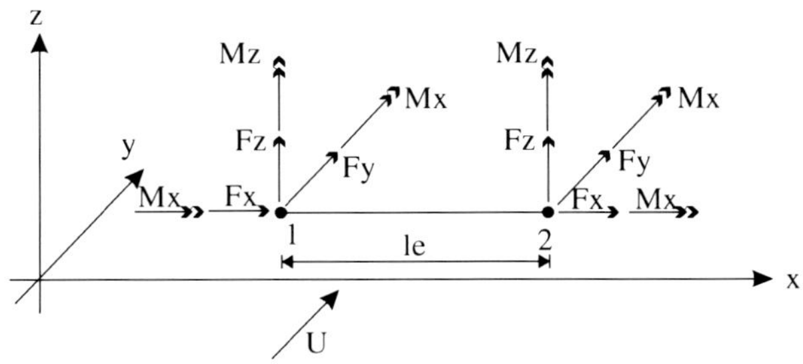

Figure 3.4.4: Forces acting on bar elements along the deck.

The forces acting on the element's outer nodes are related to its movements and speed within equation (3.4.4). As it is shown in figure 3.4.4, each node carries half the loads distributed along the element. Thus, for either end i = 1,2 one can write the following matrix expression:

$$\mathbf{f}_{ai} = \mathbf{K}_{a,i}\mathbf{u}_i + \mathbf{C}_{a,i}\dot{\mathbf{u}}_i \tag{3.4.5}$$

which, once developed, takes on the following form:

$$\begin{pmatrix} F_x \\ F_y \\ F_z \\ M_x \\ M_y \\ M_z \end{pmatrix} = \frac{l_e}{2}\begin{pmatrix} 0 \\ p_y \\ p_z \\ m_x \\ 0 \\ 0 \end{pmatrix} = \frac{l_e}{2}\begin{pmatrix} 0 & 0 & 0 & 0 & 0 & 0 \\ 0 & pK^2P_4^* & pK^2P_6^* & pBK^2P_3^* & 0 & 0 \\ 0 & pK^2H_6^* & pK^2H_4^* & pBK^2H_3^* & 0 & 0 \\ 0 & pBK^2A_6^* & pBK^2A_4^* & pB^2K^2A_3^* & 0 & 0 \\ 0 & 0 & 0 & 0 & 0 & 0 \\ 0 & 0 & 0 & 0 & 0 & 0 \end{pmatrix}\begin{pmatrix} u_i \\ v_i \\ w_i \\ \varphi_{xi} \\ \varphi_{yi} \\ \varphi_{zi} \end{pmatrix} +$$

$$+\frac{l_e}{2}\begin{pmatrix} 0 & 0 & 0 & 0 & 0 & 0 \\ 0 & pBKP_1^*/U & pBKP_5^*/U & pB^2KP_2^*/U & 0 & 0 \\ 0 & pBKH_5^*/U & pBKH_1^*/U & pB^2KH_2^*/U & 0 & 0 \\ 0 & pB^2KA_5^*/U & pB^2KA_1^*/U & pB^3KA_2^*/U & 0 & 0 \\ 0 & 0 & 0 & 0 & 0 & 0 \\ 0 & 0 & 0 & 0 & 0 & 0 \end{pmatrix}\begin{pmatrix} \dot{u}_i \\ \dot{v}_i \\ \dot{w}_i \\ \dot{\varphi}_{xi} \\ \dot{\varphi}_{yi} \\ \dot{\varphi}_{zi} \end{pmatrix} \tag{3.4.6}$$

where l_e is the length of the element and $p = \rho U^2/2$ is the dynamic wind pressure at the speed U. Aeroelastic forces can also be written as a function of the terms of the matrices, $\mathbf{K}_{ai}$ and $\mathbf{C}_{ai}$ so that expression (3.4.3) is transformed as

$$\begin{aligned} D = p_y &= K_{a22}v + K_{a23}w + K_{a24}\varphi_x + C_{a22}\dot{v} + C_{a23}\dot{w} + C_{a24}\dot{\varphi}_x \\ L = p_z &= K_{a32}v + K_{a33}w + K_{a34}\varphi_x + C_{a32}\dot{v} + C_{a33}\dot{w} + C_{a34}\dot{\varphi}_x \\ M = m_x &= K_{a42}v + K_{a43}w + K_{a44}\varphi_x + C_{a42}\dot{v} + C_{a43}\dot{w} + C_{a44}\dot{\varphi}_x \end{aligned} \tag{3.4.7}$$

In this way it becomes much clearer which parameter is the one relating a force with a certain movement or speed. K_a refers to an aeroelastic relationship between force and movement, while C_a links force with speed. On the other hand, subscript 2 goes with v; 3, with w and 4, with φ_x. For example, if the objective was to study the parameter related the lift corresponding to axis Z ($w \rightarrow 3$) with the deck speed in the direction of the wind, associated with axis Y ($v \rightarrow 2$), it would be necessary to analyse the behaviour of the term C_{a32}. Doing so seems much more intuitive than analysing the flutter derivative H_5^*.

Following the matrix notation, the vector of the aeroelastic forces for a bar element along the deck is

$$\mathbf{f}_{a,e} = \begin{pmatrix} \mathbf{f}_{a,1} \\ \mathbf{f}_{a,2} \end{pmatrix} = \mathbf{K}_{a,e}\mathbf{u}_e + \mathbf{C}_{a,e}\dot{\mathbf{u}}_e = \begin{pmatrix} \mathbf{K}_{a,1} & \mathbf{0} \\ \mathbf{0} & \mathbf{K}_{a,2} \end{pmatrix}\begin{pmatrix} \mathbf{u}_1 \\ \mathbf{u}_2 \end{pmatrix} + \begin{pmatrix} \mathbf{C}_{a,1} & \mathbf{0} \\ \mathbf{0} & \mathbf{C}_{a,2} \end{pmatrix}\begin{pmatrix} \dot{\mathbf{u}}_1 \\ \dot{\mathbf{u}}_2 \end{pmatrix} \quad (3.4.8)$$

With 12 components remaining. Of these, only six are null. In contrast, the elemental aeroelastic matrices $\mathbf{K}_{ae}$ and $\mathbf{C}_{ae}$ have the dimension of 12×12 and in their interiors, two sub-matrices 3×3 are formed, neither of which are null. The rest of the terms are zero. The assembled whole for all the bar elements making up the deck, with all the vectors $f_{a,e}$ and all the matrices $\mathbf{K}_{ae}$ and $\mathbf{C}_{ae}$, makes it possible to construct the global vector global of aeroelastic forces $\mathbf{f}_a$, and the aeroelastic global matrices $\mathbf{K}_a$ and $\mathbf{C}_a$. The following relationship is achieved:

$$\mathbf{f}_a = \mathbf{K}_a\mathbf{u} + \mathbf{C}_a\dot{\mathbf{u}} \quad (3.4.9)$$

The dimensions of the vectors $\mathbf{u}$ and $\mathbf{f}_a$ coincide with the number of degrees of freedom n in the deck. In other words, $n = 6N$, with N the number of deck nodes. Therefore, the dimension of the aeroelastic matrices $\mathbf{K}_a$ and $\mathbf{C}_a$ is $6N \times 6N$, although it has already been stated that it is only possible to have $3N \times 3N$ different terms for zero.

The system of equations governing the deck's dynamic behaviour while withstanding forces that are of aeroelastic origin is

$$\mathbf{M}\ddot{\mathbf{u}} = \mathbf{C}\ddot{\mathbf{u}} + \mathbf{K}\mathbf{u} = \mathbf{f}_a = \mathbf{K}_a\mathbf{u} + \mathbf{C}_a\dot{\mathbf{u}} \quad (3.4.10)$$

where $\mathbf{M}$, $\mathbf{C}$ and $\mathbf{K}$ are the matrices for the deck's mass, damping and stiffness, respectively. By passing the terms of the aeroelastic forces to the left side of equality, the following remains

$$\mathbf{M}\ddot{\mathbf{u}} + (\mathbf{C} - \mathbf{C}_a)\dot{\mathbf{u}} + (\mathbf{K} - \mathbf{K}_a)\mathbf{u} = \mathbf{0} \quad (3.4.11)$$

which represents the movement of a damping system in free vibration. In this system, the stiffness and damping matrices have been modified through the aeroelastic forces exerted by the wind.

3.5 Modal analysis

A procedure for resolving dynamic problems like the one defined in equation (3.4.11) is provided by modal analysis. The first task is to calculate the frequencies and natural vibration modes for the bridge so that it is possible to use one of the many structural calculation programs available on the market. These programs can solve the following eigenvalue problem:

$$(\mathbf{K} - \omega^2\mathbf{M})\boldsymbol{\phi} = \mathbf{0} \quad (3.5.1)$$

where ω is a natural frequency value, the eigenvalue ϕ is the vibration mode associated with this frequency, $\mathbf{K}$ is the elastic stiffness matrix and $\mathbf{M}$ is the mass matrix. In general, one will obtain the same number of natural frequencies and natural modes as the degrees of freedom within the system.

In the case of a suspension bridge, the frequencies and natural modes obtained in this way are very different from the real values. This discrepancy is due to the fact that, in equation (3.5.1), what has been overlooked is how the stiffness produced in the structure by some of its elements, specifically the overhead cables, are subjected to high tension forces. This effect can be taken into account when the calculation is done in second-order theory: that is, when considering the balance of forces over the deformed geometry.

For this kind of analysis, it is first necessary to carry out a static calculation to obtain the deformed geometry of the structure, as well as the axial forces that support the elements with the loads from its own weight. With these results and by means of the structure's geometric stiffness matrix it is possible to evaluate the increase in stiffness produced in elements, like the cables, which are subjected to extreme tensional forces. To obtain the frequencies and natural vibration modes, the problem can be written by considering the second-order theory as

$$\left(\mathbf{K}+\mathbf{K}_G-\omega^2\mathbf{M}\right)\boldsymbol{\phi}=\mathbf{0} \tag{3.5.2}$$

where $\mathbf{K}_G$ is the geometric stiffness matrix that depends on the axial forces mentioned earlier. Table 3.5.1 provides a comparison between the frequencies and vibration modes of the old Tacoma Bridge. These are based on the first-order theory, whereby only the elastic rigidity is considered, and second-order theory when the geometric stiffness matrix is incorporated.

Both analyses show very different results. The first-order theory analysis cannot be deemed valid. Very low frequencies are obtained because it overlooks the stiffness provided to the structure by the cables withstanding traction. Nor are the vibration modes correct; it is noticeable that all of the first-order modes arising from deck movement are torsional. The second-order analysis is a much closer approximation of the bridge's real behaviour. According to Faquharson[30], it had a torsional frequency of 0.2 Hz, which corresponded to 1.257 rad/s, very similar to the one obtained for torsional mode 5.

The movements of the bridge, the solution to the problem defined in equation (3.5.11), can be approximated with a linear combination of the most significant vibration modes m

Table 3.5.1: Comparison of dynamic analyses for the Tacoma Bridge.

First-order analysis		Second-order analysis	
Mode	ω (rad/s)	Mode	ω (rad/s)
(1) TS	0.051	(1) LS	0.568
(6) TA	0.082	(2) VA	0.817
(11) VA	0.107	(3) VS	1.189
(12) TS	0.117	(4) LA	1.296
(13) TA	0.145	(5) TA	1.505
(18) TA	0.213	(6) TS	1.608
(19) TS	0.215	(7) VA	1.705
(20) VS	0.231	(8) VS	1.792

$$\mathbf{u} = \sum_{r=1}^{m} \phi_r q_r = \mathbf{\Phi q} \tag{3.5.3}$$

where $\mathbf{\Phi}$ is the modal matrix to the order of $n \times m$

$$\mathbf{\Phi} = \begin{pmatrix} \phi_{1,1} & \cdot & \cdot & \cdot & \phi_{m,1} \\ \cdot & & & & \\ \cdot & & & & \\ \cdot & & & & \\ \phi_{1,n} & & & & \phi_{m,n} \end{pmatrix} \tag{3.5.4}$$

and each element of the vector $\mathbf{q}$ represents the participation of each one of the vibration modes in the movement vector $\mathbf{u}$. Substituting (3.5.3) in (3.4.11) and pre-multiplying by $\mathbf{\Phi}^{\mathrm{T}}$ the following is reached:

$$\mathbf{\Phi}^{\mathrm{T}}\mathbf{M\Phi}\ddot{\mathbf{q}} + \mathbf{\Phi}^{\mathrm{T}}(\mathbf{C}-\mathbf{C}_a)\mathbf{\Phi}\dot{\mathbf{q}} + \mathbf{\Phi}^{\mathrm{T}}(\mathbf{K}-\mathbf{K}_a)\mathbf{\Phi q} = \mathbf{0} \tag{3.5.5}$$

Equation (3.5.5) governs the behaviour of the deck, although it has been necessary to calculate the vibration modes for the entire bridge, including the cables and towers. Therefore, when forming the matrix $\mathbf{\Phi}$, only the n components related to the deck's degrees of freedom should be included in each column. The matrices reduced by the vibration modes m remain with the dimension $m \times m$. Defining the following matrices:

$$\mathbf{M}_R = \mathbf{\Phi}^{\mathrm{T}}\mathbf{M\Phi} \quad \mathbf{C}_R = \mathbf{\Phi}^{\mathrm{T}}(\mathbf{C}-\mathbf{C}_a)\mathbf{\Phi} \quad \mathbf{K}_R = \mathbf{\Phi}^{\mathrm{T}}(\mathbf{K}-\mathbf{K}_a)\mathbf{\Phi} \tag{3.5.6}$$

and bearing in mind that the vibration modes are normalized in terms of the mass so that the reduced mass matrix coincides with the unitary matrix $\mathbf{M}_R = \mathbf{I}$, expression (3.5.5) becomes

$$\mathbf{I}\ddot{\mathbf{q}} + \mathbf{C}_R\dot{\mathbf{q}} + \mathbf{K}_R\mathbf{q} = \mathbf{0} \tag{3.5.7}$$

where the matrix components $\mathbf{C}_R$ and $\mathbf{K}_R$ verify for $i, j = 1, \ldots, m$

$$C_{R,ij} = -\sum_{r=1}^{n} \phi_{ri} \left(\sum_{s=1}^{n} C_{a,rs}\phi_{sj} \right) \text{ with } i \neq j \tag{3.5.8}$$

$$C_{R,ii} = 2\xi_i\omega_i - \sum_{r=1}^{n} \phi_{ri} \left(\sum_{s=1}^{n} C_{a,rs}\phi_{si} \right) \tag{3.5.9}$$

$$K_{R,ij} = -\sum_{r=1}^{n} \phi_{ri} \left(\sum_{s=1}^{n} K_{a,rs}\phi_{sj} \right) \text{ with } i \neq j \tag{3.5.10}$$

$$K_{R,ii} = \omega_i^2 - \sum_{r=1}^{n} \phi_{ri} \left(\sum_{s=1}^{n} K_{a,rs}\phi_{si} \right) \tag{3.5.11}$$

ω_i and ξ_i are, respectively, the natural vibration frequency and damping coefficient for the infinitesimal mode ϕ_i. For the aeroelastic response of the deck, it is supposed that one oscillation is dampened over time so that

$$\mathbf{q}(t) = \mathbf{w}e^{\mu t} \tag{3.5.12}$$

and with this, it is possible to write (3.5.7)

$$\left(\mu^2\mathbf{I}\mathbf{w} + \mu\mathbf{C}_R\mathbf{w} + \mathbf{K}_R\mathbf{w}\right)e^{\mu t} = \mathbf{0} \tag{3.5.13}$$

The following identity provides help in obtaining the solution:

$$-\mu\mathbf{I}\mathbf{w} + \mu\mathbf{I}\mathbf{w} = \mathbf{0} \tag{3.5.14}$$

Combining both expressions

$$\left[\begin{pmatrix} \mathbf{I} & \mathbf{0} \\ \mathbf{0} & \mathbf{I} \end{pmatrix}\begin{pmatrix} \mu^2\mathbf{w} \\ \mu\mathbf{w} \end{pmatrix} + \begin{pmatrix} \mathbf{C}_R & \mathbf{K}_R \\ -\mathbf{I} & \mathbf{0} \end{pmatrix}\begin{pmatrix} \mu\mathbf{w} \\ \mathbf{w} \end{pmatrix}\right]e^{\mu t} = \mathbf{0} \tag{3.5.15}$$

and defining

$$\mathbf{w}_\mu = \begin{pmatrix} \mu\mathbf{w} \\ \mathbf{w} \end{pmatrix} \quad \text{and} \quad \mathbf{A} = \begin{pmatrix} -\mathbf{C}_R & -\mathbf{K}_R \\ \mathbf{I} & \mathbf{0} \end{pmatrix} \tag{3.5.16}$$

it can be written as

$$\left(\mathbf{A} - \mu\mathbf{I}\right)\mathbf{w}_\mu e^{\mu t} = \mathbf{0} \tag{3.5.17}$$

which is a non-linear problem of values and eigenvectors.

3.6 Aeroelastic response of a bridge

The solution for the eigenvector problem expressed in (3.5.17), the result of analysing the aeroelastic behaviour of a bridge affected by the wind with a generic speed U, provides in general terms $2m$ complex eigenvalues so that

$$\mu_j = \alpha_j + i\beta_j \qquad j = 1,\dots,2m \tag{3.6.1}$$

If an eigenvalue is complex, its complex conjugate will also be an eigenvalue

$$\bar{\mu}_j = \alpha_j - i\beta_j \tag{3.6.2}$$

For each pair of complex conjugate eigenvalues there is a corresponding pair of complex conjugate eigenvalues between them, whose dimension is $2m$.

$$\begin{aligned} \mathbf{w}_\mu &= \mathbf{w}_{\mu R} + i\mathbf{w}_{\mu I} \\ \bar{\mathbf{w}}_\mu &= \mathbf{w}_{\mu R} - i\mathbf{w}_{\mu I} \end{aligned} \tag{3.6.3}$$

Each pair of conjugate eigenvalues along with their corresponding conjugate eigenvalues define a dampened oscillation shape of the deck, which is expressed as

$$\mathbf{u} = \mathbf{\Phi}\mathbf{w}e^{\mu_j t} = \mathbf{\Phi}\left(\mathbf{w}_R + i\mathbf{w}_I\right)e^{(\alpha_j + i\beta_j)t} \tag{3.6.4}$$

where $\mathbf{w}$ is a vector whose dimension is m obtained by defining w_μ

$$\mathbf{w}_{\mu} = \begin{pmatrix} \mu \mathbf{w} \\ \mathbf{w} \end{pmatrix} \tag{3.6.5}$$

Separating the real and imaginary parts

$$\mathbf{u} = \mathbf{\Phi}\left[\mathbf{w}_R \cos\left(\beta_j t\right) - \mathbf{w}_I \sin\left(\beta_j t\right)\right] e^{\alpha_j t} + i\mathbf{\Phi}\left[\mathbf{w}_R \sin\left(\beta_j t\right) + i\mathbf{w}_I \cos\left(\beta_j t\right)\right] e^{\alpha_j t} \tag{3.6.6}$$

the real part of this expression can be compared with the free reign oscillations of a linear oscillator with one degree of freedom

$$u(t) = e^{-\xi_a \omega t}\left[A\cos\left(\omega_a t\right) + B\sin\left(\omega_a t\right)\right] \tag{3.6.7}$$

where A and B are constants that depend on the initial conditions of the oscillation, ω is the natural frequency, $\omega_a = \omega\sqrt{1-\xi_a^2}$ is the damping response frequency response and ξ_a the damping itself. Observing expressions (3.6.6) and (3.6.7) it is possible to establish the following parallel:

The imaginary part of the eigenvalues β_j takes on the role of the damping frequency ω_a of the linear oscillator, and the real part of the eigenvalues α_j is associated with the product $-\xi_a\omega$. Therefore, in a deck excited by aeroelastic loads, each pair of values α_j, β_j makes it possible to identify the characteristic parameters of the system's response:

a) The j response frequency is

$$\omega_{aj} = \beta_j \tag{3.6.8}$$

b) The j damping coefficient is

$$\xi_{aj} = \frac{-\alpha_j}{\sqrt{\alpha_j^2 + \beta_j^2}} \tag{3.6.9}$$

To resolve the eigenvalue problem $(\mathbf{A} - \mu\mathbf{I})\mathbf{w}_\mu = 0$ and obtain m pairs of complex conjugate eigenvalues $\mu_j = \alpha_j \pm \beta_j$ it is essential to follow an iterative procedure. The reason for this is that, when the matrix $\mathbf{A}$ is constructed, flutter coefficients intervene; these coefficients are a function of the reduced frequency from the system $K = Bw_d/U$. However, the frequency associated with each mode $\omega_{aj} \equiv \beta_j$ remains unknown until the eigenvalue problem has been solved. The solution to this non-linear problem rests in the procedure outlined in the flowchart of Figure 3.6.1. It consists of the following steps:

1) For the vibration mode j a value is initially taken from the β_{jp} frequency which, following iterative calculation, should coincide with the imaginary part of the eigenvalue $\omega_{aj} = b_{jp}$.
2) The eigenvalue problem is resolved $\left[\mathbf{A}(\beta_{jp}) - \mu\mathbf{I}\right]\mathbf{w}_\mu = 0$, where it is indicated that $\mathbf{A}$ depends on β_{jp}.
3) Of the $2m$ eigenvalues ($a_l \pm i\beta_l$ $l = 1,\dots,m$) obtained, it is necessary to work with the one which reduces to a minimum the difference between the value and what has been constructed $\mathbf{A}(\beta_{jp})$. This value is denoted as β_{md} and it makes minimal the difference $\left|\beta_{jp} - \beta_l\right|$.

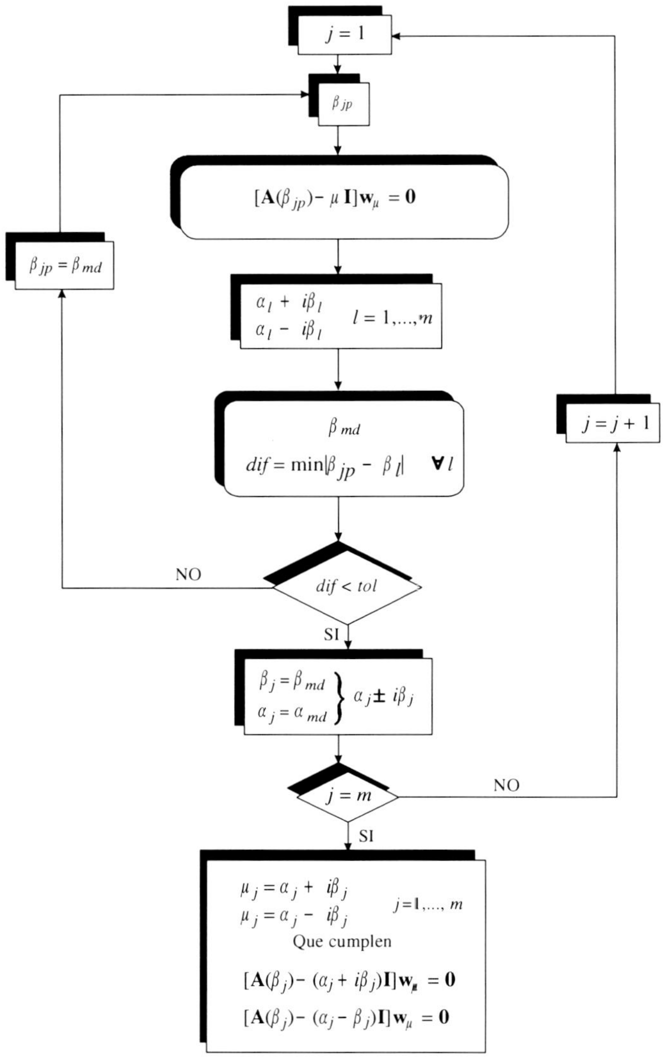

Figure 3.6.1: Flowchart about obtaining the aeroelastic response.

4) If this difference falls below the established tolerance, it may be considered that both this eigenvalue and its complex conjugate correspond to the mode j $\mu_j = a_j \pm i\beta_j$ and verify the equations $\left[\mathbf{A}(\beta_j) - (a_j \pm i\beta_j)\mathbf{I}\right]\mathbf{w}_\mu = \mathbf{0}$. If the opposite is the case and the difference is above the tolerance, Step (2) is repeated. β_{md} is taken as the initial frequency and the process is repeated until there is convergence.
5) Steps (2) to (4) are repeated for each vibration mode until m pairs of complex conjugate eigenvalues have been calculated $\mu_j = a_j \pm i\beta_j$ con $j = 1,...,m$.

The different meanings used for the β values are:

β_{jp} : Initial value for the vibration frequency of the mode j.

β_l : Imaginary part of any of the $2m$ eigenvalues obtained with $[\mathbf{A}(\beta_{jp}) - \mu\mathbf{I}]\mathbf{w}_\mu = \mathbf{0}$

β_{md} : Imaginary part of the eigenvalue reducing to a minimum the difference $\left|\beta_{jp} - \beta_l\right|$.

β_j : Frequency response for the mode j obtained after the convergence. This value complies with $\left[\mathbf{A}(\beta_j) - (\alpha_j \pm i\beta_j)\mathbf{I}\right]\mathbf{w}_\mu = 0$.

3.7 Wind speed and frequency at the outset of flutter

If, for a given wind speed, all the real parts α_j for each one of the complex eigenvalues obtained from $(\mathbf{A} - \mu\mathbf{I})\mathbf{w}_\mu e^{\mu t} = \mathbf{0}$ are negative, the effective damping of the deck – according to (3.6.9) – will always be positive. Moreover, there will be an aeroelastic response that is exponentially reduced moving towards stability The critical condition for the flutter phenomenon corresponds with the lowest wind speed. Thus, a complex eigenvalue is obtained whose real part is null; in other words it is at the transition between positive damping (decreasing oscillations) and a negative one (oscillations of growing amplitude). This trend occurs when the real part of a complex eigenvalue goes from negative to positive. Therefore

$$\alpha_j = 0 \tag{3.7.1}$$

identifies the critical condition for flutter. The process for obtaining the critical wind speed for flutter in a bridge U_f is as follows:

1) The initial speed U is taken while it is low enough to make flutter instability unlikely.
2) The non-linear eigenvalue problem is resolved $(\mathbf{A} - \mu\mathbf{I})\mathbf{w}_\mu = \mathbf{0}$, as explained in (3.5), to obtain pairs of complex conjugate eigenvalue along the lines of $\mu_j = \alpha_j \pm i\beta_j$ $j = 1, ..., m$.
3) A search is made for the eigenvalue with the least damping $\xi_{\min} = \min(\xi_{aj})$, $j = 1, ..., m$ with $\xi_{aj} = -\alpha_j \big/ \sqrt{\alpha_j^2 + \beta_j^2}$
4) If $\xi_{\min}$ is positive, there is no flutter instability. Steps (1) to (3) are repeated with a higher wind speed $U = U + \Delta U$, until $\xi_{\min}$ cancels itself out, the point at which the critical wind speed is reached, U_f. Moreover, the imaginary part of this eigenvalue β_j is the vibration frequency with which oscillations are produced in flutter.

Figure 3.7.1 has a flowchart demonstrating how to obtain flutter speed for a bridge. A similar scheme is used by Katsuchi *et al.*[24]. Because the latter is a little less explicit, special care has been taken in this text to describe each step.

In this process, care must be taken when choosing the initial frequency values β_{jp} of every mode for each wind speed U. For the first calculation, it is a good idea to take the natural vibration frequency for each mode when the wind speed U is low.

$$\beta_{jp,U} = \omega_j \tag{3.7.2}$$

Then, as the wind speeds gradually increase $U = U + \Delta U$, values can be taken based on the evolution of frequencies from previous steps

$$\beta_{jp,U} = 2\omega_{aj,U-\Delta U} - \omega_{aj,U-2\Delta U} \tag{3.7.3}$$

Figures 3.7.2 and 3.7.3 show the types of graphical representation that can be produced with the evolution of the real and imaginary parts of the eigen values – α and β

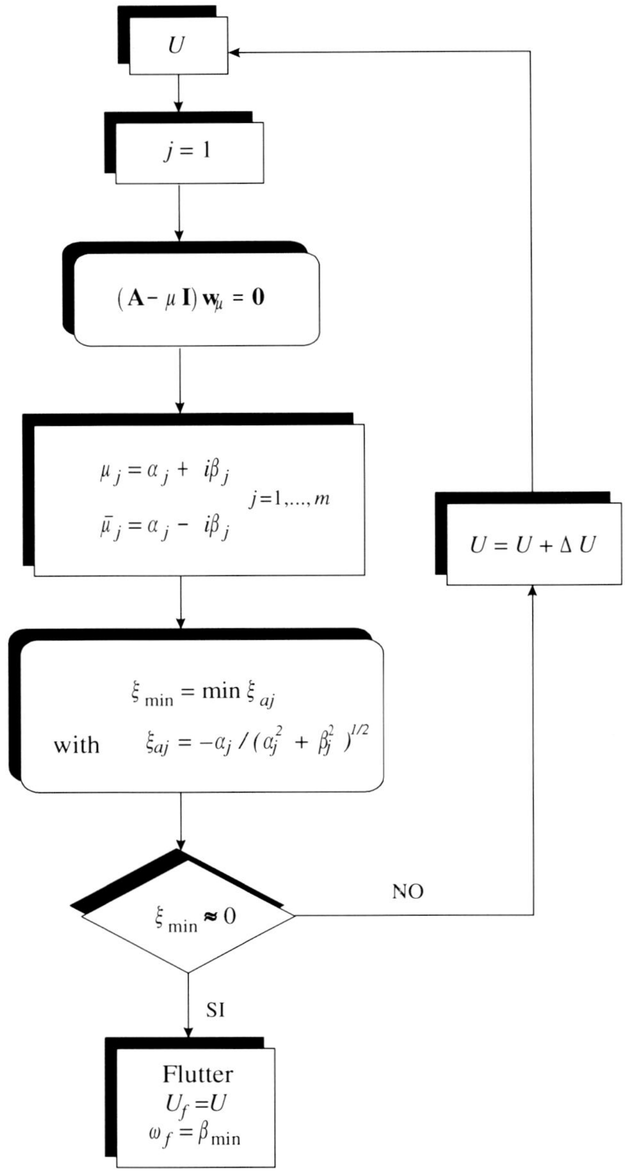

Figure 3.7.1: Flowchart about obtaining flutter speed.

respectively – in function of the wind speed U. The critical value U_f, due to which the phenomenon of flutter begins, coincides with the lower wind value and therefore one of the curves in Figure 3.7.2 dissects the horizontal axis. Representations that are similar to the parameters determining flutter in a bridge can be found in articles published by Boonyapinyo *et al.*[31].

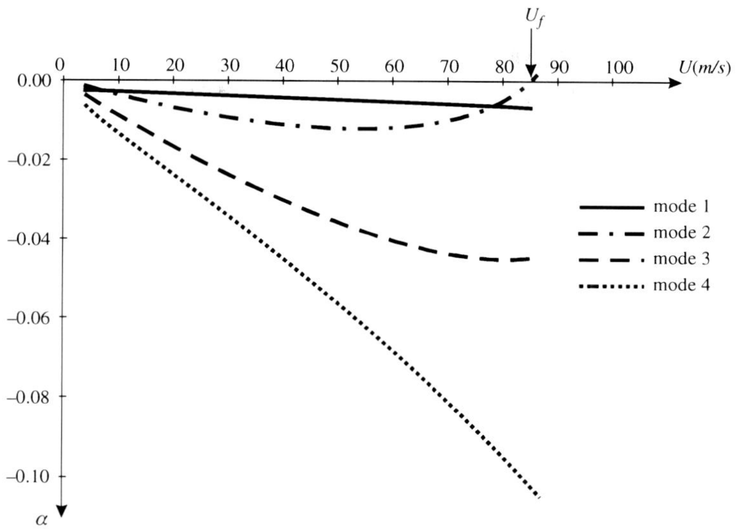

Figure 3.7.2: How the real part α of the eigenvalues evolves in an example. This entailed working with four modes.

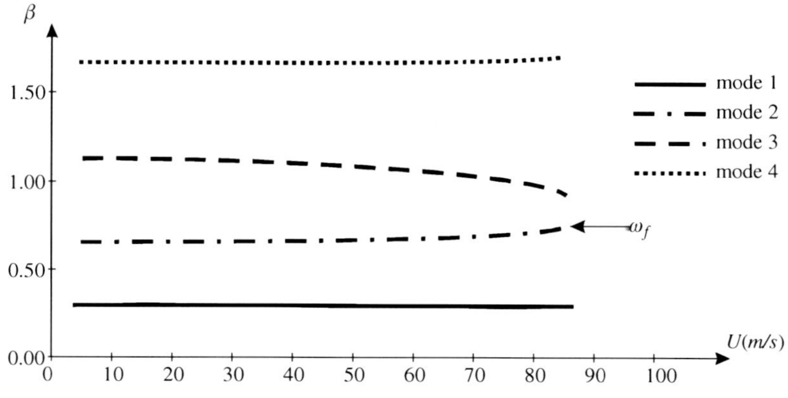

Figure 3.7.3: How the imaginary part β of the eigenvalues evolves in an example. This entailed working with four modes.

3.8 Existence of simultaneous flutter frequencies

One result becomes apparent when the flutter speed of a bridge is calculated. Several of the curves representing the real part of the eigenvalues α_j (related to damping) simultaneously dissect the horizontal axis. Figure 3.8.1 reveals how two distinct eigenvalues can appear with real parts that are simultaneously annulled at the same wind speed. Nevertheless, the imaginary part of those values would be different, hin ing at the existence of

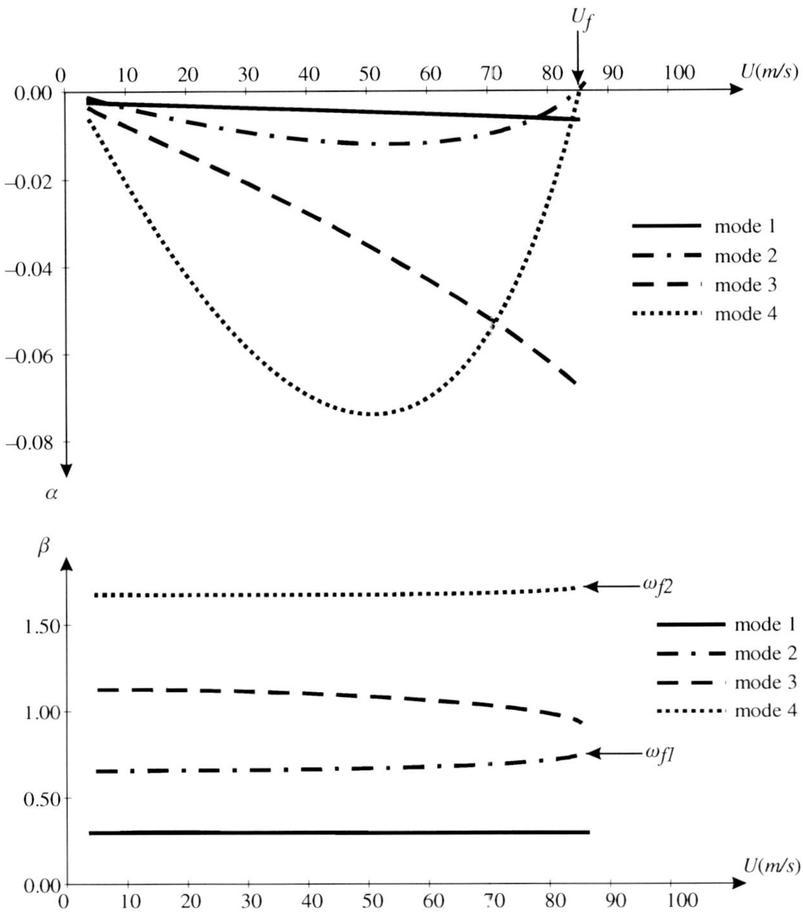

Figure 3.8.1: Example in which simultaneous flutter frequencies appear.

two possible flutter response frequencies. Later, in Chapter 5, the flutter speed for various well-known, long-span bridges is calculated. With some of these, there are other instances in which flutter can be reached with different vibration frequencies.

This phenomenon will recall the dynamic problems for obtaining frequencies and natural vibration modes. In these problems, different vibration modes correspond to the same natural frequency. The term for these is repeated vibration frequencies. They occur in the corresponding eigenvalue problem $(\mathbf{K} - \omega^2\mathbf{M})\phi = \mathbf{0}$. When the determiner of the coefficient matrix equates to zero $|\mathbf{K} - \omega^2\mathbf{M}| = 0$, multiple roots for the values w^2 are obtained. An aeroelastic problem in which there are simultaneous flutter frequencies does not necessarily involve repeated eigenvalues. Rather, these would be produced when, for the same flutter speed, there are several complex eigenvalues $\mu_j = a_j + i\beta_j$ with null real parts α_j and different imaginary ones β_j.

The same can be said of the problem through which the loads and buckling modes of a structure are obtained. These are also defined by an eigenvalue problem and, in certain cases, may lead to multiple buckling modes, as demonstrated in Khot[32]. Obtaining flutter speed entails resolving an eigenvalue problem. Moreover, there are precedents for calculating vibration modes and buckling. It is therefore surprising that no previous references are made about research on simultaneous multiple flutter modes. Undoubtedly, these situations present difficulties when it comes to carrying out sensitivity analyses, given that the derivability conditions within the design constraints in this kind of problem are very special. These circumstances are dealt with in Chapter 5 in an overview about the methods of sensitivity analysis as applied to eigenvalue problems. Chapter 6 then presents an effective way of calculating the flutter-induced phenomenon sensitivity in bridges.

3.9 References

[1] Jones N. P., Scanlan R. H., Jain A., Katsuchi H. [1998] *Advanced (and challenges) in the prediction of long-span bridge response to wind.* Proceedings of the International Symposium on Advances in Bridge Aerodynamics, Copenhagen, Denmark, 10–13 May, pp. 59–85.

[2] Scanlan R. H. [1993] *Problematics in formulation of wind-force models for bridge decks.* Journal of Engineering Mechanics, Vol. 119, No. 7, July, pp. 1353–1372.

[3] Dyrbye C., Hansen S.O. [1996] *Wind Loads on Structures.* John Wiley & Sons, Inc., New York.

[4] William J.R. Pope A. [1984] *Low-Speed Wind Tunnel Testing.* (2nd edn.) John Wiley & Sons, Inc., New York.

[5] Irwin P. A., Xie J., Stone G. K. [1997] *Controlling factors in the aerodinamic stability of cable stayed bridges.* Proceedings of Structures Congress XV, ASCE, Porland, Oregon USA, April 13–16, pp. 170–174

[6] Simiu E., Scanlan R. H. [1986] *Wind Effect on Structures.* (2nd edn.). Wiley, New York.

[7] Gamble S. L., Irwin P. A. [1985] *The action of wind on cable stayed bridges during construction.* Proceedings of the Fifth US National Conference on Wind Engineering. Lubbock, TX, USA.

[8] Saito T., Matsumoto M., Kitazawa M [1994] *Rain-wind excitation of cables on the cable-stayed Higashi-Kobe Bridge and cable vibration control.* Proceedings of the Conference on Cable-Stayed Bridges, Deauville, France, 1–15 October, pp. 507–514.

[9] Yamaguchi H., Fujino Y. [1998] *Stayed cable dynamics and its vibration control.* Bridge Aerodynamics, Larsen & Soren (eds.), Balkema, Rotterdam.

[10] Humphreys N. D., Baker C. J. [1992] *Forces on vehicles in cross–winds from moving model tests.* J. Wind Engineering and Industrial Aerodynamics, Vol. 41–44, pp. 2673–2684.

[11] King J. P. C., Mikitiuk M. J., Davenport A. G., Isyumov N. [1994] *A study of wind effects for the Nothumberland Strait crossing.* Faculty of Engineering Science, University of Western Ontario Report. BLWT-SS8-1994. September, London, Ontario.

[12] Davenport A. G., Isyumov N. [1968] *The application of the boundary-layer wind tunnel to the prediction of wind loading.* Proceedings of the International Research Seminar on Wind Effects on Buildings and Structures. University of Toronto Press, Toronto, pp. 201–230.

[13] Scanlan R. H., Tomko J. J. [1971] *Airfoil and Bridge Deck Flutter Derivatives.* Journal of Engineering Mechanics Division, ASCE, 97, No. EM6, December, pp. 1717–1737.

[14] Ibrahim S. R. Mikulcik E. C. [1977] *A method for the direct identification of vibration parameters from the free response*. The Shock and Vibration Bulletin, Vol. 47, part 4, pp. 183–198.
[15] Sarkar P. P., Jones N. P., Scanlan R. H. [1994] *Identification of aeroelastic parameter of flexible bridges*. Journal of Engineering Mechanics Division Vol. 120, No. 8, pp. 1718–1742.
[16] Scanlan R. H., Sabzevari A. [1969] *Experimental aerodynamic coefficients in the analytical study of suspension bridge flutter*. Journal of Engineering Mechanics Division, Vol. 11, No. 3, pp. 234–242.
[17] Scanlan R. H., Gade R. H. [1977] *Motion of suspended bridge spans under gusty wind*. Journal of Structural Division, ASCE. Vol. 103, No. 9, pp. 1867–1883.
[18] Scanlan R. H., Lin W. H. [1978] *Effects of turbulence on bridge flutter derivatives*. Journal of Engineering Mechanics Division, Vol. 104, No. 4, pp. 719–733.
[19] Scanlan R. H., Huston D. [1985] *Sensitivity of Bridge Decks to turbulent wind*. Proceedings of Asia Pacific Symposium on Wind Engineering, University of Roorkee, India, pp. 167–172.
[20] Kumarasena T. [1989] *Wind response prediction of long-span bridges*. Ph.D. thesis, The Johns Hopkins University, Baltimore, MD, USA.
[21] Theodorsen T. [1935] *General Theory of Aerodynamic Instability and the Mechanism of Flutter*. NACA Report 741.
[22] Fung Y. C. [1955] *An Introduction to the Theory of Aeroelasticity*. Dover, New York.
[23] Jain A. [1996] *Multi-Mode Aeroelastic an Aerodynamic Analysis of Long-Span Bridge*. A dissertation submitted to The Johns Hopkins University in conformity with the requirements for the degree of Doctor of Philosophy.
[24] Katsuchi H., Saeki S., Miyata T., Sato H. [1998] *Analytical assessment in wind-resistant design of long-span bridges in Japan*. Proceedings of the International Symposium on Advances in Bridge Aerodynamics, Copenhagen, Denmark, 10–13 May, pp. 87–98.
[25] Larsen A. [2000] *Aerodynamics of the Tacoma Narrows Bridge 60 years later.* Structural Engineering International. Vol. 10, No. 4, pp. 243–248.
[26] Namini A., Aldbrecht P. [1992] *Finite element-based flutter analysis of cable suspended ridges*. Journal of Structural Engineering. Vol. 118, No 6, pp. 1509–1526.
[27] Semiao C., Mendes P. [1998] *Avaliazao das condizoes de estabilidade aerodinâmica em pontes*. Jornadas portuguesas de engenharia de estructuras. Lisboa, Portugal.
[28] Astiz M. A. [1999] *Flutter stability of very long suspension bridges*. Journal of Bridge Engineering, Vol. 3, No. 3, pp. 132–139.
[29] Mendes P. [1997] *Estabilidade Aerodinâmica de Pontes. Cálculo da Velocidade Crítica de Flutter.* Relatorio ICIST, DT No. 19/97.
[30] Farquharson F. B. [1949] *Aerodynamic Stability of Suspension Bridges*. Parts I–V, Bulletin No. 116, University of Washington Engineering Experiment Station, Seattle, pp. 1949–1954.
[31] Boonyapinyo V., Miyata T., Yamada H. [1999] *Advanced aerodynamic analysis of suspension bridges by state-space approach*. Journal of Structural Engineering, Vol. 125, No. 12, pp. 1357–1366.
[32] Khot N. S. [1984] *Optimal Design of a Structure for System Stability for a Specified Eigenvalue Distribution*. New Direction in Optimum Structural Design. John Wiley & Sons Ltd, Chicester, UK, pp. 75–87.

Chapter 4

Flutter analysis of suspension bridges during construction

4.1 Introduction

Over the years, researchers have focused on the stability of suspension bridges withstanding wind action during construction, when the deck is being built over several stages. Their studies have involved both experimentation and computation. The most outstanding of these were carried out by Tanaka[1], Ge and Tanaka[2], Cobo[3], Brancaleoni[4,5] and Larsen[6]. It is now common knowledge that deck behaviour in the face of aeroelastic instability is worse when the bridge is being constructed than when it has already been completed. This problem can be overcome in bridges with truss decks. Stretches of the deck are laid down in stages so that its upper section is not connected to the road surface until the entire length of the deck has been completed, as described in Larsen and Gimsing[6]. However, this solution does not work with bridges that have closed aerodynamic box girders, as is the case with the following examples.

This chapter provides an application of the multimodal flutter methodology described in the previous unit. Two suspension bridges at different stages of construction will serve as examples: the Great Belt and Höga Kusten Bridges. To carry out this task so that these construction stages can be studied, the FLAS program, developed by Jurado[7], has been used. This program's most outstanding feature is the way it graphically represents the process through which the flutter speed is calculated. Figure 4.4.1 shows the screen on which, as the wind speed U is increased, and as advances are made in the search for the critical speed, points are being drawn. These points represent the development of the real parts α_j of the eigenvalues, related to damping and imaginary parts β_j, which represent frequencies. These suspension bridges were chosen because both have been subject to aeroelastic studies during their construction phase by researchers such as Tanaka[1], Ge[2] or Cobo[3]. In both examples, and for each construction phase, the following tasks are carried out:

- Dynamic calculation of the structural response in free vibration without damping. The natural frequencies and vibration eigenmodes in the second-order theory are obtained with the ADISNOL3D program, developed by Mosquera[8].
- Aeroelastic calculation of the flutter speed, with the two vibration modes, the first vertical and the first torsional one, as well as the multimodal analysis, taking into account those modes most active in aeroelastic instability. This analysis is done through the FLAS[7] program.

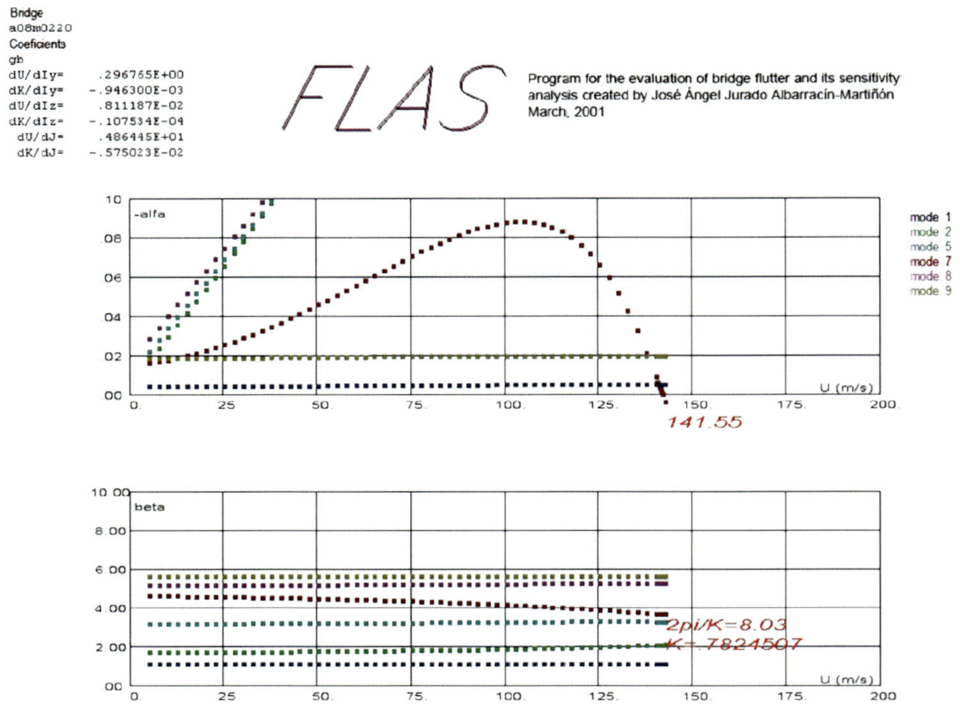

Figure 4.4.1: Graphical interface of the FLAS program.

Finally, for each bridges, a study is undertaken to show how the flutter parameters develop throughout the construction process of laying down the deck over various stages until it is in place.

Because of the scant information available, it is difficult and time-consuming to obtain data for the aeroelastic calculation of suspension bridges in the construction phase. In some cases, the construction process itself is meant; in others, the structural model is being referred to. Nevertheless, this chapter presents the results for the aeroelastic studies of two very relevant bridges at different project phases.

4.2 Höga Kusten Bridge in its construction phase

The Swedish Höga Kusten Bridge, opened in 1997, is a suspension bridge whose main span is 1210 m and side ones are 310 and 280 m, respectively. Figure 4.2.1 shows a side elevation of it, as well as a cross section of its aerodynamic box girder deck. An up-to-date photo of the bridge is found in Figure 4.2.2.

The design and construction processes took into account the location of the Höga Kusten Bridge, in a wind-swept zone. A major concern was to guarantee the bridge's security given that it was exposed to wind-produced loads during its erection. This is especially the case because, as mentioned earlier, the critical flutter speed is lower at crucial stages in the deck's construction than it is once the work has finished. A number of researchers have studied this bridge's stability under wind loads, including Tanaka[1], Cobo[3] or Ge[2]. For this reason it was chosen as an example. Table 4.2.1 provides the main geometric and mechanical characteristic of this bridge: these are from Cobo[3].

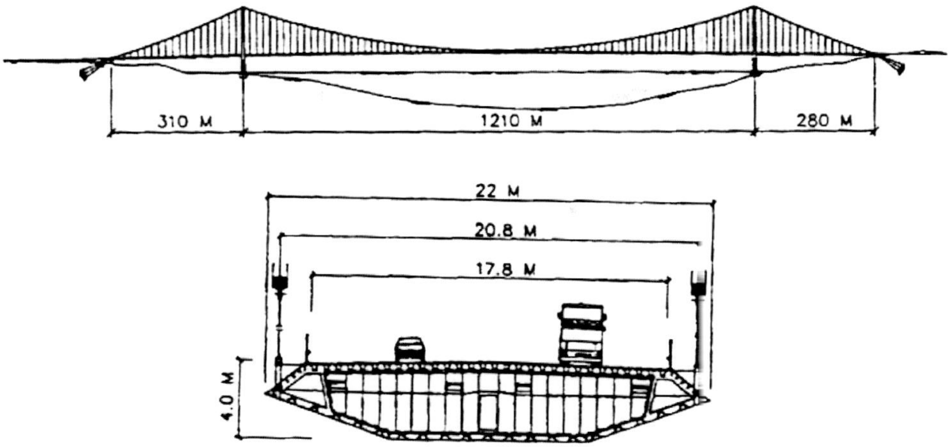

Figure 4.2.1: Elevation and cross section of the Höga Kusten Bridge.

Figure 4.2.2: Höga Kusten Bridge.

The deck's construction process was divided in 10 phases; the deck was to be laid down starting with the central span and then it would go outwards towards the towers. Figure 4.2.3 provides the deck segment order for the construction process.

4.2.1 Construction phases of the Höga Kusten Bridge

For this example, the phases described in Table 4.2.2 were chosen; these correspond to the real stages of the construction process. As is usually the case with this kind of bridge and as described in Brancaleoni[1] or Cobo[3], the prefabricated stretches of the deck's box girder are hoisted and then suspended from the main cables. Later they are joined to the next stretch by means of temporary connectors that transfer both the torsional and lateral

Table 4.2.1: Geometric and mechanical properties of the Höga Kusten Bridge.

Main span length (m)	1210
Side span lengths (m)	310–280
Distance between main cables (m)	20.8
Total deck width (m)	22
Deck depth (m)	4.0
Section of each main cable (m^2)	0.234
Inertia moment under lateral bending for the deck I_z (m^4)	31.57
Torsional inertia moment for the deck J (m^4)	1.1
Deck mass (segments) (t/m)	
25 (L = 25 m)	11.0
24, 26 (L = 16 m)	8.94
23, 27 (L = 40 m)	8.033
12–22, 28–38 (L = 40 m)	7.615
11, 39 (L = 40 m)	7.788
Polar moment for deck inertia (segments) (tm^2/m)	
25 (L = 25 m)	518.3
24, 26 (L = 16 m)	383.6
23, 27 (L = 40 m)	332.1
12–22, 28–38 (L = 40 m)	306.9
11, 39 (L = 40 m)	310.1

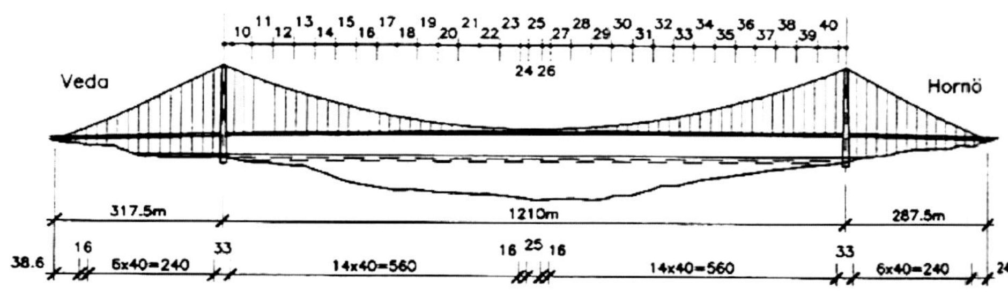

Figure 4.2.3: Defining the deck segment for the Höga Kusten Bridge.

bending inertia. At the same time, the connectors allow for vertical bending until the deck has been completed.

4.2.1.1 Phase 1: 18% of the main span

In this first phase, the length of the deck hung from cables was 217 m: 18% of the 1210 m making up the total length of the main span. At this point, the erection of the side spans had yet to begin and the configuration was symmetrical. The dynamic analysis was achieved through the ADISNOL3D program by obtaining the natural frequencies vibration eigenmodes and taking the geometric nonlinearity into account. The bar structure – in graphical form in Figure 4.2.4 – was done, with 140 bar elements and 134 nodes.

Table 4.2.3 provides the natural frequencies obtained for the significant modes along the deck until the second symmetrical torsional mode appeared. It also shows those obtained for the first symmetrical vertical bending mode and the first symmetrical torsional mode

Table 4.2.2: Defining the construction phases studied for the Höga Kusten Bridge.

	Segments laid	Deck laid (m)	Percentage of central span laid (%)	Configuration
Phase 1	22–28	217	18	Symmetrical
Phase 2	19–35	617	51	Asymmetrical
Phase 3	14–35	817	68	Symmetrical
Phase 4	10–40	1177	97	Symmetrical

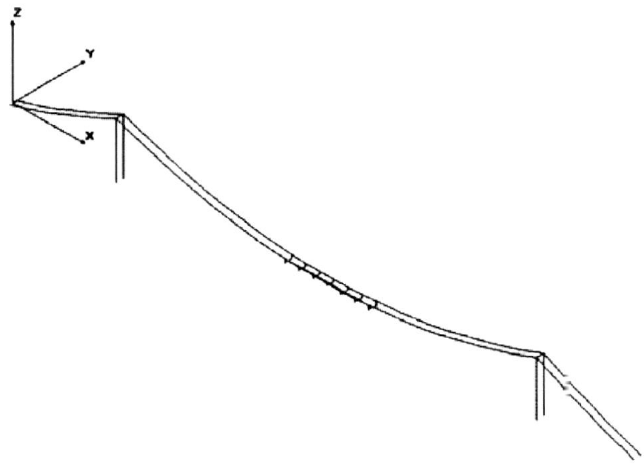

Figure 4.2.4: Structural model of Phase 1 for the Höga Kusten Bridge.

for this same phase, attributed Cobo[3] and Tanaka[1]. The modes are described in the following way:

V: vertical vibration mode
L: lateral vibration mode
T: torsional vibration mode
S: symmetric vibration mode
A: asymmetric vibration mode

Table 4.2.4 contrasts the ratios between the first symmetrical torsional and vertical bending modes obtained with those found in Cobo[3] and Tanaka[1]

It may be considered positive that these different results coincide, bearing in mind that, with the lack of data, reasonable if not real values were taken on for some of the bridge's mechanical characteristics, such as its towers and hangers.

The FLAS code also played a role in the aeroelastic analysis. In keeping with the work done by other researchers and with the aim of contrasting the results, a decision was made to use the flutter derivatives of the Great Belt. The literature offered insufficient data for the flutter derivatives along the Höga Kusten Bridge's deck section, as argued in Cobo[3]. Consequently the flutter parameters obtained here are not directly comparable with those produced in wind-tunnel testing by Tanaka[1] and Ge[2].

In terms of structural damping, a viscous type of damping was considered for all modes whose value $\xi = 0.00318$ reached a 2% level of logarithmic decrease.

The flutter parameters obtained are shown in Table 4.2.5. These include those from the aeroelastic analysis with the two classic flutter modes, (4)VS and (10)TS, as well as

Table 4.2.3: Natural frequencies (rad/s) in Phase 1 of the Höga Kusten Bridge.

Mode type	Natural frequencies (rad/s)		
	ADISNOL3D	Cobo[3]	Tanaka[1]
(1) LS	0.297		
(2) VA	0.727		
(3) LA	0.734		
(4) VS	0.850	0.92	0.94
(7) LTA	0.984		
(9) LA	1.052		
(10) TS	1.061	1.1	1.15
(12) VA	1.366		
(15) VS	1.451		
(21) LS	1.731		
(22) LA	1.758		
(23) VS	1.817		
(24) TA	1.970		
(25) VA	2.089		
(26) TS	2.097		

Table 4.2.4: Ratios (ω_a/ω_h) in Phase 1 of Höga Kusten Bridge.

	Ratio (ω_a/ω_h)
ADISNOL3D	1.25
Cobo[3]	1.19
Tanaka[1]	1.26

Table 4.2.5: Results of flutter analysis at Phase 1 for the Höga Kusten Bridge.

	Cobo[3] 2 modes	Cobo[3] multimodal	Tanaka[1] T.V.	FLAS 2 modes	FLAS 15 modes
U_f (m/s)	34.75	35.0	38.2	32.17	32.23
K_f				0.63406	0.63250

the ones for multimodal aeroelastic analysis with the 15 modes calculated. All the modes were considered until the deck's second symmetrical torsional mode appeared.

A perusal of Table 4.2.5 reveals that, for this phase, the aeroelastic analysis with 2 modes brings almost identical results to those produced in the multimodal analysis. This happens because, in this early stage of the bridge's construction, there is only one symmetrical vertical bending vibration mode with a frequency lower than the first symmetrical torsional vibration mode.

Thus the way in which these various results coincide can be viewed in a favourable light given the lack of data, already mentioned, to make a structural model of the bridge's exact shape. Moreover, as done by other researchers, the Great Belt Bridge's flutter derivatives were used; these would not have perfectly corresponded with those for the other bridge's deck.

4.2.1.2 Phase 2: 51% of the central span

By the second phase, 617 m of deck had already been suspended: 51% of the main span's total length of 1210 m. As mentioned in Cobo[3], the side spans were already in place in this phase and, as shown in Table 4.2.2, the configuration is not symmetrical. Figure 4.2.5 offers a graphical representation of the structural bar model, comprising 190 bar elements and 164 nodes. Tables 4.2.6–4.2.8 contrast the natural frequencies and flutter speeds obtained for this phase with those found in other studies.

By studying Table 4.2.8 one notices how, in this phase, aeroelastic analysis with 2 modes produces results that are superior to the flutter speed of the multimodal analysis. This point is also confirmed by Cobo[3], without actually specifying the number of modes included in his multimodal analysis. As for the data from wind-tunnel testing mentioned in the literature, there seems to be two kinds of results in various publications. Once again, one may view the concordance of these results in a favourable light.

4.2.1.3 Phase 3: 68% of the central span

Along with the entire lateral spans, 817 m of the deck's length were hanging from the cables by this third stage: 68% of the main span's length of 1210 m. Referring back to Table 4.2.2, it may be seen that the configuration was symmetrical. Figure 4.2.6 has the new structural bar model: 220 bar elements and 179 nodes. The results obtained for this phase, operating in an analogue form to the previous ones shown in Tables 4.2.9–4.2.11, can be compared with the results found in other publications. Table 4.2.11 shows how the aeroelastic analysis with 2 modes brings out results once again superior to those of the multimodal analysis.

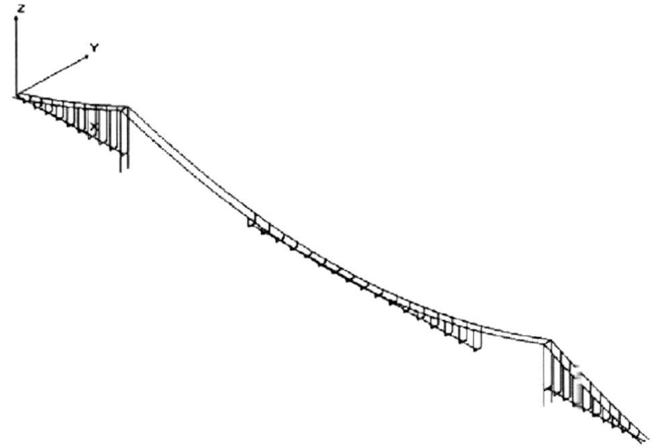

Figure 4.2.5: Structural model for Phase 2 of the Höga Kusten Bridge.

Table 4.2.6: Natural frequencies (rad/s) in Phase 2 for the Höga Kusten Bridge.

Mode type	Natural frequencies (rad/s)		
	ADISNOL3D	Cobo[3]	Tanaka[1]
(1) LS	0.290		
(2) LA	0.497		
(3) VA	0.630		
(4) VS	0.885	0.92	1.05
(7) LA	1.220		
(8) VS	1.268		
(10) LTS	1.318		
(12) VA	1.477		
(16) TS	1.760	1.82	1.87
(17) VS	1.828		
(21) LA	1.979		
(22) VA	2.114		
(23) LS	2.128		
(26) VS	2.412		
(27) TA	2.436		
(28) VA	2.667		
(30) TS	2.710		

Table 4.2.7: Ratios (ω_a/ω_h) in Phase 2 of the Höga Kusten Bridge.

	Ratio (ω_a/ω_h)
ADISNOL3D	1.98
Cobo[3]	1.94
Tanaka[1]	1.78

Table 4.2.8: Results of flutter analysis at Phase 2 for the Höga Kusten Bridge.

	Cobo[3] 2 modes	Cobo[3] multimodal	Tanaka[1]	Ge and Tanaka[2] T.V.	FLAS 2 modes	FLAS 17 modes
U_f (m/s)	70.25	65.75	70.00[1]	64.54[2]	67.45	62.63
K_f					0.44051	0.42492

4.2.1.4 Phase 4: 97% of the main span

In this fourth phase, the length of deck already suspended from the cables is 1177 m: that is, 97% of the total made up by the main span and the two side ones, or a total of 1210 m. There is a symmetrical configuration, as seen in Table 4.2.2. The structural bar model, found in Figure 4.2.7 consists of 250 bar elements and 197 nodes. Table 4.2.14 reveals

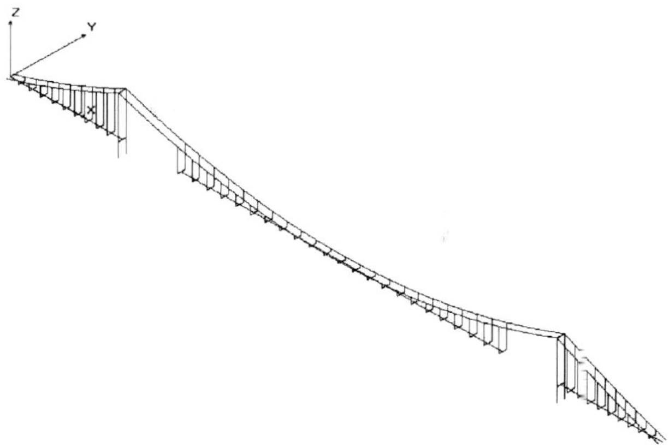

Figure 4.2.6: Structural model for Phase 3 of the Höga Kusten Bridge.

Table 4.2.9: Natural frequencies (rad/s) in Phase 3 of the Höga Kusten Bridge.

Mode type	Natural frequencies (rad/s)		
	ADISNOL3D	Cobo[3]	Tanaka[1]
(1) LS	0.291		
(2) LA	0.428		
(3) VA	0.598		
(4) VS	0.870	0.91	0.95
(5) LS	0.895		
(7) VS	1.189		
(9) VA	1.344		
(13) LTA	1.575		
(14) VS	1.738		
(17) TS	2.024	1.84	1.90
(22) VA	2.100		
(23) LA	2.168		
(24) LTS	2.179		
(26) LS	2.396		
(27) LA	2.498		
(28) VS	2.594		
(29) TS	2.653		

Table 4.2.10: Ratios (ω_a/ω_h) in Phase 3 of the Höga Kusten Bridge.

	Ratio (ω_a/ω_h)
ADISNOL3D	2.32
Cobo[3]	2.03
Tanaka[1]	2.00

Table 4.2.11: Results of flutter analysis at Phase 3 for the Höga Kusten Bridge.

	Cobo[3] 2 modes	Cobo[3] multimodal	FLAS 2 modes	FLAS 17 modes
U_f (m/s)	69.50	62.00	72.94	61.30
K_f			0.46963	0.44332

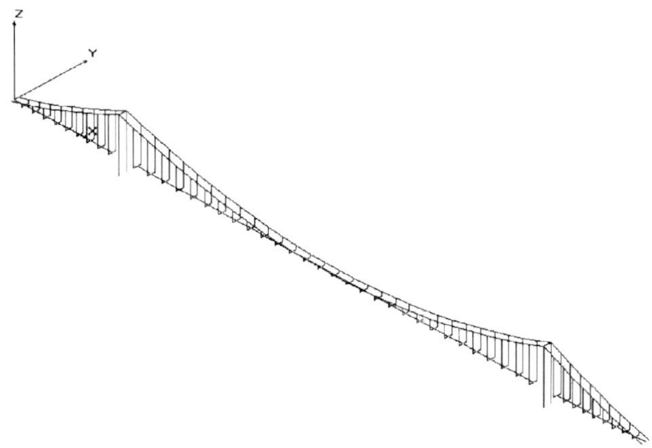

Figure 4.2.7: Structural model for Phase 4 of the Höga Kusten Bridge.

Table 4.2.12: Natural frequencies (rad/s) in Phase 4 of the Höga Kusten Bridge.

Mode type	Natural frequencies (rad/s)		
	ADISNOL3D	Cobo[3]	Tanaka[1]
(1) LS	0.292		
(2) LA	0.377		
(3) VA	0.584		
(4) LS	0.614		
(5) VA	0.840		
(6) VS	0.865	0.88	0.88
(7) VS	1.182		
(8) VA	1.281		
(9) LA	1.289		
(14) VS	1.620		
(15) TA	1.754		
(18) VA	1.919		
(19) TS	2.076	2.10	2.06
(24) LTS	2.152		
(25) VS	2.238		
(28) VA	2.548		
(30) TS	2.677		

Table 4.2.13: Ratios (ω_a/ω_h) in Phase 4 of the Höga Kusten Bridge.

	Ratio (ω_a/ω_h)
ADISNOL3D	2.40
Cobo[3]	2.45
Tanaka[1]	2.34

Table 4.2.14: Results of flutter analysis at Phase 4 for the Höga Kusten Bridge.

	Cobo[3] 2 modes	Cobo[3] multimodal	Tanaka[1] T.V.	FLAS 2 modes	FLAS 17 modes
U_f (m/s)	78.25	77.75	70.00	80.40	60.45
K_f				0.46718	0.52880

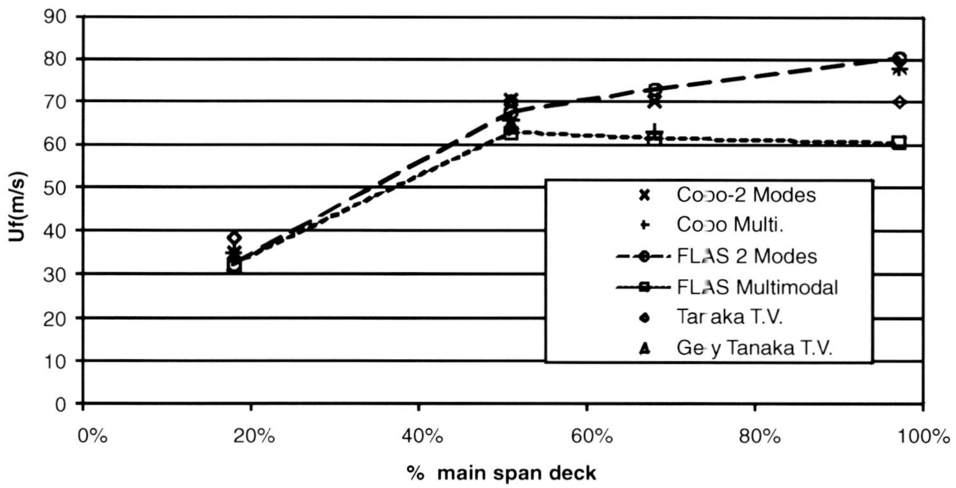

Figure 4.2.8: Flutter speed evolution U_f as the Höga Kusten deck was erected.

that, for this phase, the aeroelastic analysis with 2 modes gives results that are once again superior to those of the multimodal analysis. This conclusion is confirmed in Cobo[3]. Tables 4.2.12–4.2.14 offer the comparative results for this construction phase.

4.2.2 Flutter parameter evolution in the construction phase of the Höga Kusten Bridge

Figures 4.2.8 and 4.2.9 show how the results of the flutter analysis develop as advances are made in the erection of the deck throughout the phases analysed earlier. The results obtained are provided. Moreover, the two vibration modes are considered, along with the

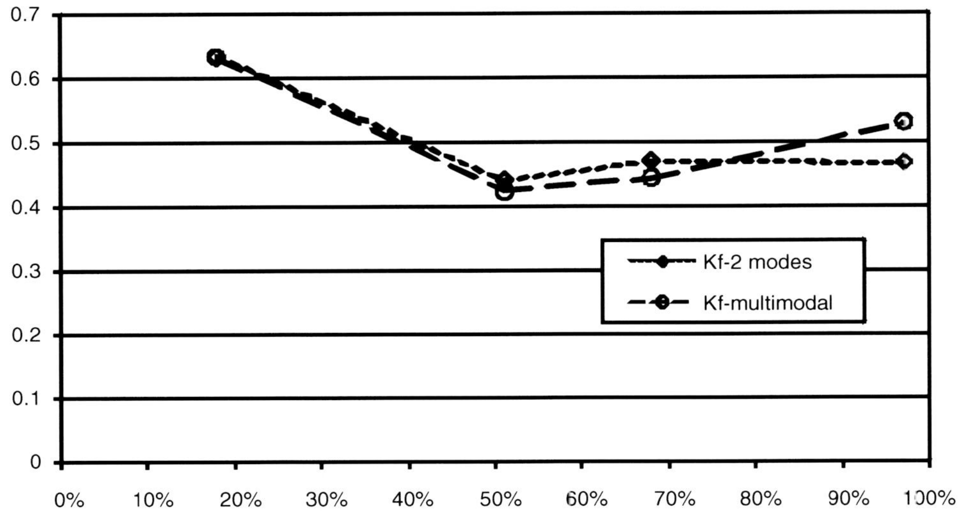

Figure 4.2.9: Reduced frequency evolution K_f as the Höga Kusten deck was erected.

Table 4.2.15: Summary of flutter speeds (m/s) obtained by various researchers for the Höga Kusten Bridge in the construction phase.

	Percentage of main span (%)	Cobo[3] 2 modes	Cobo[3] Multimodal	Tanaka[1]	Ge[2] T.V.	FLAS 2 modes	FLAS multimodal
Phase 1	18	34.75	35.0	38.2[ ]		32.17	32.23
Phase 2	51	70.25	65.75	70.00[1]	64.54[2]	67.45	62.63
Phase 3	68	69.50	62.00			72.94	61.30
Phase 4	97	78.25	77.75	70.00[1]		80.40	60.45

multimodal analyses; the results from Cobo[3], for which the Great Belt Bridge's derivatives were used; and the results from Tanaka[1], who used complete reduced models in wind tunnels. Table 4.2.15 has the numerical values for the flutter speed.

After studying the figures, it can be concluded that one must consider a sufficient number of vibration modes in the aeroelastic analysis. Greater differences can be observed between bimodal and multimodal analyses as the deck length increases across the different phases.

Also noteworthy is how the flutter speed is minimal at the early construction phase under consideration (18% of the central span). It remains almost steady with few variations once 50% of the main span has been constructed. Once again the experimental results of Tanaka[1] corroborate this point. Similar types of behaviour have also been detected by Brancaleoni[1] for the Humber Bridge in its construction phase.

4.3 The Great Belt Bridge in its construction phase

As can be appreciated in Figure 4.3.1, the construction process in the bridge over the Great Belt in Denmark has followed a sequence of phases starting from the main span and working out towards the towers. Table 4.3.1 provides the main geometric and mechanical characteristics of this suspension bridge. As mentioned earlier, the prefabricated box girder sections for the deck are hoisted and then hung from the main cables. Later, provisional hangers are used to connect these cables to the contingent deck span, transferring both the torsional and lateral bending inertia. However, vertical bending is permitted until the entire deck is constructed. Brancaleoni[1] and Cobo[3] mention that this is a common practice with this type of bridge. The elevation and two cross sections of this magnificent bridge are found in Figure 4.3.2a and b.

Figure 4.3.1: Laying the deck of the Great Belt Bridge.

Table 4.3.1: Geometrical and mechanical properties of the Great Belt Bridge.

Main span length (m)	1624
Side span length (m)	535
Main cable sag (m)	180
Distance between the main cables (m)	27
Total deck width (m)	31
Deck depth (m)	4.4
Cross section of each main cable (m^2)	0.44
Mass of each main cable (t/m)	3.74
Inertial moment under lateral bending in deck I_z (m^4)	100.0
Inertial moment under torsion in deck J (m^4)	7.6
Deck mass (t/m)	15.26
Deck's polar inertial moment (tm^2/m)	1106.8

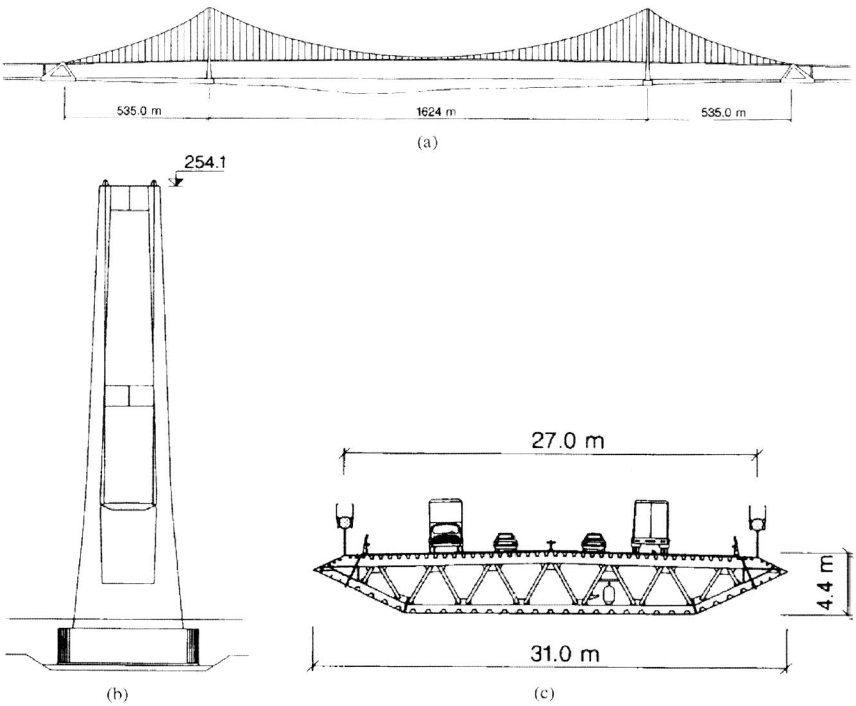

Figure 4.3.2: (a) Elevation of Great Belt Bridge, (b) Tower and (c) cross section of Great Belt Bridge.

Table 4.3.2: Construction phases of Great Belt Bridge being studied.

	Central span length (m)	Percentage of central span (%)
Phase 1	150	9.25
Phase 2	250	15.4
Phase 3	350	21.6
Phase 4	450	27.8
Phase 5	649.6	40
Phase 6	974.4	60
Phase 7	1299.2	80
Phase 8	1461.6	90

4.3.1 Construction phases of the Great Belt Bridge

The only experimental data related to the construction of this bridge which are available have been taken from Tanaka[1]. These correspond with testing carried on a complete reduced model of the bridge on a scale of 1/200 and only at the early phases of its execution, with 150, 250, 350 and 450 m of deck laid, or 9.25%, 15.4%, 21.6% and 27.8% of the main deck complete. For this example, these construction phases, as well as more advanced ones, were considered. Taking the above data into account, a study was carried out on the phases outlined in Table 4.3.2, all of which with a symmetrical

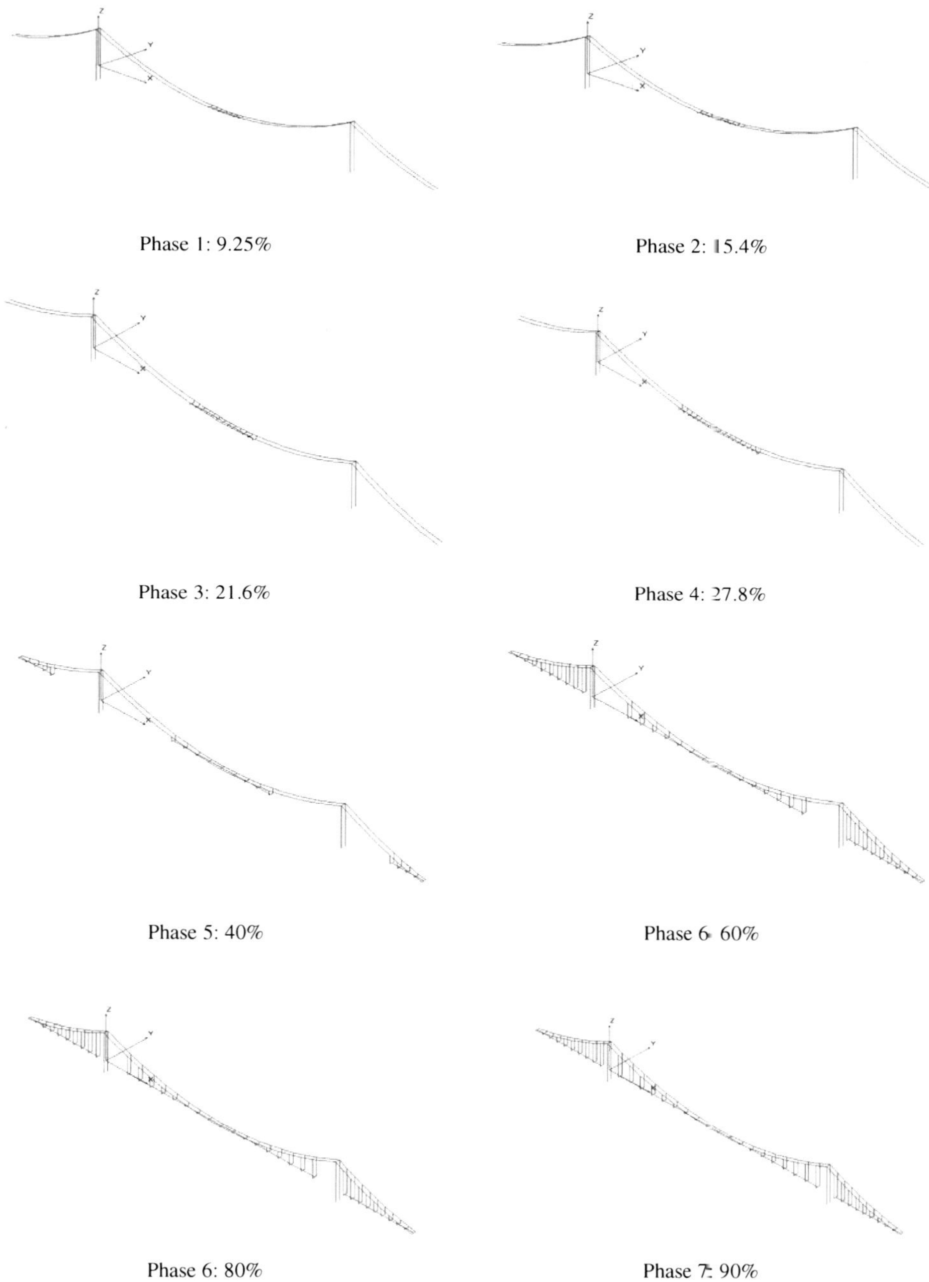

Phase 1: 9.25% — Phase 2: 15.4%

Phase 3: 21.6% — Phase 4: 27.8%

Phase 5: 40% — Phase 6: 60%

Phase 6: 80% — Phase 7: 90%

Figure 4.3.3: Structural models of the construction phases of the Great Belt Bridge.

configuration. In Figure 4.3.3 there is a structural bar model for each phase, while in Figure 4.3.3 information about each of these structural models is detailed.

With these data and through the ADISNOL3D program, it was possible to calculate the natural frequencies and eigenmodes of vibration in second-order theory. Table 4.3.4

outlines the natural frequencies obtained for the most significant modes along the deck until the second symmetrical torsion mode.

The aeroelastic analysis was carried out using the FLAS program. Figures 4.3.4 and 4.3.5 show the flutter derivatives for this bridge's deck, provided by Cobo[3] and Jurado[7].

As for structural damping, the viscous variety was considered for those modes whose value was $\xi = 0.00318$ corresponding to a logarithmic decrease δ of 2%

Table 4.3.3: Nodes and elements from the structural models at each construction phase.

	Nodes	Bar elements
Phase 1	171	180
Phase 2	177	190
Phase 3	189	210
Phase 4	201	230
Phase 5	142	174
Phase 6	184	244
Phase 7	196	264
Phase 8	202	274

Table 4.3.4(a): Vibration modes and natural frequencies throughout the Great Belt phases.

Phase 1: 9.25%		Phase 2: 15.4%		Phase 3: 21.6%		Phase 4: 27.8%	
Modes	ω_n (rad/s)	Modes	ω_n (rad/s)	Modes	ω_n (rad/s)	Modes	ω_n (rad/s)
(1) LS	0.241	(1) LS	0.238	(1) LS	0.233	(1) LS	0.231
(2) VA	0.548	(2) VS	0.537	(2) VS	0.532	(2) VS	0.538
(3) VS	0.549	(3) VA	0.557	(3) VA	0.560	(3) VA	0.539
(4) LA	0.581	(4) LA	0.592	(4) LA	0.583	(4) LA	0.544
(7) LS	0.715	(7) LA	0.772	(7) LTS	0.831	(7) LTS	0.894
(9) TS	0.764	(9) TS	0.789	(8) TS	0.862	(8) TS	0.927
(10) VS	0.853	(10) VS	0.881	(9) LA	0.877	(9) LA	0.961
(16) LA	1.102	(15) VA	0.929	(10) VS	0.935	(10) VS	0.970
(17) VA	1.116	(16) LA	1.037	(11) LA	0.938	(11) LA	0.975
(20) LS	1.322	(17) VA	1.097	(16) VA	0.959	(16) LA	1.019
(22) VS	1.346	(20) LS	1.387	(17) VA	1.074	(17) VA	1.108
(24) LA	1.451	(21) VS	1.395	(18) VS	1.426	(18) VA	1.366
(25) VA	1.600	(22) LA	1.432	(21) LS	1.524	(19) VS	1.588
(37) VS	1.869	(23) LA	1.439	(22) LA	1.530	(22) LS	1.646
(41) TA	2.015	(25) VA	1.598	(24) VA	1.681	(24) LA	1.731
(43) TS	2.057	(34) TS	1.875	(25) VA	1.742	(25) VA	1.895
		(35) VS	1.913	(26) TS	1.754	(26) VA	1.938
		(36) VA	1.937	(35) VS	1.906	(29) TS	1.983
		(42) TS	2.015	(40) VS	2.203	(31) VS	1.989
		(43) TA	2.164	(45) TA	2.511	(36) TS	2.086

Table 4.3.4(b): Vibration modes and natural frequencies throughout the Great Belt phases.

Phase 5: 40%		Phase 6: 60%		Phase 7: 80%		Phase 8: 90%	
Modes	ω_n (rad/s)	Modes	ω_n (rad/s)	Modes	ω_n (rad/s)	Modes	ω_n (rad/s)
(1) LS	0.234	(1) LS	0.234	(1) LS	0.234	(1) LS	0.234
(2) LA	0.515	(2) LA	0.343	(2) LA	0.343	(2) LA	0.318
(3) VA	0.539	(3) VA	0.491	(3) VA	0.491	(3) VA	0.488
(4) VS	0.545	(4) VS	0.574	(4) VS	0.574	(4) LS	0.499
(5) VA	0.856	(7) VA	0.596	(5) LS	0.596	(5) VS	0.578
(6) VS	0.908	(8) VS	0.777	(8) VA	0.777	(8) VA	0.782
(9) LS	0.973	(9) LS	0.791	(9) VS	0.791	(9) VS	0.784
(10) LA	0.979	(12) LA	1.054	(10) VA	1.054	(10) VA	1.032
(11) TS	1.117	(13) LS	1.120	(11) LA	1.120	(11) LA	1.034
(12) VA	1.138	(14) VA	1.221	(14) LS	1.221	(14) LS	1.237
(17) VS	1.300	(15) VS	1.284	(15) VS	1.284	(15) VS	1.262
(18) VS	1.637	(22) TS	1.390	(16) LA	1.390	(16) LA	1.298
(19) VA	1.657	(24) VA	1.447	(20) LTA	1.447	(18) LA	1.412
(20) VA	1.720	(25) LA	1.505	(21) VS	1.505	(21) VS	1.496
(23) LA	1.743	(28) LS	1.589	(24) VA	1.589	(22) VA	1.533
(24) LS	1.774	(29) VS	1.752	(25) TS	1.752	(25) VS	1.797
(25) TA	1.805	(30) TA	1.882	(26) VS	1.882	(26) TS	1.801
(27) VA	1.956	(31) VA	2.010	(29) LTA	2.010	(27) LS	1.956
(28) TS	2.010	(32) VA	2.056	(30) LS	2.056	(28) VA	2.017
		(33) VS	2.100	(31) VA	2.100	(29) LA	2.022
		(34) TA	2.139	(32) TA	2.139	(32) TA	2.128
		(35) TS	2.274	(33) VA	2.274	(33) LS	2.130
				(34) VS	2.310	(34) VS	2.247
				(35) VS	2.357	(35) VA	2.294
				(36) LTS	2.445	(36) VS	2.341
				(37) TS	2.497	(37) TS	2.444

$$\delta = \frac{2\pi\xi}{\sqrt{1-\xi^2}} \approx 2\pi\xi = 0.02 \tag{4.3.1}$$

In Table 4.3.5, the flutter speeds are obtained. These refer to both the aeroelastic analysis with the two classic flutter modes, vertical and torsional of the first order, and aeroelastic mulitmodal analysis with all the calculated modes. Also added are the experimental data about critical wind speed obtained by Tanaka[1].

By analysing Table 4.3.5, one can deduce that:

- In phases 1, 2, 3 and 4, the 2 mode aeroelastic and multimodal analyses lead to very similar results. The former is put in such a way that there is only one vertical and symmetrical vibration mode with a frequency lower than the first symmetrical torsional mode. There is no coupling between vertical modes that would produce variations in the flutter speed. This flutter speed is similar to the one obtained by Tanaka[1] through wind-tunnel testing of a reduced scale complete model of the bridge in its construction phase.

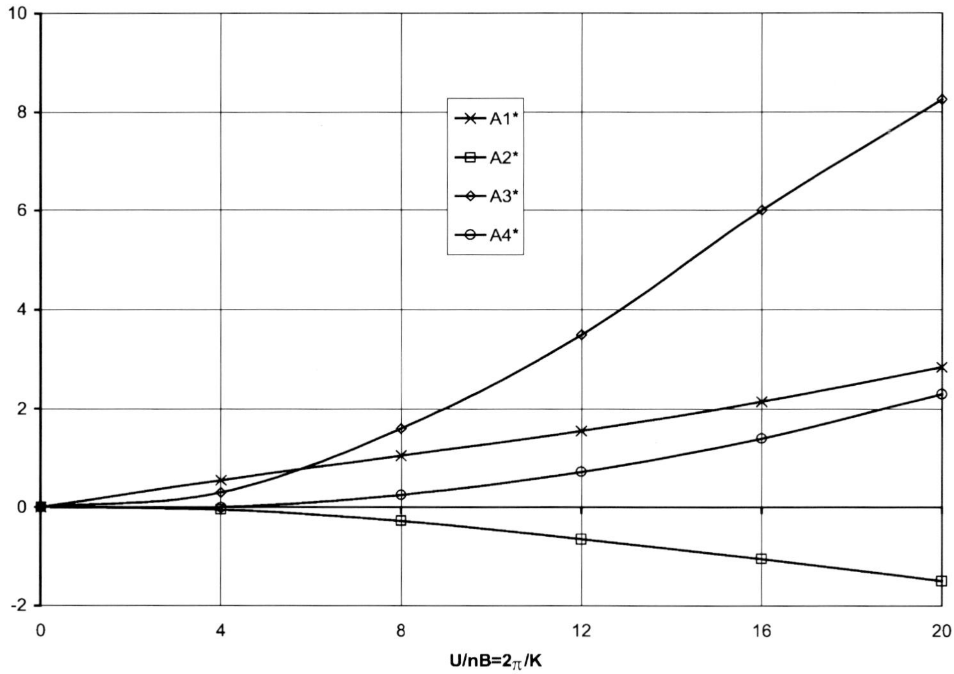

Figure 4.3.4: Flutter derivatives A_i* for the Great Belt Bridge.

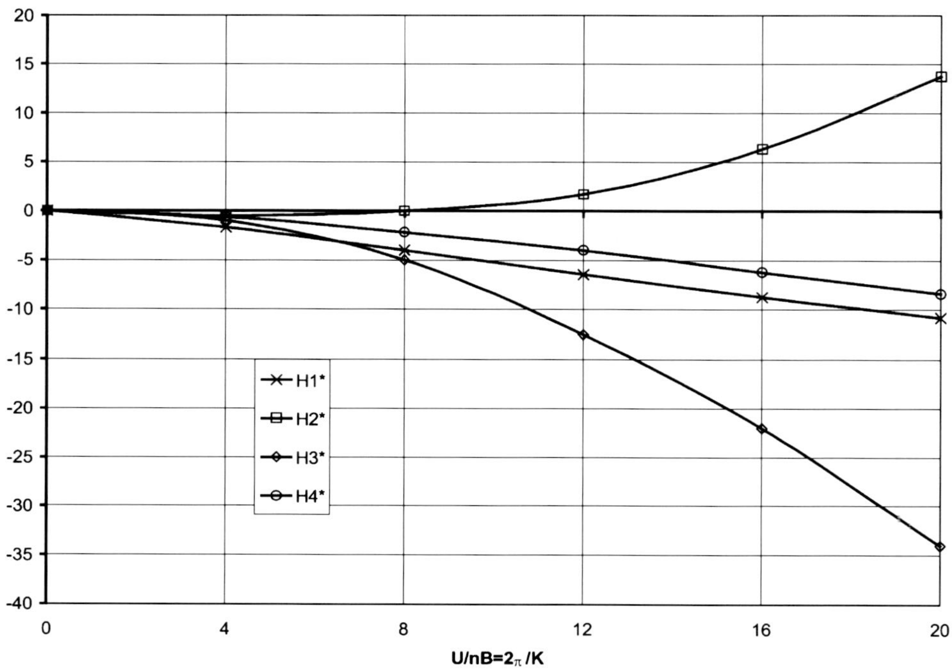

Figure 4.3.5: Flutter derivatives H_i* for the Great Belt Bridge.

Table 4.3.5: Numerical results of flutter analysis for the Great Belt construction phases.

		FLAS 2 modes	FLAS multimodal	Tanaka[1] (wind tunnel)
Phase 1	U_f (m/s)	33.83	35.00	32.5
9.25%	K_f	0.64312	0.61620	
Phase 2	U_f (m/s)	37.25	38.12	41.0
15.4%	K_f	0.52244	0.50664	
Phase 3	U_f (m/s)	42.14	42.04	52
21.6%	K_f	0.49182	0.48633	
Phase 4	U_f (m/s)	46.33	46.28	57
27.8%	K_f	0.48903	0.48393	
Phase 5	U_f (m/s)	52.95	56.15	
40%	K_f	0.53887	0.47888	
Phase 6	U_f (m/s)	75.62	58.88	
60%	K_f	0.44825	0.63306	
Phase 7	U_f (m/s)	87.26	54.54	
80%	K_f	0.42573	0.92167	
Phase 8	U_f (m/s)	90.26	58.79	
90%	K_f	0.426013	0.868412	

- Tanaka's[1] results seem excessively high for Phases 3 and 4 if one takes into account that the critical flutter speed for the whole bridge obtained in this example is 62.02 m/s. Furthermore, these results vastly differ from those provided by Larsen[6] for the phase in which 20% of the main span had been constructed. The results given for this example fall between the other two.
- In Phase 5, with 40% of the span in place, multimodal aeroelastic analysis gives a somewhat higher flutter speed than the one emerging from the analysis with two classic flutter modes.
- It can be seen in Table 4.3.5 that, for Phases 6, 7 and 8, the 2 mode aeroelastic analysis results contrast greatly with those for the multimodal analysis. The former analysis leads to a flutter speed value that becomes more unstable as more of the span is constructed and its length is increased.

4.3.2 Flutter parameter evolution in the construction phase of the Great Belt Bridge

Figures 4.3.6 and 4.3.7 illustrate how the flutter parameters develop as more and more of the deck is put up throughout the phases considered earlier. They give the results obtained for this example, for both the bimodal vibration and multimodal analyses. Moreover, there are the results of other researchers, whose approaches differ: Tanaka[1], through wind-tunnel testing of a reduced complete model of the bridge in its construction phase, and the only datum produced by Larsen[6], whose calculations were done by computer using the flutter coefficients for the phase in which 20% of the deck was raised.

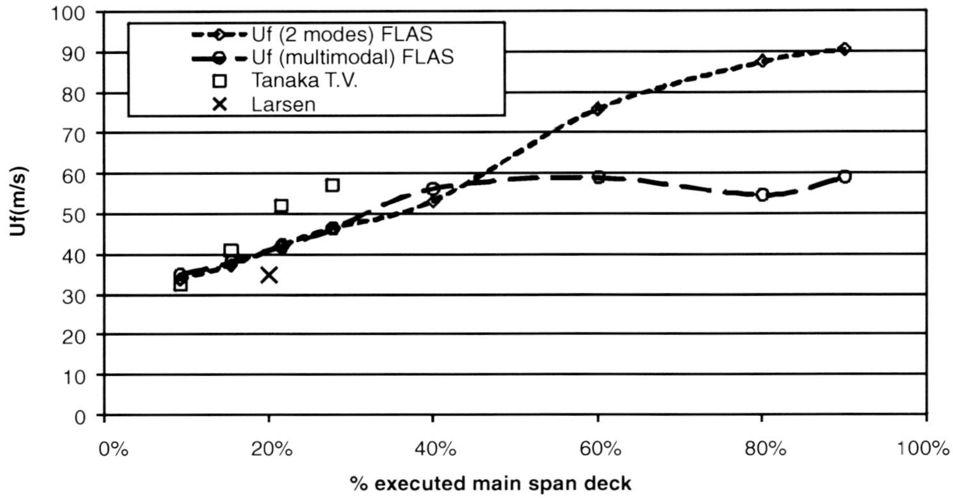

Figure 4.3.6: Flutter speed evolution U_f as the Great Belt Bridge Construction.

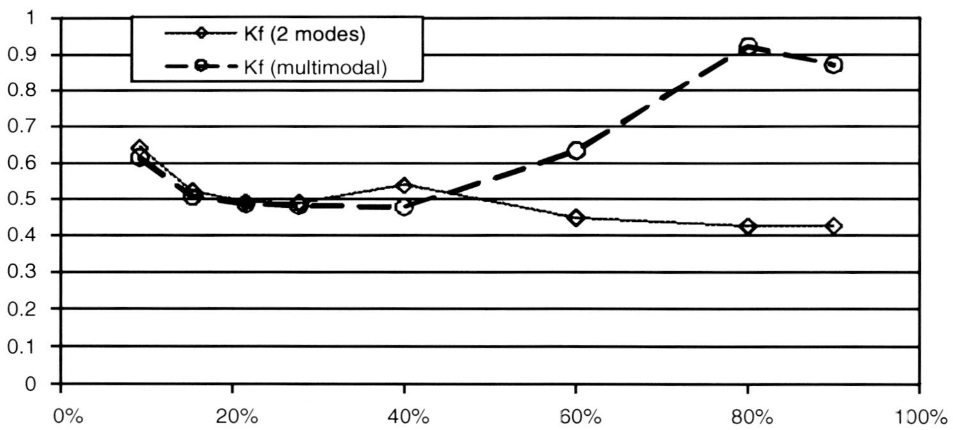

Figure 4.3.7: Reduced frequency evolution K_f during Great Belt Bridge Construction.

Referring to Figure 4.3.6, it is possible to glean that wind speeds U_f emerging from the bimodal and multimodal aeroelastic analysis for the initial construction phases (with up to 40% of the central span completed) are very close. For the more advanced phases, there are significant differences of up to 50%. This same phenomenon has already been confirmed in the Höga Kusten Bridge analysed earlier; once again demonstrating the need to include a sufficient number of modes to carry out a correct analysis of aeroelastic response in this kind of bridge. Another noteworthy point is how the flutter speed U_f (in multimodal analysis) increases from minimal values in early construction phases (10% of the central span) until 40% of the central span is raised. It then stabilises 60 m/s in the remaining phases. This behaviour has been confirmed in the previous example of the Höga Kusten Bridge, as well as in studies by Brancaleoni[1] for the Humber Bridge at the construction phase.

4.4 References

[1] Tanaka H. [1998] *Aeroelastic stability of suspension bridges during erection.* Structural Engineering International. Journal of the International Association for Bridge and Structural Engineering (IABSE). Vol. 8, No. 2, pp. 118–123.

[2] Ge Y.J., Tanaka H. [2000] *Aerodynamic stability of long-span suspension bridges under erection.* Journal of Structural Engineering. Vol. 126, No. 12, pp. 1404–1412.

[3] Cobo del Arco D. [1998] *An analysis of wind stability. Improvements to the response of suspension bridges.* Ph.D. Thesis from the Polytechnic University of Catalonia, School of Civil Engineering.

[4] Brancaleoni F. [1992] The construction phase and its aerodynamic issues. *Aerodynamics of Large Bridges*, A. Larsen (ed.) Balkema, Rotterdam.

[5] Brancaleoni F. [1988] *Behaviour of long span suspension bridges under construction.* 13th IABSE Congress Report. Helsinki, pp. 905–910.

[6] Larsen A. [1993] *Aerodynamic aspects of the final design of the 1624 m. suspension bridge across the Great Belt.* Journal of Wind Engineering and Industrial Aerodynamics, Vol. 48, pp. 261–285.

[7] Jurado J.A. [2001] *Análisis Aeroelástico y de Sensibilidad del Fenómeno del Flameo en Puentes Soportados por Cables*. Ph.D. Thesis, University of A Coruña. School of Civil Engineering. (in Spanish).

[8] Mosquera A. [2002] *Sensibilidad analítica de la vibración libre y del flameo en puentes colgantes de gran vano en servicio y en fase constructiva.* Ph.D. Thesis, University of A Coruña. School of Civil Engineering. (In Spanish).

Chapter 5

Flutter analysis of completed cable-supported bridges

5.1 Introduction

As a way of demonstrating the aeroelastic and sensitivity analysis methodology within the context of flutter speed for fully constructed cable-supported bridges, this chapter will provide an in-depth study of four examples. These include the bridges over the Great Belt and Akashi Strait, as well as the Vasco de Gama and the old Tacoma Narrows Bridges. Among these examples are the two suspension bridges which currently hold the record for span length and the bridge whose spectacular collapse due to the phenomenon of flutter won it world fame. Finally, a cable-stayed bridge has been chosen to show how it is also possible to apply the methodology to this typology.

Thus, it will be possible to discern differences in aeroelastic behaviour among the various deck types: the spaced truss of the Akashi-Kaikyo; the aerodynamic box girder cross section of the Great Belt; and the twin-beamed construction of the Vasco de Gama and original Tacoma Bridges. For all of these the flutter derivatives are available. Yet it is a paradox that within the scientific publications covering the wind studies of these bridges, there is great inconsistency among the values for the decks' mechanical parameters; they vary significantly from one author to the next.

An analysis of the flutter speed will be carried out taking two vibration modes into account: the first vertical mode and the first torsional one. Mulitmodal analysis will also be performed to consider any modes related to aeroelastic instability.

5.2 Great Belt Bridge

On 14 June, 1998, this line of communication over the Great Belt in Denmark was opened to traffic. Its main structure is the suspension bridge with an aerodynamic box deck cross section; it has a 1624-m span, which is shown in Figure 5.2.1.

The bridge's design, developed by the consultancy firm COWI, is characterised by the continuity of the deck beneath the towers, the artful relationship between the cable system's sag and the main span's length (1/9), the scrupulously calculated geometry of the deck's aerodynamic box deck cross section, and the great care given to the overall aesthetic and environmental impact of the bridge. Among the most noteworthy of the studies carried out on this bridge are those by Larsen[1] in 1993, preceding the deck's final design.

In Table 5.2.1 and Figure 5.2.2, one finds the geometrical characteristics and mechanical parameters. These properties were obtained from Larsen[1] and Cobo[2].

Figure 5.2.1: Great Belt Bridge.

Table 5.2.1: Geometrical and mechanical properties of Great Belt Bridge.

Main span length (m)	1624
Side span length (m)	535
Cable sag (m)	180
Distance between cables (m)	27
Deck's total width (m)	31
Deck depth (m)	4.4
Section of each main cable (m^2)	0.44
Mass of each main cable (t/m)	3.74
Inertia moment under vertical bending I_y (m^4)	4.0
Inertia moment under lateral bending I_z (m^4)	100.0
Inertia moment under torsion J (m^4)	7.6
Deck mass (t/m)	15.26
Polar inertia moment of deck (tm^2/m)	1106.8

5.2.1 Frequencies and natural modes for the Great Belt Bridge

The dynamic analysis of the bridge is a fundamental step before studying flutter instability. Natural frequencies and vibration modes for the bridge were obtained from a bar model which is shown in Figure 5.2.3 with 208 nodes and 323 elements calculated through the ADISNOL3D program. This model uses bar elements and concentrated masses so that the bridge's characteristics can be modelled with sufficient accuracy. The deck is modelled with bars running lengthwise along the X-axis; to these the properties outlined in Table 5.2.1 have been attributed. To make the deck movements compatible with those of the main hangers and cables, it is necessary to define a series of imaginary bars perpendicular to the deck and with a high level of rigidity. These would connect each end of the deck to the hangers.

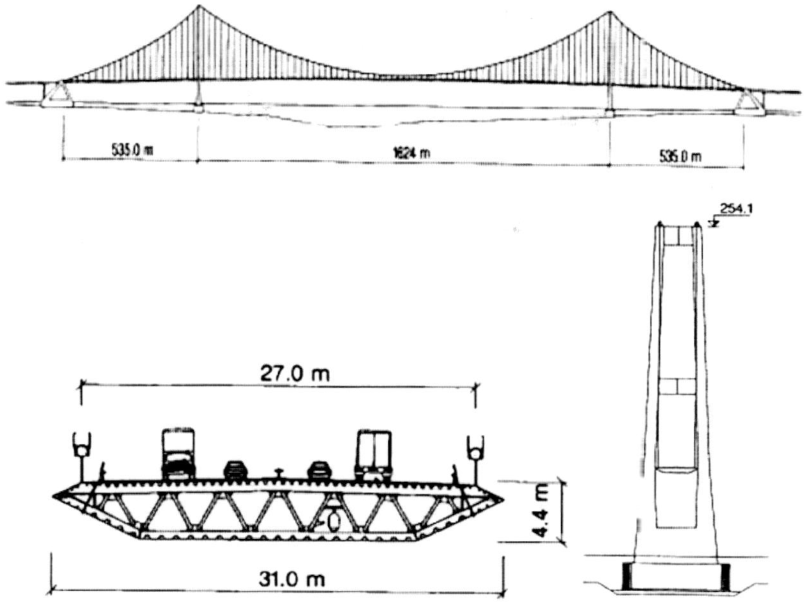

Figure 5.2.2: Principal dimensions of the Great Belt Bridge.

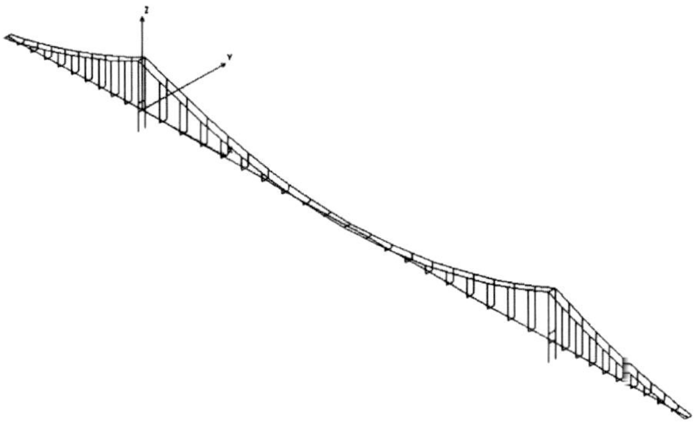

Figure 5.2.3: Structural model of the Great Belt Bridge.

The dynamic calculation made with ADISNOL3D takes into account the increasing stiffness produced in the structure, especially along the main cables, when the second-order effects are considered. Before this kind of dynamic analysis takes place, a preliminary static analysis has to be carried out because a deformed geometry is being worked with. Therefore, a calculation of the structure at its own weight has been performed. When second-order theory is considered in suspension bridges, the structure's stiffness grows considerably. Already the cables are supporting great tension loads.

Basically, instead of the traditional elastic rigidity matrix $\mathbf{K}_E$, this kind of dynamic calculation uses a rigidity matrix modified by the geometric rigidity matrix $\mathbf{K}_G$. The last matrix depends on the forces produced by the static load case. Hence the eigenvalue and vector problem, that in the linear case, is

$$(\mathbf{K}_E - \omega^2\mathbf{M})\boldsymbol{\phi} = \mathbf{0} \tag{5.2.1}$$

which becomes

$$\left(\mathbf{K}_E + \mathbf{K}_G - \omega^2\mathbf{M}\right)\boldsymbol{\phi} = \mathbf{0} \tag{5.2.2}$$

whereby ω^2 is an eigenvalue that coincides with the square of a natural vibration frequency value and ϕ is an eigen vector or vibration mode. Table 5.2.2 compares the natural frequencies obtained, expressed in (rad/s), with those provided by Larsen[1] and Cobo[2]. The following terminology is used to indicate the type of vibration mode associated with each frequency.

V: vertical vibration mode
L: lateral vibration mode
T: torsional vibration mode
S: symmetrical mode
A: asymmetrical mode

Thirty-seven vibration modes were calculated so that at least three of each kind would be available for the aeroelastic analyses. Thus mode 37 is the third pure mode of torsion, although 10 vertical modes and 4 lateral ones had preceded it. Only the modes that presented movements in the main span of the deck were dealt with, considering that the others (for instance, modes that are only deformed by the cables) are unrelated to deck flutter.

By perusing Table 5.2.2, one can see that the frequency values coincide well. They do so even though some of the mechanical properties of the bridge are estimates and, as

Table 5.2.2: Comparison of the natural frequencies in (rad/s) for the Great Belt.

Mode type	ADISNOL 3D	LARSEN	COBO
(1) LS	0.326	0.326	0.328
(2) VS	0.606	0.628	0.622
(3) VA	0.693	0.723	0.712
(4) LA	0.781	0.773	0.765
(5) VS	0.813	0.848	0.832
(6) VA	1.104	–	1.123
(11) VA	1.284	–	1.189
(12) VS	1.360	–	1.351
(13) LS	1.382	1.175	1.251
(18) VS	1.559	–	1.527
(19) VA	1.747	–	–
(21) VA	1.821	–	–
(22) TS	1.855	1.747	1.730
(27) VS	2.115	–	–
(28) LTA	2.207	–	–
(29) LS	2.434	–	–
(30) TA	2.461	2.406	2.216
(37) TS	2.816	3.154	–

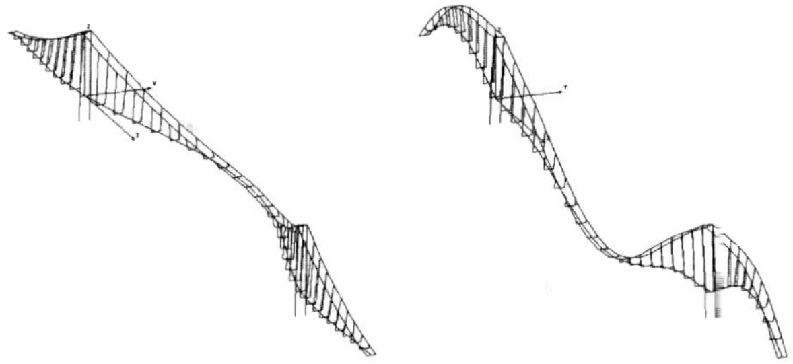

Figure 5.2.4: LS mode 1 and VS mode 2 of the Great Belt.

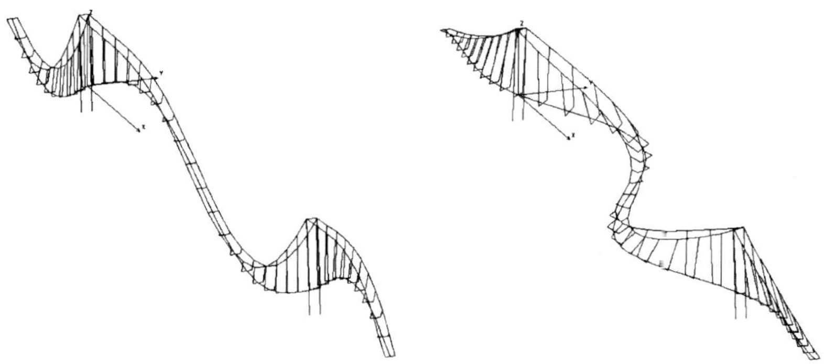

Figure 5.2.5: VA mode 3 and LA mode 4 of the Great Belt.

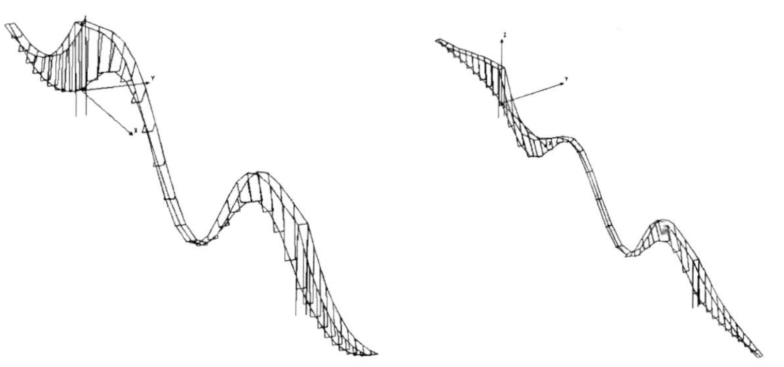

Figure 5.2.6: VS mode 5 and VA mode 6 of the Great Belt.

such, they are reasonable, but not real, values. Figures 5.2.4–5.2.12 show the shapes of these vibration modes.

As for the structural damping of the Great Belt Bridge, it has been assumed that with all the modes, there is a viscose damping $\xi = 0.00318$ corresponding with a 2% level of logarithmic decrease.

$$\delta = \frac{2\pi\xi}{\sqrt{1-\xi^2}} \approx 2\pi\xi = 0.02 \tag{5.2.3}$$

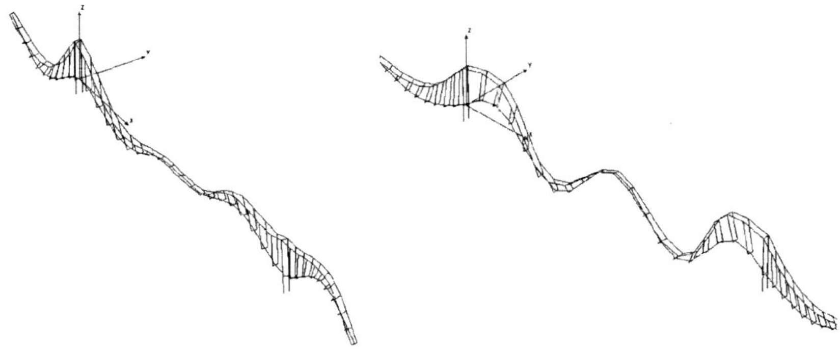

Figure 5.2.7: VA mode 11 and VS mode 12 of the Great Belt.

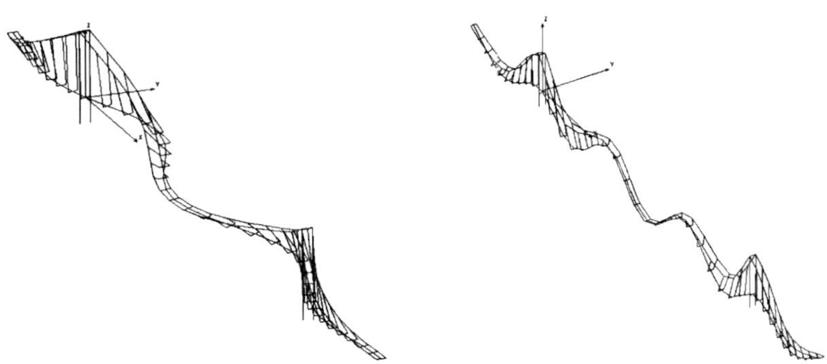

Figure 5.2.8: LS mode 13 and VS mode 18 of the Great Belt.

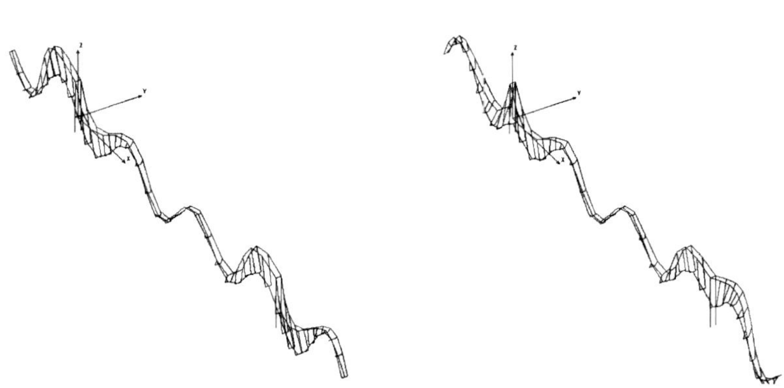

Figure 5.2.9: VA mode 19 and VA mode 21 of the Great Belt.

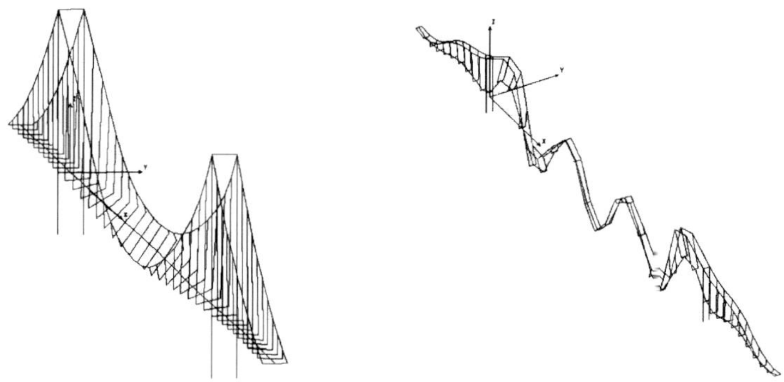

Figure 5.2.10: TS mode 22 and VS mode 27 of the Great Belt.

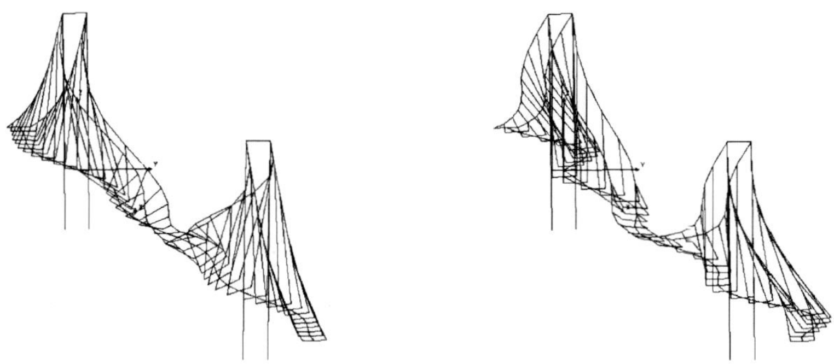

Figure 5.2.11: LTA mode 28 and LS mode 29 of the Great Belt.

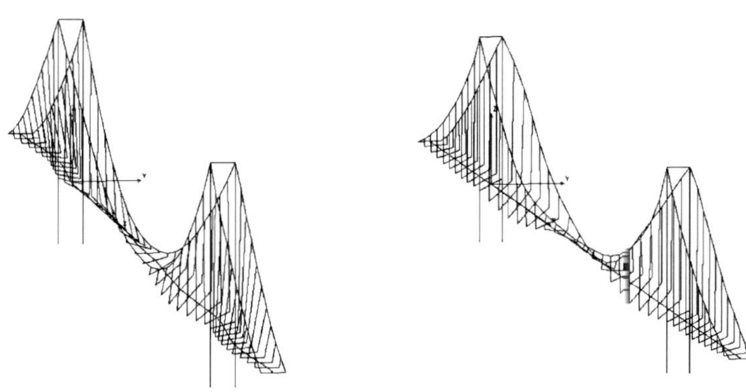

Figure 5.2.12: TA mode 30 and TS mode 37 of the Great Belt.

5.2.2 Aeroelastic analysis of the Great Belt Bridge

The modal analysis linked to the aeroelastic analysis of this bridge uses the 18 vibration modes represented in the previous figures. As mentioned in Chapter 3, to begin the multi-modal analysis, it is necessary to construct the aeroelastic matrices of the deck, $\mathbf{K}_a$ and $\mathbf{C}_a$, which are functions of the average wind speed U and reduced frequency K. They modify the structure's damping and rigidity. The system is reduced by means of the modal matrix Φ constructed with the selected modes

$$\boldsymbol{\Phi}^T\mathbf{M}\boldsymbol{\Phi}\ddot{\mathbf{q}}+\boldsymbol{\Phi}^T\left(\mathbf{C}-\mathbf{C}_a\right)\boldsymbol{\Phi}\dot{\mathbf{q}}+\boldsymbol{\Phi}^T\left(\mathbf{K}-\mathbf{K}_a\right)\boldsymbol{\Phi}\mathbf{q}=\mathbf{0} \tag{5.2.4}$$

Supposing there are damping solutions like $\mathbf{q}(t) = \mathbf{w}e^{\mu t}$, the equation system could be written

$$\left(\mu^2\mathbf{I}\mathbf{w}+\mu\mathbf{C}_R\mathbf{w}+\mathbf{K}_R\mathbf{w}\right)e^{\mu t}=\mathbf{0} \tag{5.2.5}$$

leading to a problem with eigenvalues and vectors.

$$\left(\mathbf{A}-\mu\mathbf{I}\right)\mathbf{w}_\mu e^{\mu t}=\mathbf{0} \tag{5.2.6}$$

where

$$\mathbf{w}_\mu=\begin{pmatrix}\mu\mathbf{w}\\ \mathbf{w}\end{pmatrix}\ \text{and}\ \mathbf{A}=\begin{pmatrix}-\mathbf{C}_R & -\mathbf{K}_R\\ \mathbf{I} & \mathbf{0}\end{pmatrix} \tag{5.2.7}$$

Because the matrix **A** is not symmetrical, complex eigenvalues are obtained. These are conjugated like this

$$\mu_j=\alpha_j+i\beta_j\ \text{ y }\ \bar{\mu}_j=\alpha_j-i\beta_j\ \text{ with }\ j=1,\ldots,m \tag{5.2.8}$$

The imaginary part of the eigenvalues obtained coincide with the damping frequencies of the response

$$\omega_{aj}=\beta_j \tag{5.2.9}$$

In contrast, the real part is associated with the damping through the equation

$$\xi_{aj}=\frac{-\alpha_j}{\sqrt{\alpha_j^2+\beta_j^2}} \tag{5.2.10}$$

The critical condition for the flutter phenomenon corresponds with the lowest wind speed, which produces a complex eigenvalue whose real part is null. That is, there is a transition between positive damping with diminishing oscillations and negative damping, whose oscillations have increasing amplitude. This occurs when the real part of a complex eigenvalue goes from negative to positive.

This formulation for fully constructed bridges was applied using the FLAS program. More detailed information about this method can be found in Chapter 3.

When the aeroelastic matrices are built, flutter derivatives intervene. The latter should be obtained with experimental tests in an aerodynamic wind tunnel. The graphs in Figures 5.2.13–5.2.20 represent the values for the flutter derivative of the Great Belt Bridge (GBT) used by Cobo[2], alongside those for a flange shape following Theodorsen's

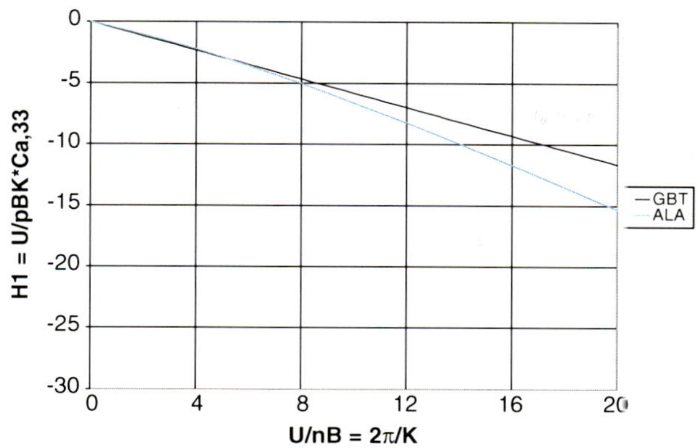

Figure 5.2.13: Flutter derivatives H_1*.

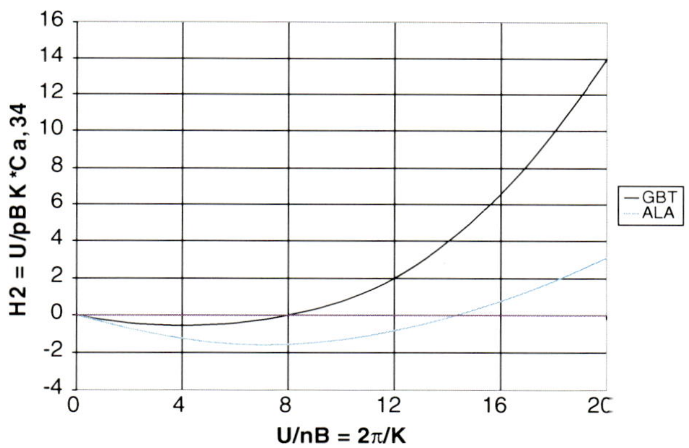

Figure 5.2.14: Flutter derivatives H_2*.

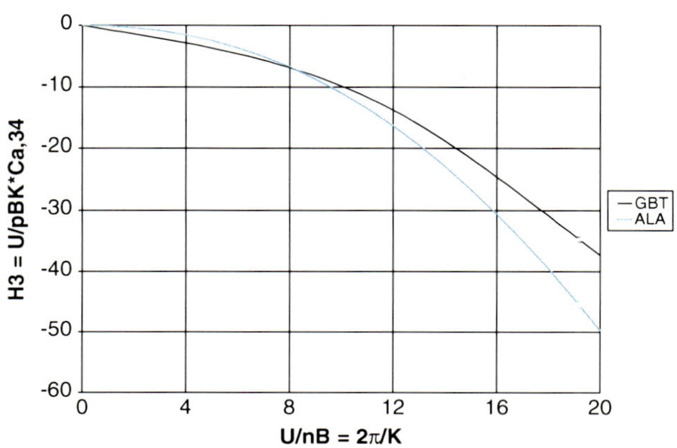

Figure 5.3.15: Flutter derivatives H_3*.

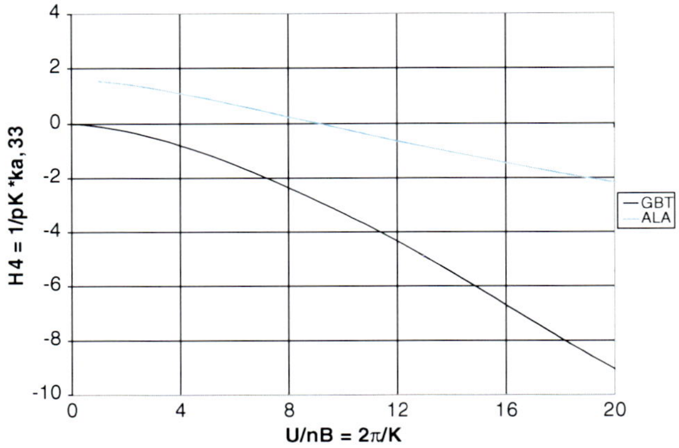

Figure 5.2.16: Flutter derivatives H_4*.

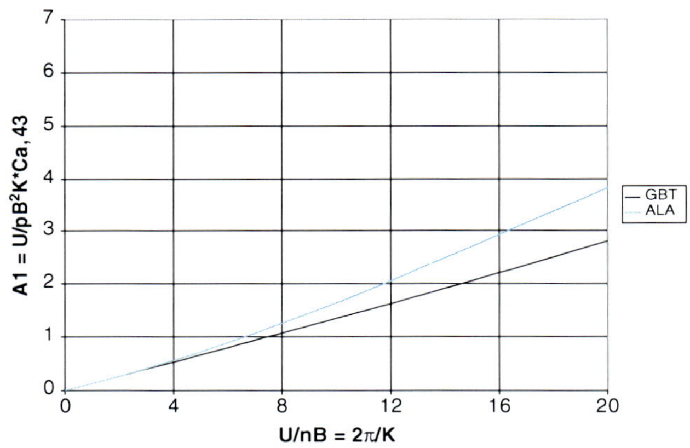

Figure 5.2.17: Flutter derivatives A_1*.

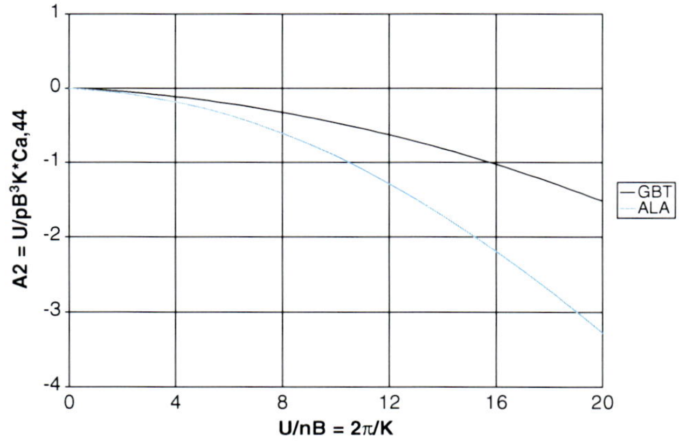

Figure 5.2.18: Flutter derivatives A_2*.

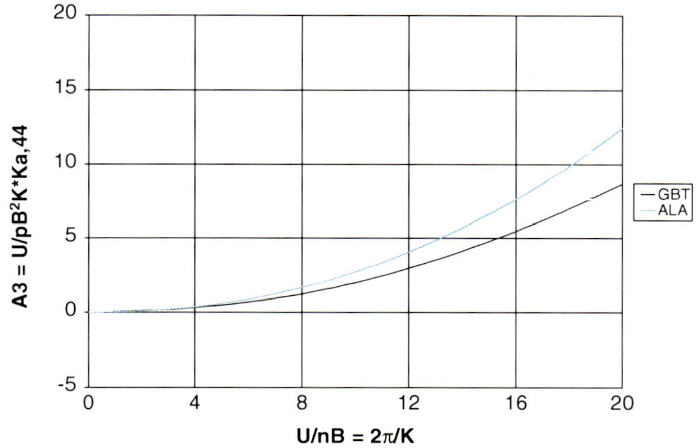

Figure 5.2.19: Flutter derivatives A_3*.

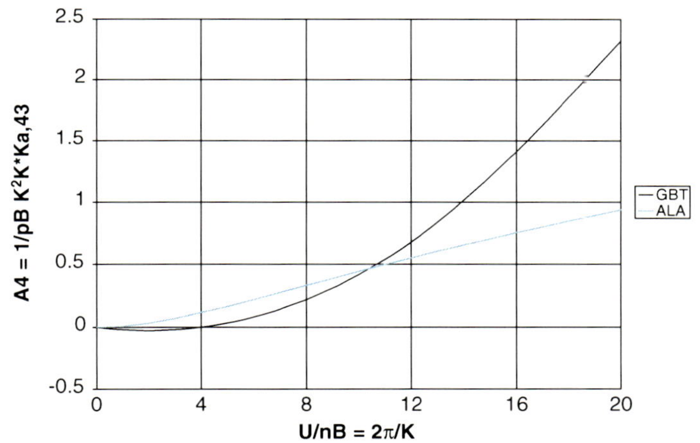

Figure 5.2.20: Flutter derivatives A_4*.

theory[3] (ALA). For the Great Belt, there was no information available for the coefficients $P_i^*(i = 1...6)$, which influence the aeroelastic drag force D.

Aeroelastic analysis with two vibration modes

The first analysis carried out on the Great Belt Bridge considers the coupling of the two classic modes found in the phenomenon of flutter: the symmetrical mode 2 of vertical bending and the symmetrical mode of torsion (22). The graphs in Figures 5.2.21 and 5.2.22 show how the real and imaginary parts of the eigenvalues, obtained for growing wind speeds, evolve. Also represented are the values for the critical wind speed and its associated reduced frequency.

As can be perceived in the figures, the torsional mode is responsible for the flutter-induced instability reached for $U_f = 89.91$ m/s and with a reduced frequency of $K_f = 0.4494$, corresponding to $w = 1.3$ rad/s.

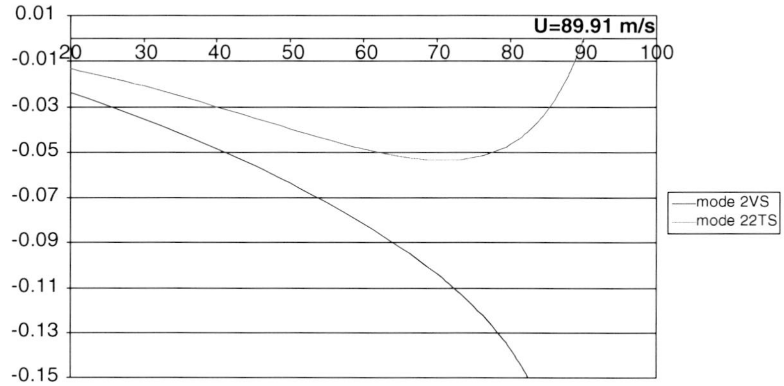

Figure 5.2.21: Evolution of the real part versus U using two modes for the Great Belt.

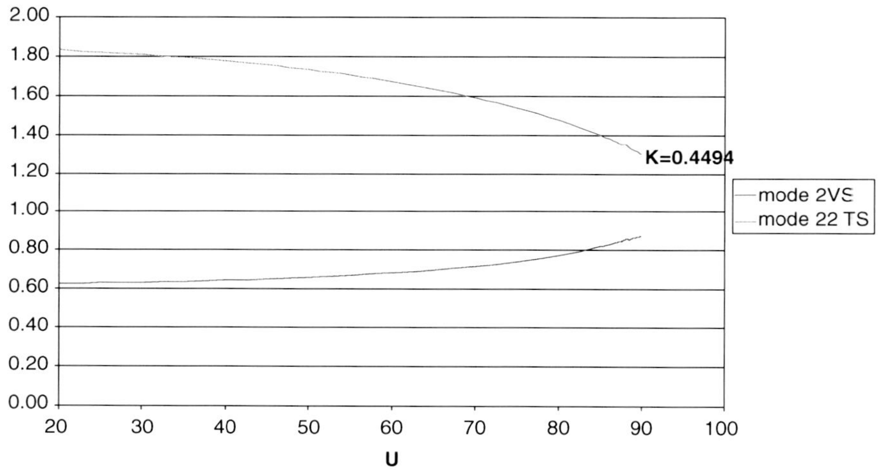

Figure 5.2.22: Evolution of the imaginary part β versus U using two modes for the Great Belt.

By perusing the graphs in Figure 5.2.21, which detail the evolution of a, one sees that the damping for mode 2 always increases within the range of analysed wind speeds. However, the one for the torsional mode 22 groes up until about $U = 70$ m/s and sharply decreases once this value is reached, becoming null and inducing flutter. Such behaviour would indicate that up to values of 80% of flutter speed, the wind's effect on the bridge is positive, given that the damping for both modes rises. Nevertheless, from that point on, the situation changes radically; U just needs to increase to 20 m/s in order to bring on instability in the bridge.

The graphs in Figure 5.2.22 provide information about the damping response frequencies. With low wind speeds, these frequencies almost coincide with the natural frequencies for the chosen modes and there is a detectable change due to an increase

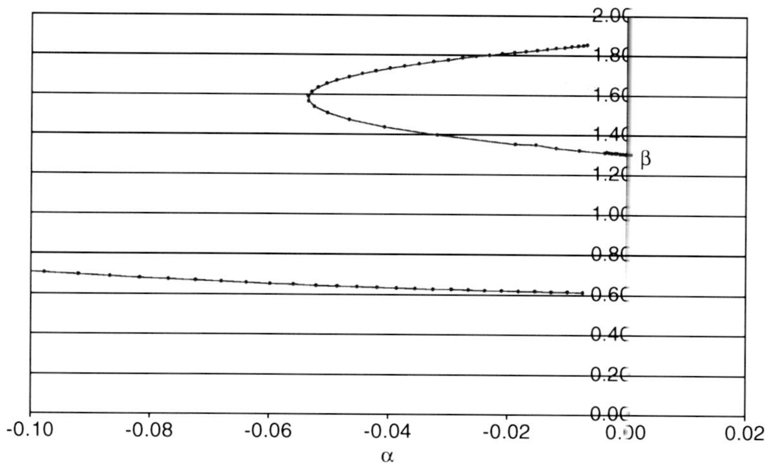

Figure 5.2.23: Evolution of β versus α using two modes for the Great Belt Bridge.

in U. Thus, up to the point of flutter, the frequency for the vertical mode goes up to around 0.3 rad/s, a 50% rise. Similarly, the torsional mode decreases by approximately 0.5–1.3 rad/s, or by about 30%. Also, the frequencies vary more noticeably when speeds approach flutter.

Another kind of representation, the Argand diagram in Figure 5.2.23, outlines the joint evolution of both the real and imaginary parts of the eigenvalues. Each point in the graphic represents an eigenvalue for a specific speed. Flutter is produced when a point lies to the right of vertical axis β.

In this graphic, the C-shaped curve corresponding with torsional mode 22 clearly shows that as the wind speed rises, an inversion of the damping values is produced. At the same time, there is a drop in the frequency β.

Aeroelastic analysis with 18 vibration modes

This multimodal analysis entailed working with all the vibration modes previously calculated. In other words, the study included all the significant modes that appeared, even the third mode of pure torsion 36. In this way, various vibration modes of each kind intervene. It is therefore possible to confirm that the results of the aeroelastic study provide a more precise representation of how the bridge would really respond to flutter.

The same kind of graphical representation obtained for the bimodal analysis can be found in Figures 5.2.24–2.26.

The resulting critical flutter speed is $U_f = 62.41$ m/s, about 30% below the one calculated through the bimodal method, which has produced an unreliable result. The marked difference means that multimodal calculation, using a sufficient quantity of vibration modes, proves reliable for analysing large-span bridges. The reduced frequency of response in this analysis is $K_f = 0.8371$, which, referring to Figure 5.2.25, corresponds to 1.68 rad/s.

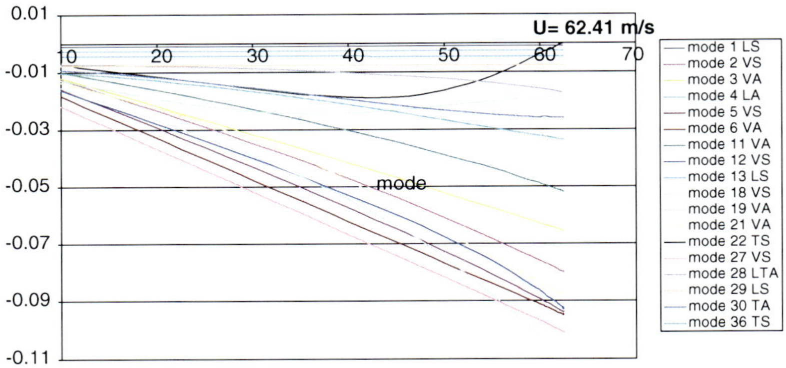

Figure 5.2.24: Real part's evolution versus U using 18 modes for the Great Bell Bridge.

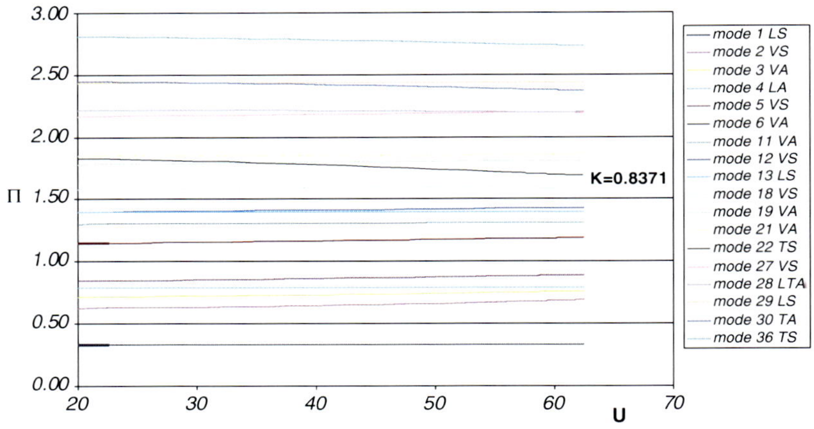

Figure 5.2.25: Imaginary part's evolution β versus U using 18 modes for the Great Belt Bridge.

In Figure 5.2.24, the line corresponding to mode 22 crosses the horizontal axis $\alpha = 0$, indicating that a null damping value is reached and, as a result, there is flutter-induced instability. A glance at this figure will also reveal how the damping of the symmetrical torsional modes (22 and 36) tends to diminish after a certain wind speed. However, the damping in their vertical counterparts always goes up and, in the lateral ones, it remains constant. Figure 5.2.25 shows decreases in the frequency values related to torsion and increases in the vertical bending frequencies.

Figure 5.2.26 has the graphical representation of β versus α and the C-shaped curve once again appears. It cuts across the vertical axis and has a null damping value. Therefore it belongs to torsional mode 22.

Table 5.2.3 offers a comparison of previous aeroelastic studies of the Great Belt Bridge,alongside the results obtained by Larsen[1] and Cobo[2].

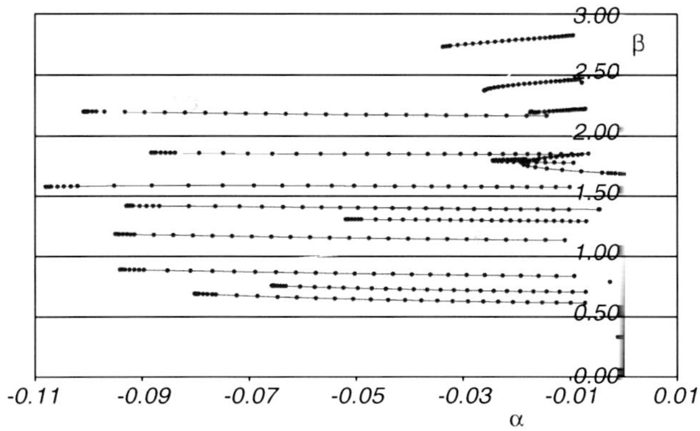

Figure 5.2.26: Evolution of β versus α using 18 modes for the Great Belt Bridge.

Table 5.2.3: Comparison of various aeroelastic studies on the Great Belt Bridge.

Analysis	Flutter speed Uf (m/s)
FLAS, 2 modes	89.91
FLAS, 18 modes	62.41
Larsen, cross-section test	70–74
Larsen, whole model test	70–75
Larsen, flutter derivative analysis, 9 modes	75
Cobo, 2 modes	82.75
Cobo, 6 modes	96.75

One can see from the table the similarity between the results for the calculations using two vibration modes: between 80 and 90 m/s. The speed of 96.75 m/s corresponding to Cobo's multimodal analysis may seem excessive alongside the experimental studies by Larsen. Nonetheless, the 62.41 m/s obtained in the multimodal analysis, using 18 modes, come closer to the experimental results (70–75 m/s). For these last results, it was difficult accurately modelling the highest frequencies for the bridge. As for Larsen's study using flutter derivatives (75 m/s), it is important to note that only nine modes were included. These were the first three vibration modes of each type (vertical, lateral and torsional). As a result, vertical symmetrical modes 12 and 18 do not come into play. Yet, alongside mode 22, which is torsional and symmetrical, these are of considerable importance in any effort to obtain a lower flutter speed.

By comparing the data found in Table 5.2.3, one concludes that it is necessary to continue working with a sufficiently high number of modes, thereby considering various vibration modes of the same kind. Doing away with some of these may mean that a dangerously high flutter speed is obtained, leading to a false sense of security in terms of the bridge design.

5.3 Bridge over the Akashi Strait

Shown in Figure 5.3.1, this bridge over the Akashi Strait is one among several situated along the Kobe-Naruto route between the Japanese islands of Honshu and Shikoku. It

Figure 5.3.1: Akashi-Kaikyo Bridge.

Table 5.3.1: Geometrical and mechanical properties of the Akashi-Kaikyo Bridge.

Main span length (m)	1990
Lateral span length (m)	960
Tower height (m)	297
Width between cables (m)	35.5
Total deck width (m)	35.5
Deck edge (m)	14
Cross section of each main cable (m^2)	0.79
Inertia moment under vertical bending I_y (m^4)	24
Inertia moment under lateral bending I_z (m^4)	130
Inertia moment under torsion J (m^4)	17.8
Deck mass (t/m)	28.7
Polar inertia moment in deck (tm^2/m)	5800

crosses the Strait at a distance of about 4 km and connects Honshu, Japan's main island, to neighbouring Awaji. The waterway is among the most navigated ones in the world so that maritime traffic is an important consideration.

For the design, a suspension bridge was chosen with an extremely rigid latticework deck so that the Strait could be crossed using the minimum number of supports. Its total length is 3910 m, with a main span of 1991 m. Its two towers are 297 m high; crossbeams have been used to reinforce these.

Because Japan is an area buffeted by typhoons, rigorous testing has been carried out to ensure an appropriate response from the bridge when exposed to wind loads. Among the most noteworthy of these studies is one by Katsuchi[4] at John Hopkins University. Others include the experimental testing on a full model of the bridge on a scale of 1/100, done by Miyata *et al.*[5]. Table 5.3.1 and Figure 5.3.2 offer some of the bridge's geometrical

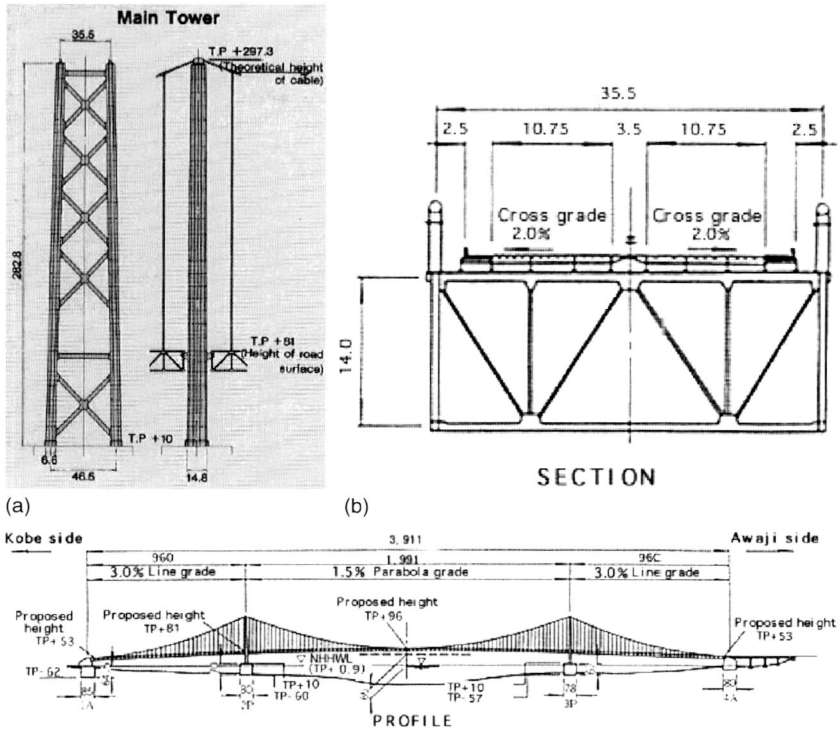

(a) (b) (c)

Figure 5.3.2: Tower, cross section, and elevation of the Akashi Strait Bridge.

features and mechanical parameters. The deck's mechanical properties have been taken from Cobo[2].

5.3.1 Natural frequencies and modes for the Akashi Strait Bridge

Using the ADISNOL3D program, a bar model was produced with 224 nodes and 359 elements. Figure 5.3.3 provides an illustration of this model in which the deck is represented with lengthwise bars that are governed by the mass and rigidity properties outlined in Table 5.3.1. In order to make the deck node and cable movements compatible, fictitious transversal bars were defined. These exceptionally rigid bars were to connect the lower ends of the hungers to the deck.

Another factor taken into account was the non-linear geometry that occurs in the behaviour of the cable system. Thus, to obtain the frequencies and vibration modes, the approach behind the formula (7.1.2) was taken. Table 5.3.2 compares the natural frequencies obtained with the ones used by Katsuchi[4] and those indicated by Miyata *et al.*[5].

The 38 first vibration modes were calculated so that various vibration modes of each type would be available for aeroelastic analysis. However, only those that presented movement along the central deck span were dealt with, thus eliminating those related to cable or side-span deformation.

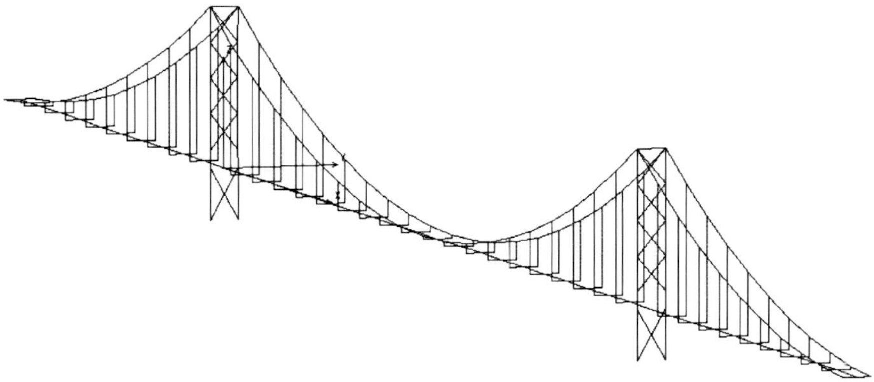

Figure 5.3.3: Finite element model for the Akashi Strait Bridge.

Table 5.3.2: Comparison of natural frequencies in (rad/s) for the Akashi Strait Bridge.

Mode type	ADISNOL3D	KATSUCHI	MIYATA
(1) LS	0.2836	0.2438	0.2431
(2) VS	0.4160	0.4095	0.4008
(3) VA	0.4623	0.4729	0.4681
(4) VA	0.5339	0.5339	0.5246
(5) LA	0.5943	0.4921	0.4869
(6) VS	0.7708	0.7646	0.7621
(10) LTS	0.9934	0.7986	0.9405
(11) VS	1.0196	1.0293	–
(12) TLS	1.0275	0.9745	–
(14) VA	1.0939	1.0768	–
(24) TA	1.3797	1.3281	1.3050
(25) VS	1.4263	1.3896	–
(26) TA	1.5019	1.6001	–
(27) LA	1.5731	1.3896	–
(28) TS	1.6401	1.8662	–
(29) VA	1.7711	–	–
(38) TS	2.0172	–	–

Sixteen modes were used in all. Until mode 38, the fourth pure torsional one was reached, others appeared: eight vertical, three lateral and two that simultaneously have both lateral and torsional components. It appears that the frequency values for the different modes coincide to a great extent. Below, Figures 5.3.4–5.3.12 provide the shapes for the vibration modes.

As for the structural damping in the Akashi Strait Bridge, it is supposed that all the damping modes are of the viscose kind at $\xi = 0.0318$ corresponding to a 2% level of logarithmic decrease.

$$\delta = \frac{2\pi\xi}{\sqrt{1-\xi^2}} \approx 2\pi\xi = 0.02 \qquad (5.3.1)$$

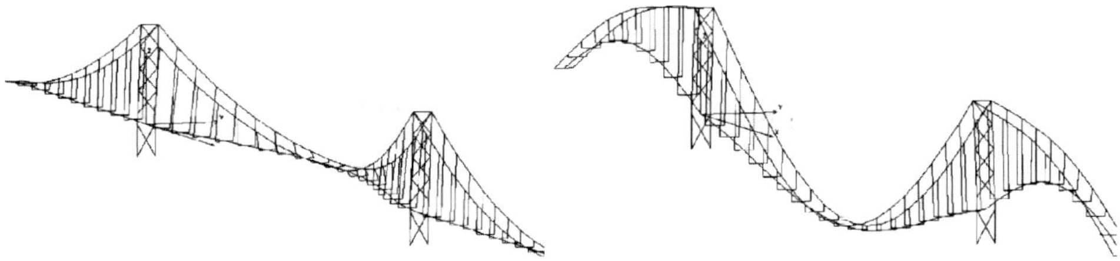

Figure 5.3.4: LS mode 1 and VS mode 2 for the Akashi Strait Bridge.

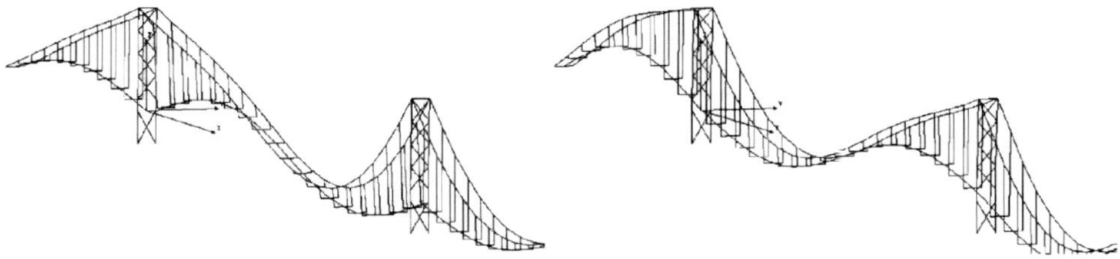

Figure 5.3.5: VA mode 3 and VA mode 4 for the Akashi Strait Bridge.

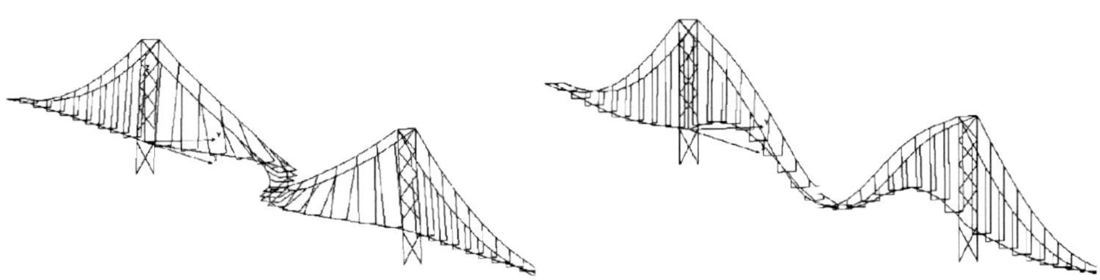

Figure 5.3.6: LA mode 5 and VS mode 6 for the Akashi Strait Bridge.

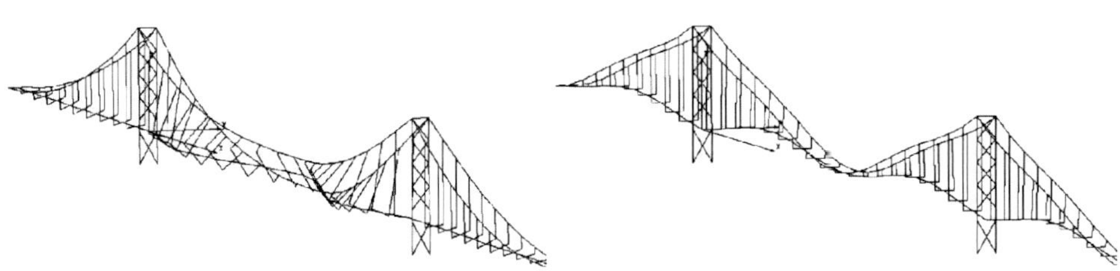

Figure 5.3.7: TLS mode 10 and VS mode 11 for the Akashi Strait Bridge.

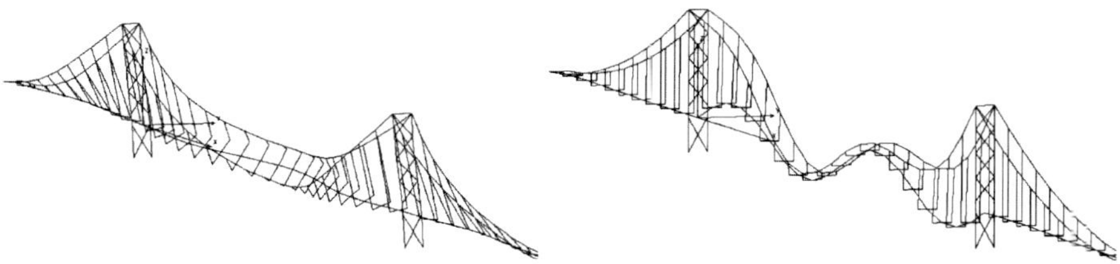

Figure 5.3.8: TLS mode 12 and VA mode 14 for the Akashi Strait Bridge.

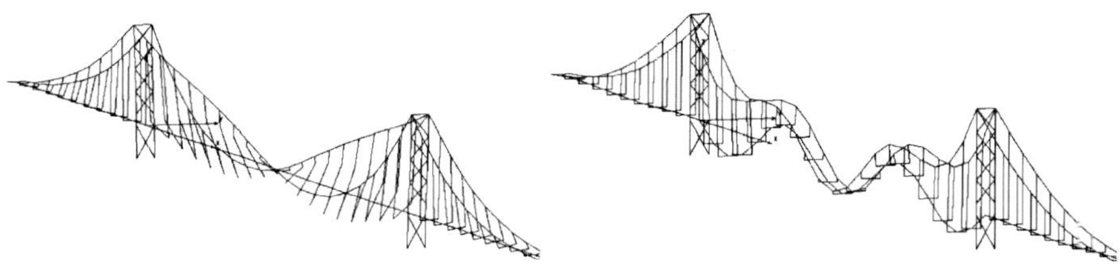

Figure 5.3.9: TA mode 24 and VS mode 25 for the Akashi Strait Bridge.

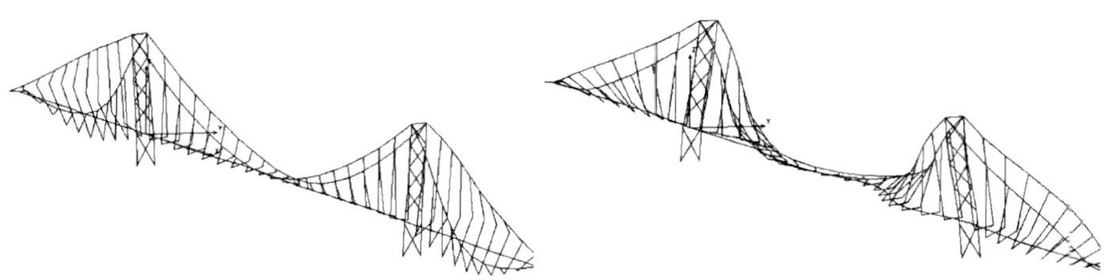

Figure 5.3.10: TA mode 26 and LA mode 27 for the Akashi Strait Bridge.

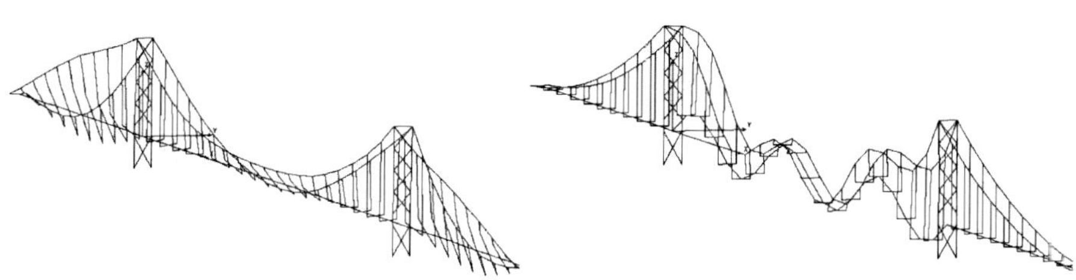

Figure 5.3.11: TS mode 28 and VA mode 2 for the Akashi Strait Bridge.

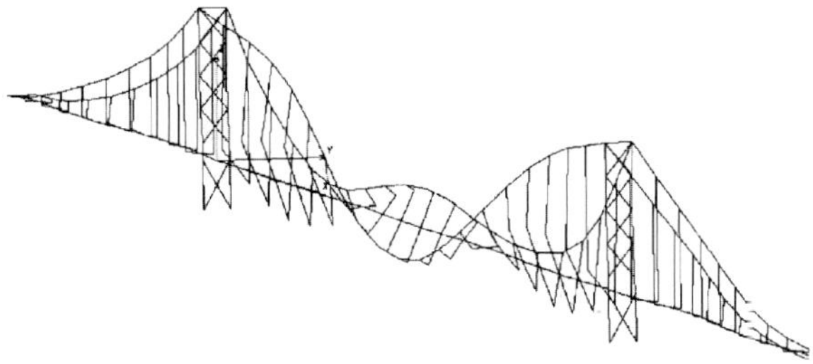

Figure 5.3.12: TS mode 38 for the Akashi Strait Bridge.

Figure 5.3.13: Wind testing on the full model of the Akashi Strait Bridge.

This is the value set for torsional modes by the Japanese code, although, for example, in the full model tests carried out by Miyata *et al.*[5] values close to 3% of logarithmic decrease were used.

5.3.2 Aeroelastic analysis of the Akashi Strait Bridge

For this bridge, an aeroelastic analysis using the 16 modes shown in the previous figures was chosen. As done in the example presented earlier, all the analyses shown here are based on multimodal theory. The same equations as before, (7.1.4)–(7.1.10), were applied. The flutter derivatives for the deck of this bridge were the ones used by Katsuchi[4] in his own study on the bridge's aeroelastic behaviour. It is worth mentioning that also used for the study of this bridge were the coefficients P_i^* ($i = 1,...,6$) relating the force of horizontal thrust D (drag) to deck movements. According to the conclusions of Katsuchi[4], these coefficients have a marked influence on the bridge's behaviour under the phenomenon of flutter. Moreover, they have a destabilizing effect when the angle of attack is not null. With these calculations, presented at a later stage, coefficients for a zero angle of attack are used, given that its influence is slightly less in this case.

Figure 5.3.13 shows a full model of the bridge tested in the wind tunnel at the Centre for Research and Development in Nagasaki. In this photograph, one can discern the lateral deformation produced by wind loads. This deformation, along with the fact that many of the first vibration modes of the bridge were lateral, leads one to think that forces in the direction of the wind should be taken into account with a bridge like this, with a main span of nearly 2000 m.

The graphs for Figures 5.3.14–5.3.20 represent the flutter derivative values H^*_i *and* A^*_i for the Akashi Strait Bridge (AKK), compared with the values for an aerodynamic shape (ALA) according to Theodorsen's theory. This theory does not consider lateral movements. Therefore, the coefficients P^*_i are shown separately in Figure 5.3.21.

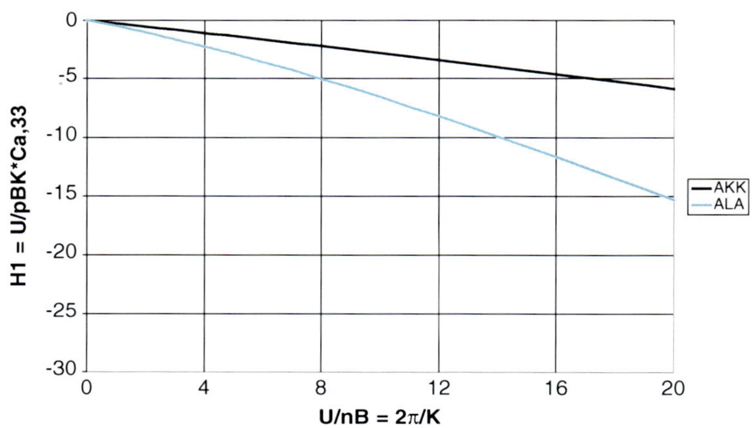

Figure 5.3.14: Flutter derivatives H_1*.

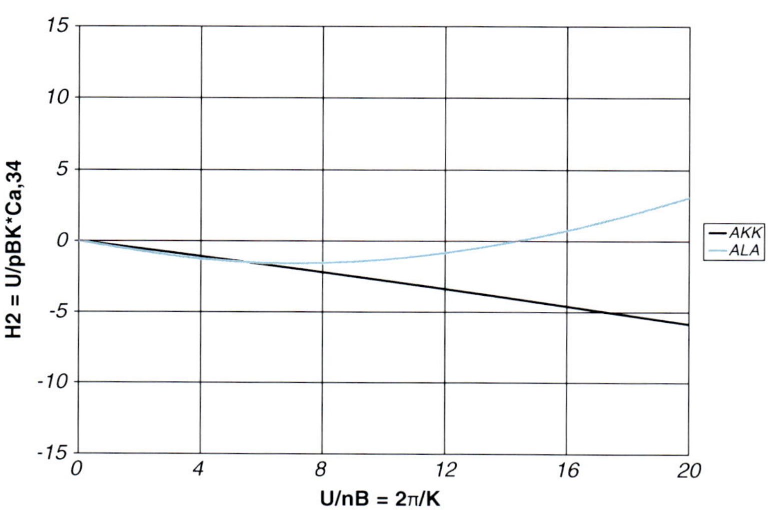

Figure 5.3.15: Flutter derivatives H_2*.

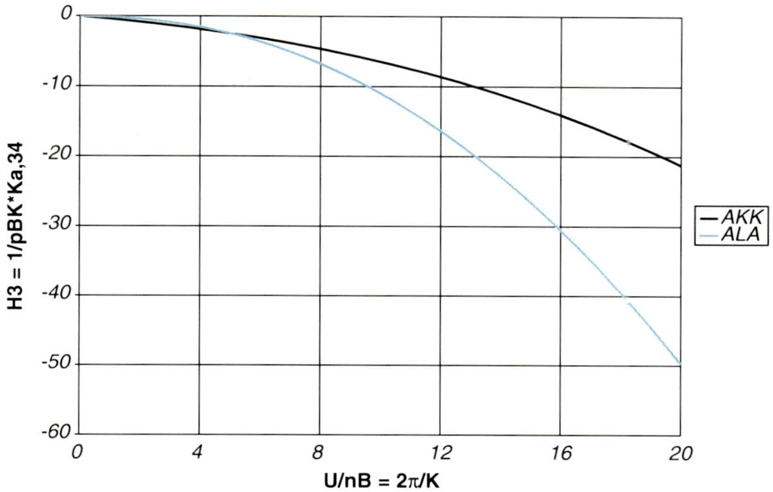

Figure 5.3.16: Flutter derivatives H_3*.

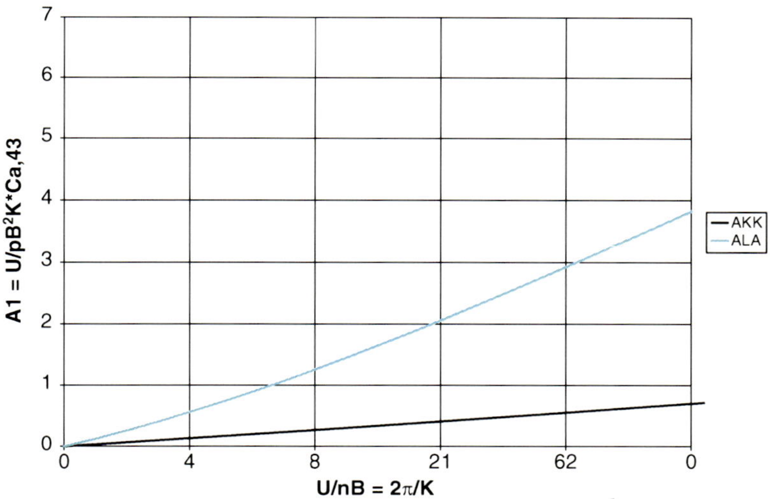

Figure 5.3.17: Flutter derivatives A_1*.

Aeroelastic analysis of the Akashi Strait Bridge using two vibration modes

As was the case in the previous example, the first step in studying the behaviour of the Akashi Strait Bridge was to consider the two classic modes in flutter phenomena: the first mode of vertical bending and the first one of torsion. It must be remembered that in this bridge the first torsional modes also have an appreciable component of lateral bending. Modes 10 and 12 are of this kind (TLS): 10 has the highest lateral component; 12, the highest torsional one.

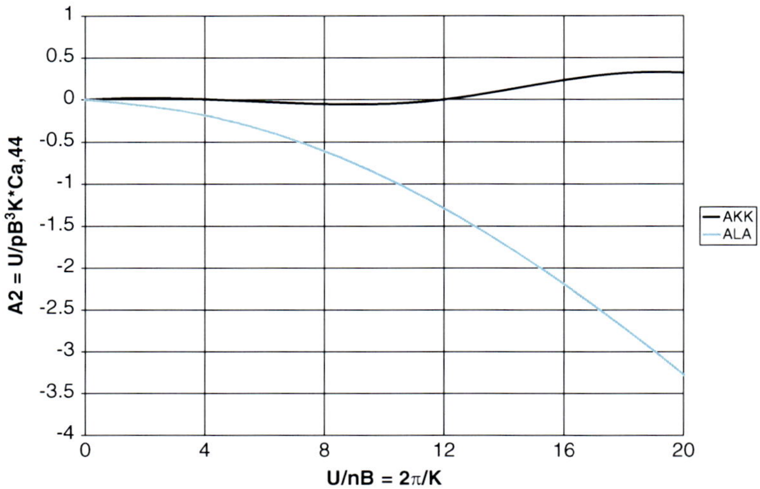

Figure 5.3.18: Flutter derivatives A_2*.

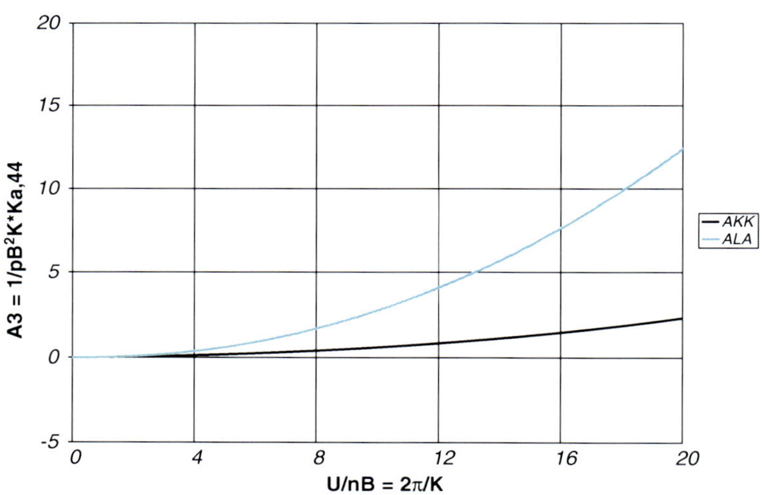

Figure 5.3.19: Flutter derivatives A_3*.

Tests were carried out on the two possibilities. The results are that when modes 2 and 10 are considered, flutter does not occur until speeds of more than 100 m/s are reached. In contrast, if modes 2 and 12 are chosen, flutter is produced at 77.69 m/s, a more probable result. The graphical representations found in Figures 5.3.22 and 5.3.23 portray the evolution of the real and imaginary parts of the eigenvalues obtained for the increasing wind speeds, as well as the critical wind speed values and the reduced frequency associated with it.

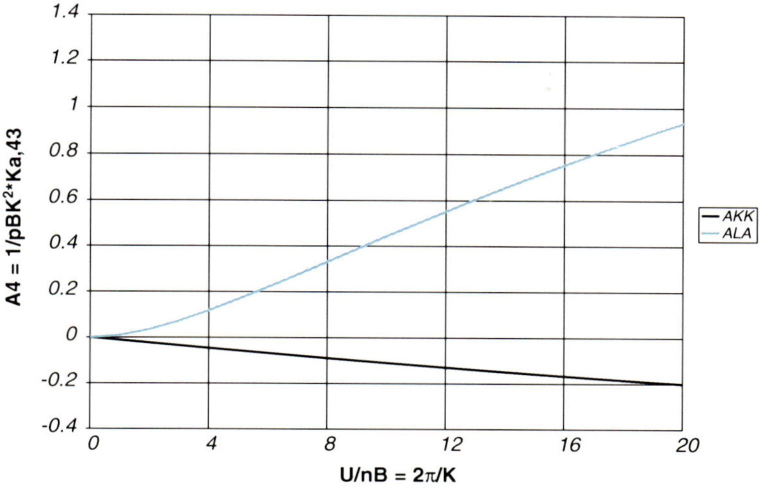

Figure 5.3.20: Flutter derivatives A_4*.

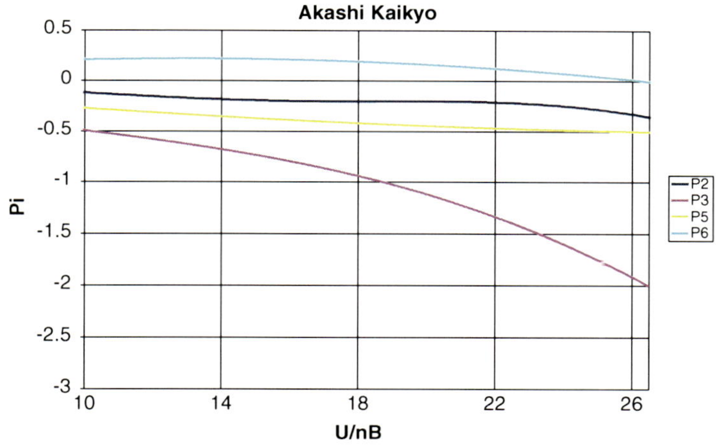

Figure 5.3.21: Flutter derivatives P_i*.

As can be detected in the figures, mode 12 of symmetrical torsion, with a noticeable lateral component, is responsible for the flutter instability, reached at a wind speed of $U = 77.69$ m/s, and a reduced frequency of $K = 0.4554$ corresponding to $\omega = 0.9967$ rad/s.

The variation of a for vertical mode 2 indicates that its damping continually grows within the range of analysed speeds. In contrast with mode 12, the trend shifts. After about 50 m/s damping lessens until it is null and flutter is induced. A similar trend for the variation of a has already been witnessed in the Great Belt Bridge example. Perhaps the only difference in the curves for these two bridges is that the

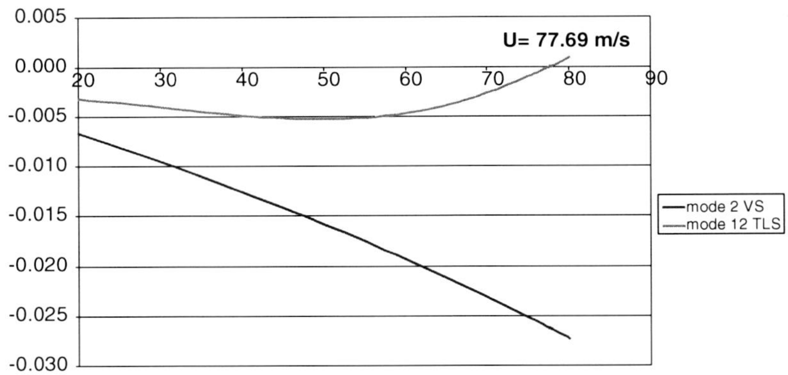

Figure 5.3.22: Evolution of the real part versus U using two modes.

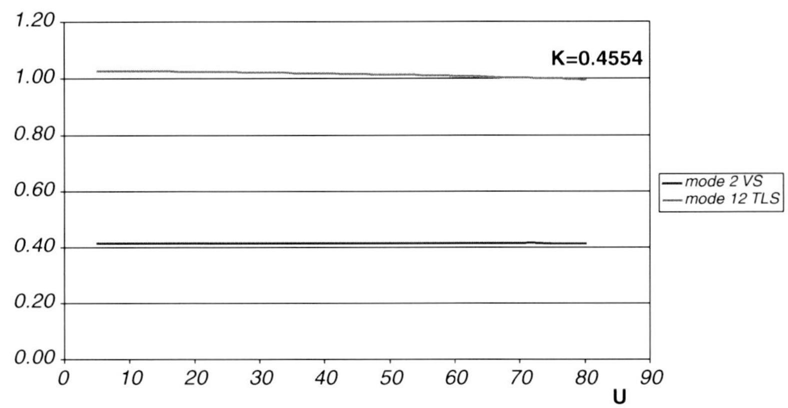

Figure 5.3.23: Evolution of the imaginary part versus U using two modes.

inversion of the damping for the torsional mode is more gradual in the case of the Akashi Strait Bridge.

For the Akashi Strait Bridge, the curves representing the evolution of β, or the frequencies, vary very little as the speed increases. Nevertheless, one can detect a slight decrease in the frequency value for the symmetrical torsional mode 12.

Figure 5.3.24 shows the evolution of both the real and imaginary parts of the eigenvalues. Each point in the graph represents an eigenvalue with a specific wind speed. Flutter is produced when a point crosses the vertical axis.

As was the case with the graphic for the previous example, there is a C-shaped curve, showing both the inversion for the damping of the torsional mode and a decrease in its frequency.

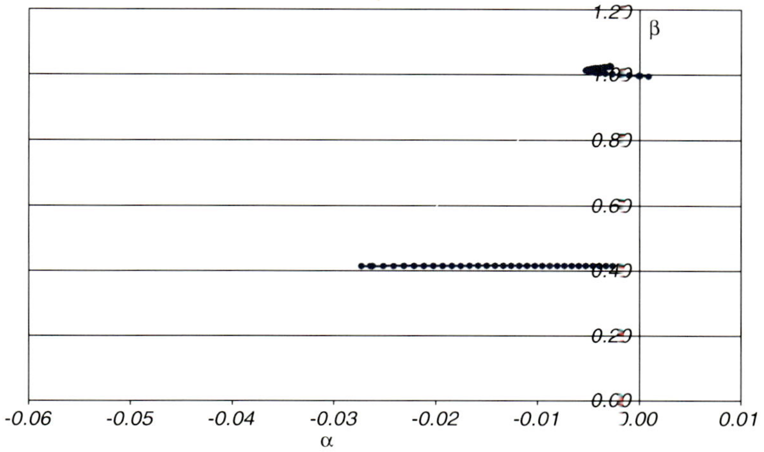

Figure 5.3.24: Evolution of β in contrast with α using two modes.

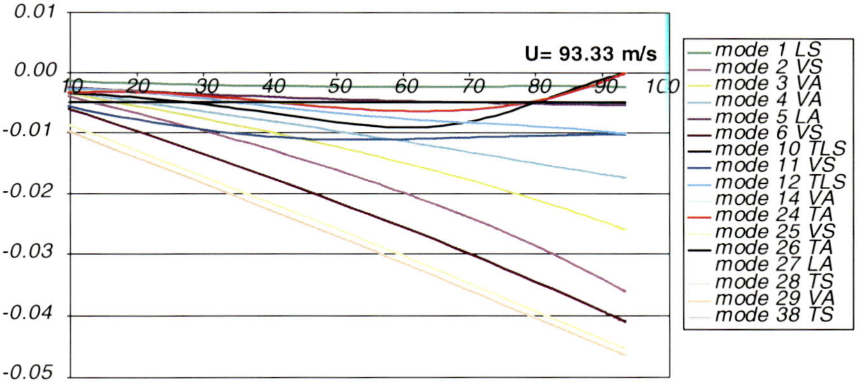

Figure 5.3.25: Evolution of the real part versus U using 17 modes.

Aeroelastic analysis for the Akashi Strait Bridge using 17 vibration modes

Figures 5.3.25–5.3.27 have the same kind of graphs used for the two-mode analysis. However, in this case, all the vibration modes previously described in the dynamic analysis are taken into account.

Once again Figure 5.3.25 shows how, with some of the torsional modes, changes appear in the curve's crest. For the lateral modes, there is hardly any variation; for the vertical ones, the damping grows in a steady fashion. On the other hand, Figure 5.3.26 indicates there are virtually no changes in the frequencies.

The critical wind speed is $U = 93.33$ m/s with $K = 0.5064$. By observing Figure 5.3.25, one can see how, in this analysis, the factor responsible for flutter is mode 24, which is the one for asymmetric torsion (AT). Therefore, the mode in which null damping is reached within the two-mode analysis undergoes a variation (*TLS* mode 12). Another considerable change takes place when the flutter speed increases more than 15 m/s. These differences

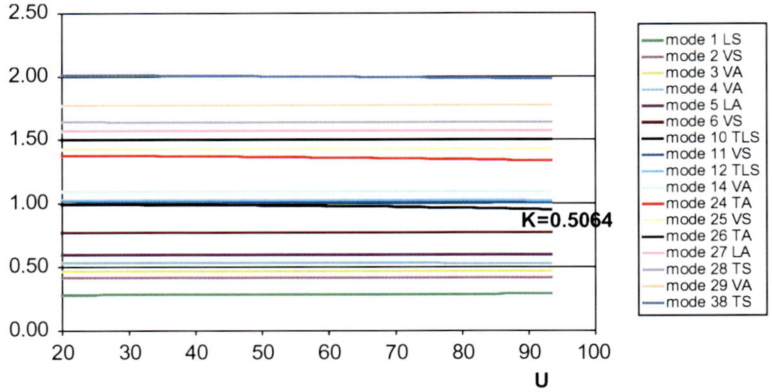

Figure 5.3.26: Evolution of the imaginary part versus U using 17 modes.

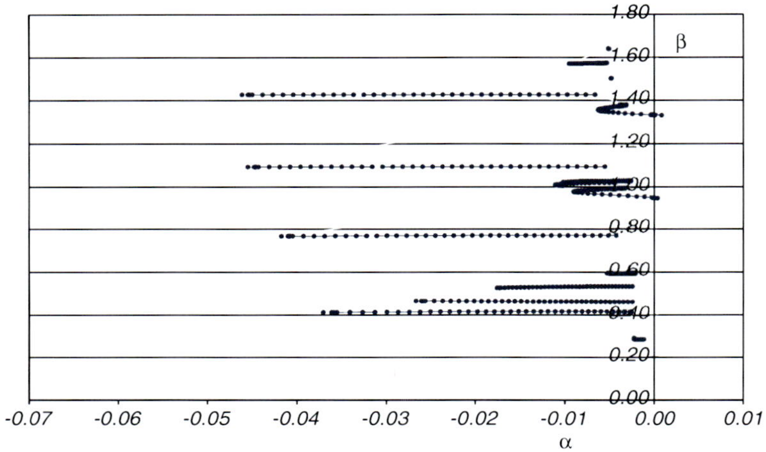

Figure 5.3.27: Evolution of β in contrast with α using17 modes.

invalidate the aeroelastic analysis carried out with two vibration modes on the Akashi Strait Bridge. Moreover, with the Great Belt Bridge example, it was possible to detect striking differences between the bimodal and multimodal analyses. Thus it seems that, with any bridge with a main span over 1500 m long, an analysis based on two modes of vibration could provide unreliable results.

Another interesting phenomenon comes to light when one studies the graph related to the real part α of the eigenvalues in Figure 5.3.25. At the point in which AT mode 24 cuts across the horizontal axis, the value for α of TLS mode 10 is almost null. This indicates that the flutter of the bridge can be reached, at practically the same wind speeds (93 m/s), with two kinds of responses. In one response, the asymmetric modes are coupled. Real part α of AT mode 24 is the one becoming null for a frequency response (imaginary part β) of 1.33 rad/s. The second kind of response takes place with the coupling of the symmetrical modes, in which TLS mode 10 reaches negative damping with a frequency different from 0.94 rad/s.

Table 5.3.3: Comparison of various aeroelastic analyses for the Akashi Strait Bridge.

Analysis	Flutter speed *U (m/s)*	Frequency ω (rad/s)
FLAS, 2 modes	77.69	0.99
FLAS, 17 modes	93.33	1.33–0.95
Katsuchi, 2 modes	58.9	–
Katsuchi, 15 modes	87–96	0.92–0.93
Miyata, reduced model	85–95	1.3

For this bridge it is possible to reach the critical flutter speed with two possible combinations of modes and also to get two different frequencies. As a result, one will later face difficulties when carrying out a sensitivity analysis around the deck's mechanical parameters. One will no longer be able to consider the frequency or flutter speed as derivable once there are changes between the two flutter shapes.

Figure 5.3.27 has the Argand diagram related to this bridge. Yet again there are various C-shaped curves representing the torsional modes. The fact that two of these curves cut across the vertical axis β confirms that, with flutter, the damping of the two modes are simultaneously annulled.

Table 5.3.3 compares the results from the aeroelastic analyses carried out on the Akashi Strait Bridge with the analytical calculations made by Katsuchi[4] and the testing done by Miyata ***et al.***[5] using a full model on a 1/100 scale.

Drawing one's attention to the multimodal analyses, much more precise than those which use only two vibration modes, it is possible to observe how all the values in the table hover around 90 m/s, returning to more or less the same point according to the suppositions one makes. As for the response frequency, the value obtained by Katsuchi approximates one of the possibilities found in the analysis for 17 modes. On the other hand, the value obtained by Miyata in the reduced scale model study of the bridge is similar to the frequency of the other flutter shape.

The chance that simultaneous flutter shapes appear has gone undetected in the aeroelastic analyses done by other research teams. However, it seems strange that the frequency results for two of these studies are markedly different, food for thought and a topic worthy of further investigation.

5.4 Original Tacoma Bridge

Obviously this bridge is of great significance as the starting point in the study of aeroelasticity for long-span bridges. For this reason, it seems appropriate to include this structure among the examples targeted for the application of the aeroelastic analysis.

The original bridge over the Tacoma Narrows was revolutionary in its design and infamous for its collapse. Seventh November, 1940, put paid to the trend in bridge engineering for increasingly lighter and more flexible constructions. From the end of the 19th century, the prevailing taste in bridges was for slenderness and elegance in the name of aesthetics. With its slender deck and flexile, stylized towers, the Tacoma Bridge was considered quintessential for its time. Figure 5.4.1 shows what happened to this daring design.

The geometric dimensions needed to produce the finite element model were taken from the plans made by the Federal Work Agency[6]. These plans appear in Figure 5.4.2.

Figure 5.4.1: Original Tacoma Bridge.

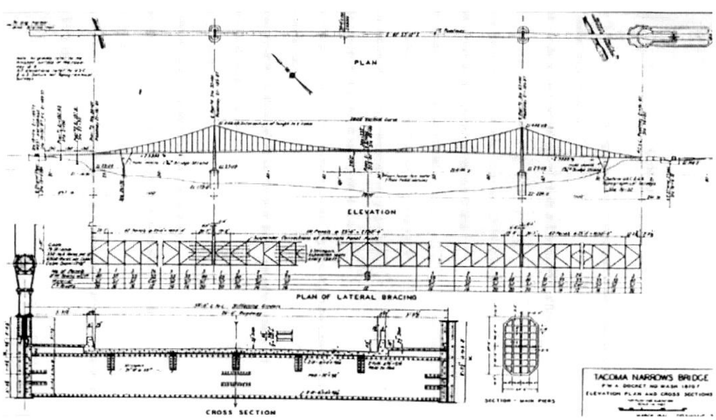

Figure 5.4.2: Plans for the original Tacoma Bridge.

On the other hand, the deck's mechanical parameters were based on the values found in an article by Herzog[7] comparing many of the dimensional features of the world's most noteworthy suspension bridges. Table 5.4.1 provides this data.

5.4.1 Frequencies and natural modes for the Tacoma Bridge

A dynamic analysis of the bridge was carried out using a bar model with 208 nodes and 323 elements calculated through the finite element program called ADISNOL3D. As seen in Figure 5.4.3, this model has parallels with the two previous examples. Among the similarities is the geometrical non-linearity that appears in the main cables when expression (6.1.2) is used to obtain the frequencies and vibration modes.

Table 5.4.2 compares the natural frequencies obtained with those provided by Cobo[2], indicating the type of vibration mode associated with each frequency: *V*, vertical; *L*, lateral; *T*, torsional; *S*, symmetrical; and *A*, asymmetrical. Frequencies are given in (rad/s). These correspond with the 10 most important vibration modes that have appreciable deck movement. Also there is the oscillation frequency that was recorded before the collapse.

Table 5.4.1: Geometrical and mechanical propert es of original Tacoma Bridge.

Main span length (m)	853.4
Side span length (m)	335.3
Tower height (m)	71
Width between cables (m)	11.9
Total deck width (m)	11.9
Deck edge (m)	2.3
Cross section of each main cable (m^2)	0.124
Mass of each main cable (t/m)	1.05
Inertia moment for vertical bending I_y (m^4)	0.154
Inertia moment for lateral bending I_z (m^4)	5.69
Inertia moment for torsion J (m^4)	6.07×10^{-6}
Deck mass (t/m)	6.22
Polar inertia moment for deck (tm^2/m)	106.5

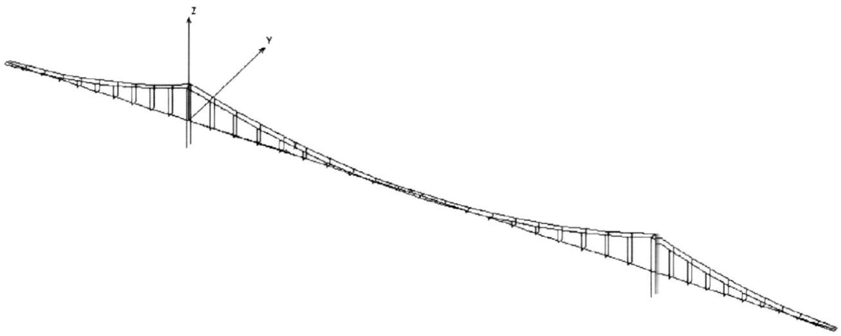

Figure 5.4.3: Finitre element model of the original Tacoma Bridge.

Table 5.4.2: Comparison of natural frequencies in (rad/s) for t1e Tacoma Bridge.

Mode type	ADISNOL3D	COBO	COLAPSE
(1) LS	0.568	0.435	–
(2) VA	0.817	0.795	–
(3) VS	1.189	0.809	–
(4) LA	1.296	0.949	–
(5) TA	1.505	1.147	1.256
(6) TS	1.608	1.165	–
(7) VA	1.705	1.055	–
(8) VS	1.792	–	–
(12) VS	2.179	–	–
(16) TA	2.321	–	–

As was the case in previous examples, the calculation was made with a sufficiently high number of modes so that at least three torsional ones were included. Here mode 16 is the third mode of pure torsion. Moreover, with lower frequencies, various vertical and lateral ones appeared.

The frequency values for the different modes coincide on a satisfactory basis. Figures 5.4.4–5.4.8 show the deformed geometry for each vibration mode.

With the structural damping for the Tacoma Bridge, it is supposed that there is a viscose type of damping $\xi = 0.00318$ in all the modes corresponding to a 2% level of logarithmic decrease.

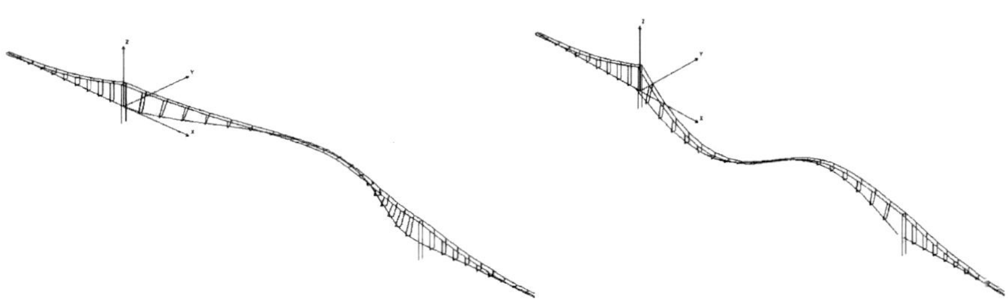

Figure 5.4.4: LS mode 1 and VA mode 2 of the Tacoma Bridge.

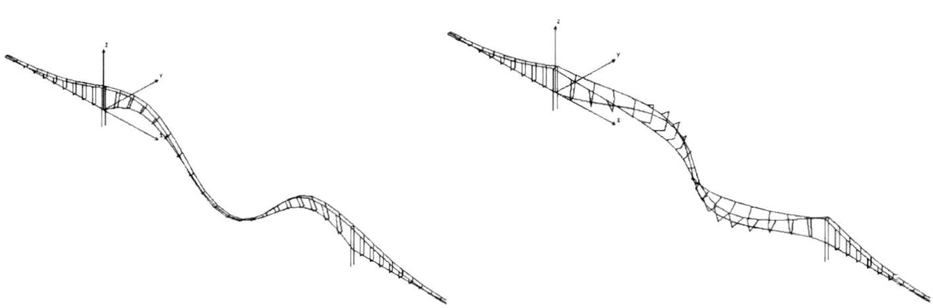

Figure 5.4.5: VS mode 3 and LA mode 4 of the Tacoma Bridge.

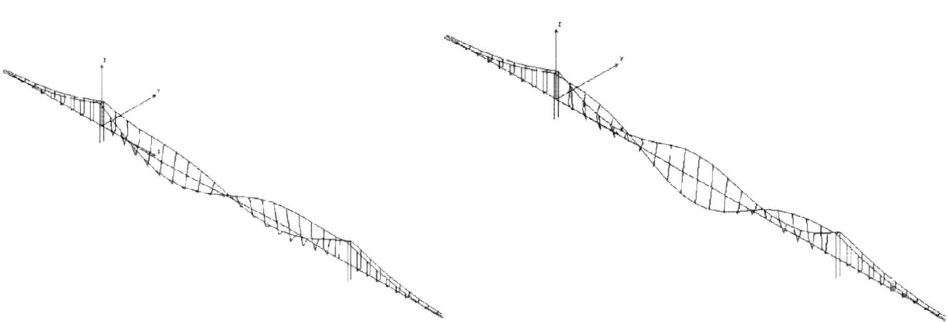

Figure 5.4.6: TA mode 5 and TS mode 6 of the Tacoma Bridge.

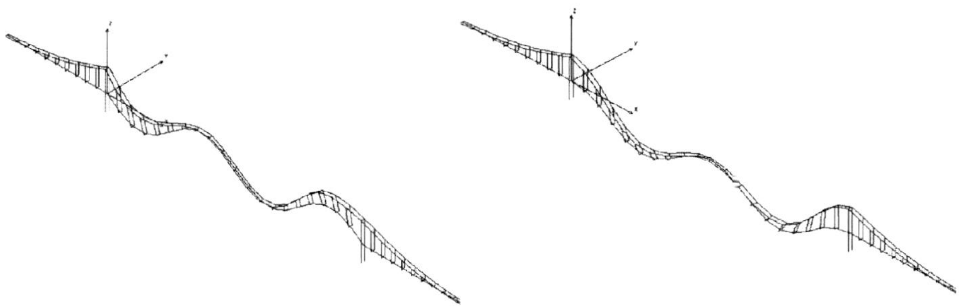

Figure 5.4.7: VA mode 7 and VS mode 8 of the Tacoma Bridge.

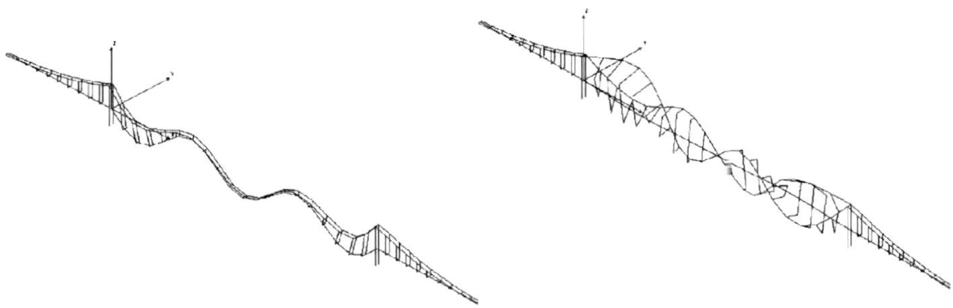

Figure 5.4.8: VS mode 12 and TS mode 16 of the Tacoma Bridge.

$$\delta = \frac{2\pi\xi}{\sqrt{1-\xi^2}} \approx 2\pi\xi = 0.02 \tag{5.4.1}$$

Because there are no data of real value about the damping found in the original Tacoma Bridge, many researchers work within a very broad range, defined by the values $\xi = 0.001$ and $\xi = 0.02$. For this study, the previously indicated damping value was used, along the lines of other suspension bridges.

5.4.2 Aeroelastic analysis of the Tacoma Bridge

For this bridge, the 10 modes represented in the figures above will be used in the aeroelastic analysis. As was the case with the bridges analysed earlier, all the aeroelastic analyses are based on a multimodal theory, an application of expressions (7.1.4)–(7.1.10).

The flutter derivatives of Scanlan and Tomko[8] will be utilized for the Tacoma Bridge's H section. These can be found in Figures 5.4.9–5.4.12; they are contrasted with what may be obtained analytically through Theodorsen's theory for aerodynamic shapes. Some coefficients, such as $H_1^*, A_1^*, A_3^*, A_4^*$, are almost null, according to Scanlan, and that is why they are absent from the figures. Also absent is information about the values taken by the P_i^* coefficients in this bridge.

Aeroelastic analysis of the Tacoma Bridge using two vibration modes

The first aeroelastic analysis carried out on the Tacoma Bridge considers the coupling of the two classic modes of flutter phenomena. These are the first mode of vertical bending

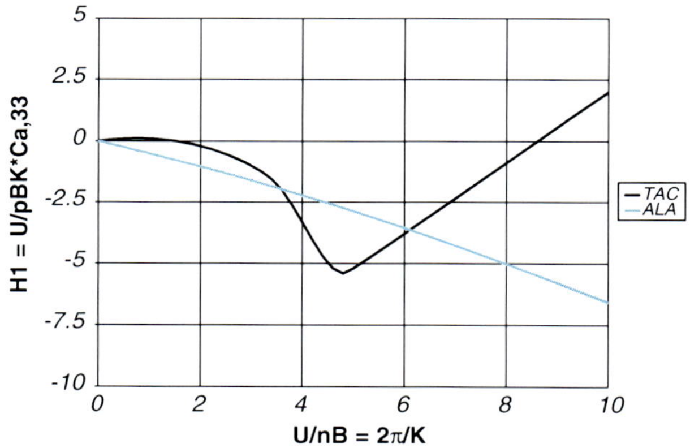

Figure 5.4.9: H_1* flutter derivatives.

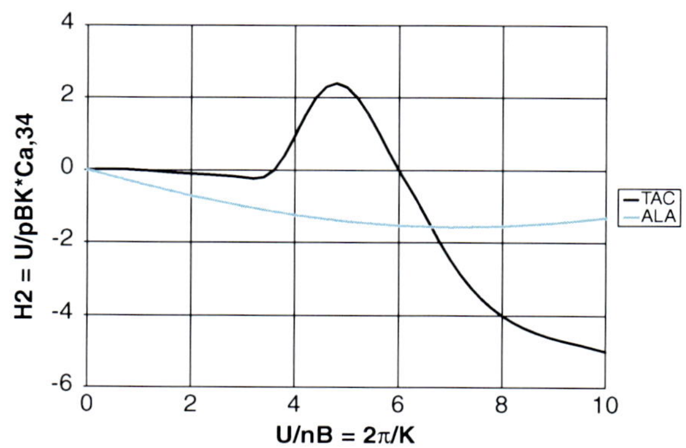

Figure 5.4.10: H_2* flutter derivatives.

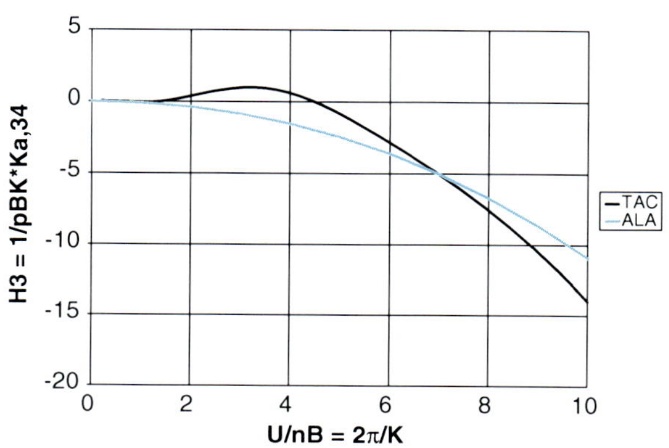

Figure 5.4.11: H_3* flutter derivatives.

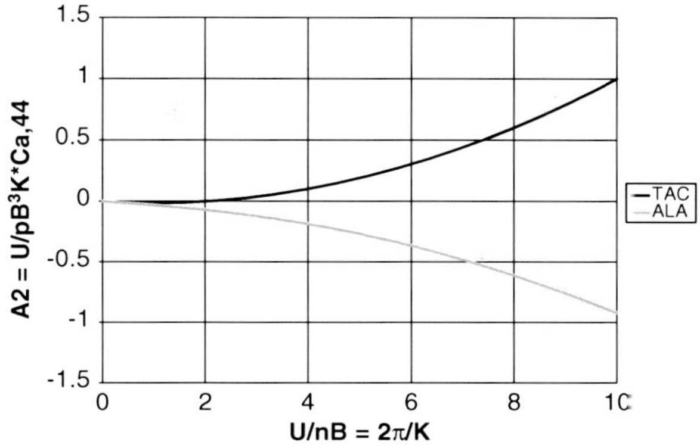

Figure 5.4.12: A_2* flutter derivatives.

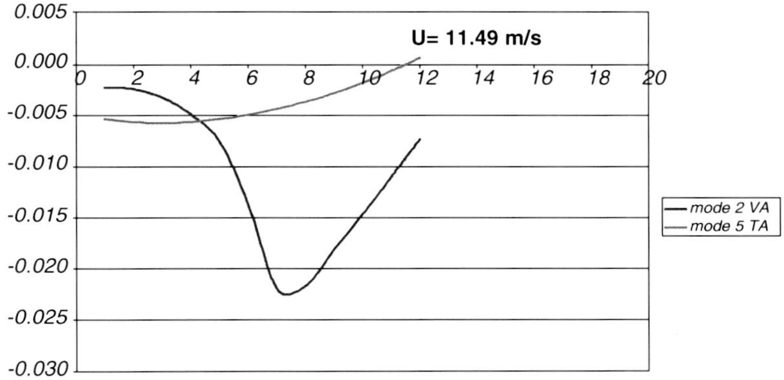

Figure 5.4.13: Evolution of a versus U using two modes.

and the first one of torsion, which correspond with modes 2 and 5 of those presented in Table 5.4.2. Here the two modes are asymmetrical, a slight variation on what is found in the Great Belt and Akashi Bridges. In both of these, the first modes of each type are symmetrical.

The graphs in Figures 5.4.13 and 5.4.14 represent the evolution of the real and imaginary parts of the eigenvalues for the increasing wind speeds. Also provided are the critical wind speeds and the reduced frequency associated with them. It should be remembered that real part a is related to damping, whereas imaginary part β coincides with the response frequency.

As revealed in the figures, the torsional mode is responsible for the flutter-induced instability reached with wind speeds of U = 11.49 m/s, and a reduced frequency of K = 1.5592, corresponding with ω = 1.505 rad/s.

Figure 5.4.13 shows how the damping for torsional mode 5 already decreases with speeds below 3 m/s. This gives one an idea of just how unstable the bridge was under torsion. The damping for the vertical mode sharply increases as the speed U rises. It then drops dramatically from 8 m/s on. This behaviour may be explained by the sharp peaks observed in the graphs related to the flutter derivatives in Figures 5.4.9–5.4.12.

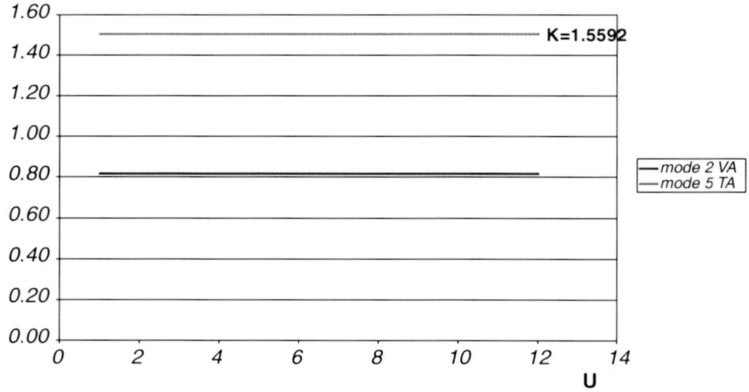

Figure 5.4.14: Evolution of β versus U using two modes.

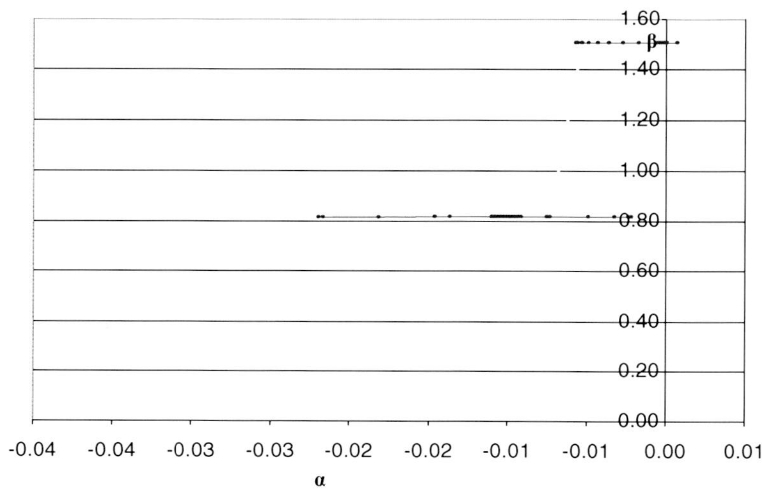

Figure 5.4.15: Evolution of β versus α using two modes for the Tacoma Bridge.

On the other hand, in Figure 5.4.14, the response frequencies practically remain unchanged as the speed grows. Thus, during flutter, the response frequency almost coincides with the natural frequency for torsional asymmetrical mode 5.

The Argand diagram in Figure 5.4.15 shows the evolution for both the real and imaginary parts of the eigenvalues. Each point along the graph represents an eigenvalue with a determined wind speed. Flutter happens when a point dissects the vertical axis.

There are no C-shaped curves in this example owing to the fact that the frequencies hardly change with the speed. Consequently, their evolution is almost horizontal.

Aeroelastic analysis of the Tacoma Bridge using 10 vibration modes

Below, in Figures 5.4.16–5.4.17, are graphs identical to the ones used for the two-mode analysis. However, the focus this time is on the first 10 modes with movements along the deck, while those that only affect cables have been eliminated.

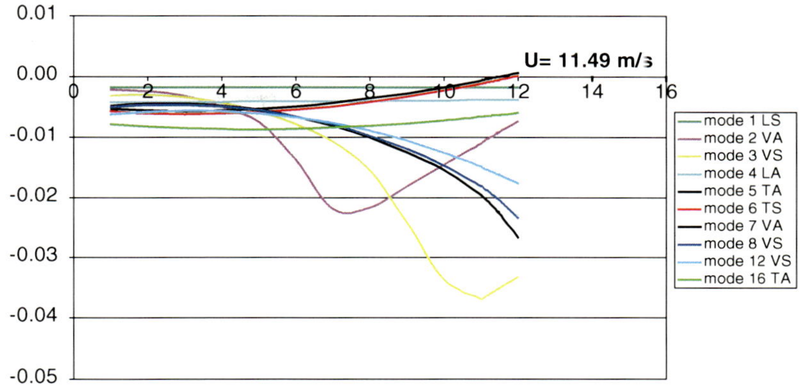

Figure 5.4.16: Evolution of α in comparison with U using 10 modes.

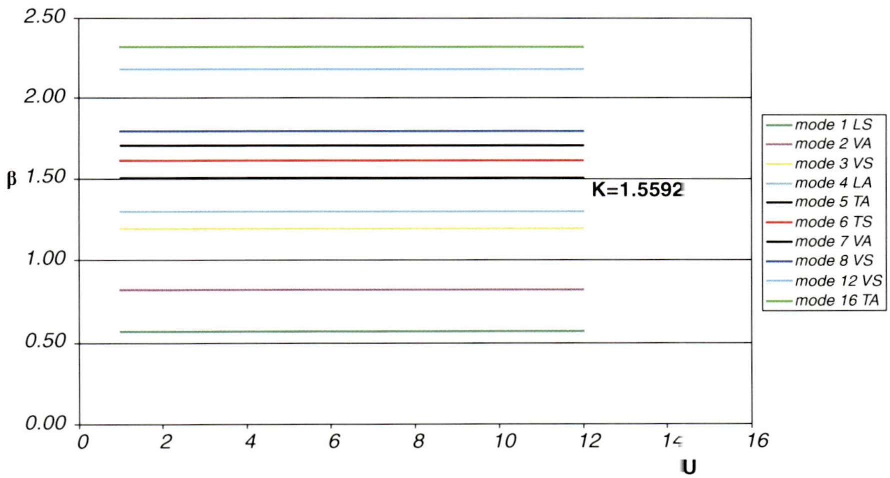

Figure 5.4.17: Evolution of β in comparison with U using 10 modes.

The critical flutter speed is $U = 11.49$ m/s, exactly the same as the one obtained with two vibration modes. The reduced frequency also coincides with $K = 1.5592$. From these results, it may be deduced that the aeroelastic analysis using two vibration modes comes very close to the bridge's behaviour under flutter. This is because with the Tacoma Bridge, the first asymmetric torsional mode chiefly causes the instability, whereas the other modes play a more modest part. With long-span suspension bridges, like the Great Belt or Akashi-Kaikyo, this is not the case as a wider variety of vibration modes participates.

Figure 5.4.16 demonstrates how, in all the torsional modes, damping decreases with the speed. In the vertical modes, it goes up sharply, reaches a peak and then later falls. Finally, in the lateral modes, the damping remains almost completely unchanged.

Another noticeable feature is how the evolution of α symmetrical torsional mode 6 has strong parallels with that of torsional asymmetrical mode 5. On the other hand,

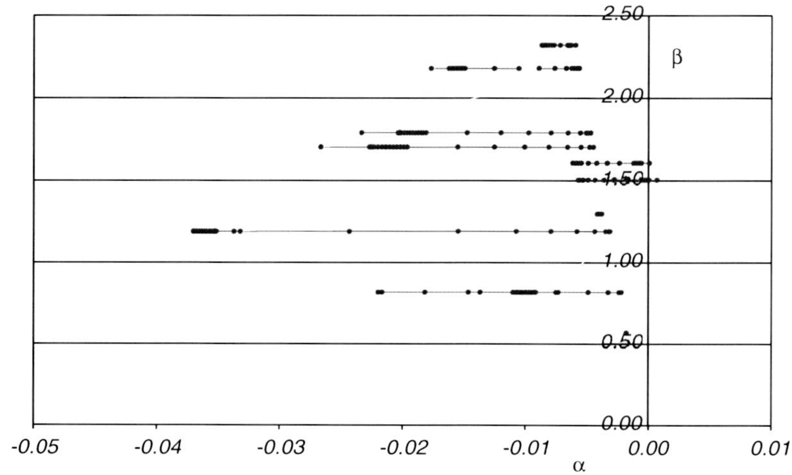

Figure 5.4.18: Evolution of β in comparison with α using 10 modes.

Table 5.4.3: Comparison of diverse aeroelastic analyses for the Tacoma Bridge.

Analyses	U (m/s)
Jurado, 2 modes $\xi = 0.00318$	11.49
Jurado, 10 modes $\xi = 0.00318$	11.49
Scanlan torsional flutter $\xi = 0.003$	7.60
Scanlan torsional flutter $\xi = 0.001$	10.23
Farquharson real collapse	18.77

Figure 5.4.18 also reveals two curves that cross the vertical axis. In these graphs, there are two distinct types of flutter at almost the same speed. Such was also the case with the Akashi Strait Bridge. Nevertheless, here the curves for these modes, related to a, never actually coincide: they become null at points that are fairly close, yet different.

Table 5.4.3 compares the aeroelastic analyses carried out on the Tacoma Bridge with calculations done by Scanlan[8], in which the torsional mode was considered, as well as the real collapse values published by Farquharson[9].

The flutter speed obtained 11.49 m/s is lower than the one reached on the day the bridge collapsed, which – according to Farquharson's report – was almost 19 m/s. Nonetheless, one may suppose that the incipient flutter could have been produced at a lower speed. The value proposed in this work does fall within the speeds determined by Scanlan, between 7.5 and 13.8 m/s.

5.5 The Vasco da Gama Bridge

The Vasco da Gama Bridge was opened to mark the International Exposition of 1998 (EXPO98), held in Lisbon that year. Crossing the Tajo river, further along from the 25 April suspension bridge, the main structure over this navigation channel comprises a

Figure 5.5.1: Vasco da Gama Bridge.

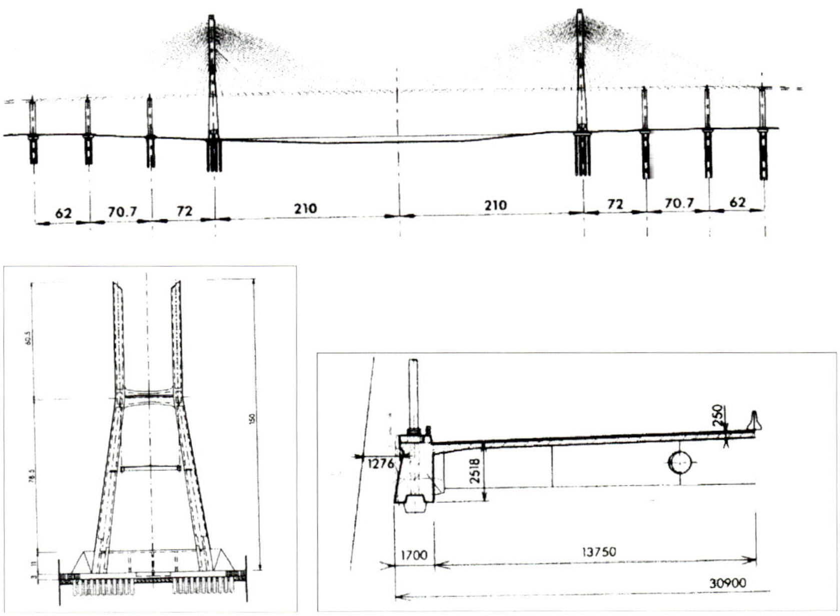

Figure 5.5.2: Elevation, tower and deck of the Vasco da Gama Bridge.

830-m long cable-stayed bridge. Its main span is 420 m, while the lateral spans are 62, 70.7 and 72 m either side. Its 31-m-long deck has two crossbeams made of reinforced concrete that are connected by a 25-cm thick concrete slab, as well as I-shaped lateral steel girder placed at 4.4 m intervals. The deck is continuous along its complete length, moving freely both vertically and sideways at tower height. To lessen these lateral displacements, there are elastoplastic dampers. The towers are 150 m tall and in the shape of an H. Figure 5.5.1 provides a photograph of the bridge.

The geometric dimensions needed to produce the finite element model, used to calculate frequencies, were taken from Capra[10], as well as the description of the bridge made by Machado and Branco[11]. Figure 5.5.2 has plans for the bridge's elevation, one of its towers and the deck cross section. Because it was impossible to find any

Table 5.5.1: Geometrical and mechanical properties of the Vasco da Gama Bridge.

Main span length (m)	420
Side span length (m)	62+70.7+70
Tower height (m)	150
Width between cables (m)	29.2
Total deck length (m)	30.9
Deck depth (m)	2.65
Cross section of each cable (m^2)	0.019
Inertia moment under vertical bending I_y (m^4)	10.5
Inertia moment under lateral bending I_z (m^4)	2300
Inertia moment under torsion J (m^4)	6.2
Deck mass (t/m)	53.9
Polar moment of inertia for deck (tm^2/m)	6800

references to the deck's mechanical parameters, they were calculated from the dimensions appearing in the figure. The mass and polar inertia for the deck were provided by Grillaud *et al.*[12] in the experimental aeroelastic study they made of the bridge. Table 5.5.1 outlines the main geometrical properties and mechanical parameters for the Vasco da Gama Bridge.

5.5.1 Frequencies and natural modes for the Vasco da Gama Bridge

The dynamic analysis of the bridge was carried out using a bar model using 2473 nodes and 2013 elements calculated through the finite element code COSMOS/M. The model, appearing in Figure 5.5.3, contains BEAM3D elements to simulate the deck, cables and towers. MASS elements were also used to provide a sufficiently accurate simulation of the mass distribution along the deck. Various elements were situated at each cable stay to obtain a suitable analysis of its deformation when second-order theory was applied. In an approach similar to the one taken for the previous examples, highly rigid, fictitious elements were added to connect the longitudinal bars representing the deck to the ends of the cable stays. The purpose of these elements was to make the movements compatible.

The dynamic calculation made takes into account the increased rigidity produced in the structure when second-order effects were considered, especially along the cable stays. A previous calculation was made of the structure at its own weight. The deformation obtained was used to get the second-order forces and thus it was possible to make an accurate assessment of the non-linear geometric behaviour of the cables.

Table 5.5.2 compares the natural frequencies obtained with those given by Mendes[13] and Branco, as well as those found in Capra and Leveille[10], indicating the type of vibration mode associated with each frequency: *V*, vertical; *L*, lateral; *T*, torsional; *S*, symmetrical; and *A*, asymmetric. Also provided are the frequencies in (rad/s) corresponding with the first 11 vibration modes in which the deck has noticeable movements. The modes exclusive to cable movement were excluded given that the aeroelastic analysis was limited to the deck. The other modes, therefore, would not provide data useful for the aeroelastic calculation.

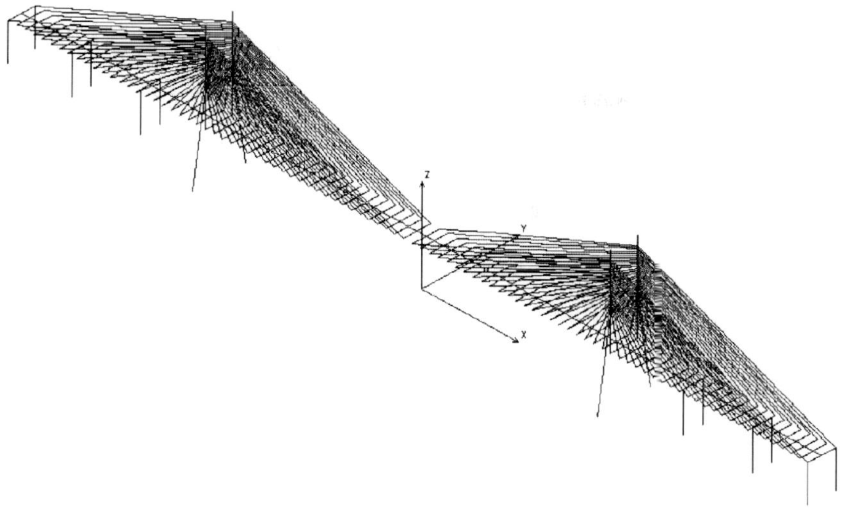

Figure 5.5.3: Finite element model of the Vasco da Gama Bridge.

Table 5.5.2: Comparison of the frequencies (rad/s) for the Vasco da Gama Bridge.

Type of mode	ADISNOL3D	SEMIAO-MENDES	CAPRA-LEVEILLE
(1) LS	1.490	1.583	1.317
(2) VS	1.704	2.086	1.933
(3) VA	1.945	2.789	–
(4) TS	2.977	2.909	–
(5) TA	3.513	3.085	–
(6) VS	3.523	3.933	–
(7) TS	4.179	4.172	–
(8) VA	4.294	–	–
(9) VS	4.877	4.882	–
(10) VA	5.387	–	–
(11) TS	5.434	5.292	–

Eleven modes were chosen because it was the necessary number to reach the third purely torsional mode. Moreover, as seen in the previous examples, flutter usually makes the damping for one of the modes of this type null.

The amount by which the frequency values for the different modes coincide may be deemed acceptable, especially if one takes into account that the mechanical properties used for the deck were calculated from plans referred to above. Figures 5.5.4–5.5.9 show the vibration mode shapes.

As for the structural damping of the Vasco da Gama Bridge, it is assumed that there is, for all modes, a viscose type of damping $\xi = 0.00477$ corresponding with a logarithmic decrease of 3%

$$\delta = \frac{2\pi\xi}{\sqrt{1-\xi^2}} \approx 2\pi\xi = 0.03 \qquad (5.5.1)$$

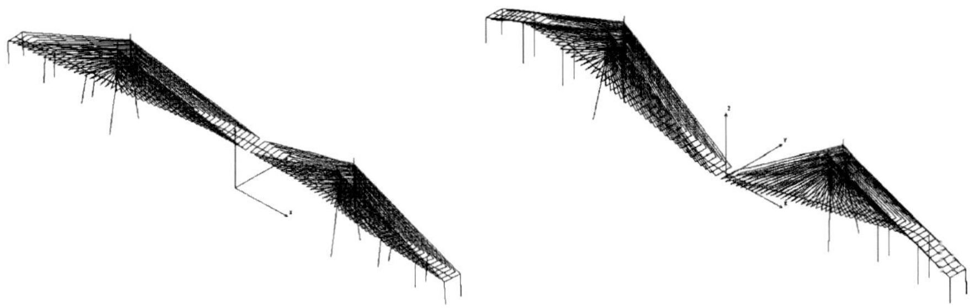

Figure 5.5.4: LS mode 1 and VS mode 2 for the Vasco da Gama Bridge.

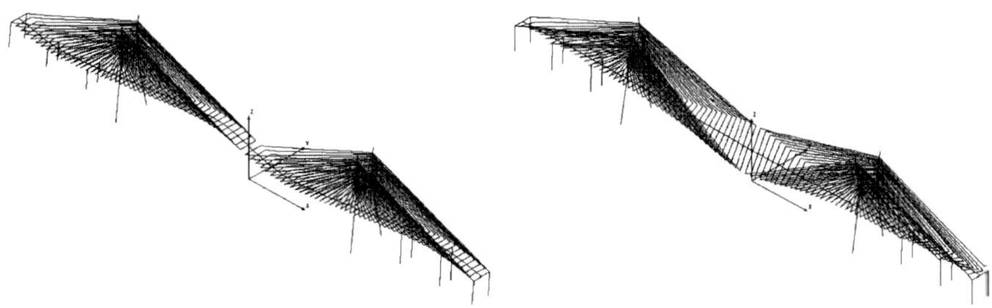

Figure 5.5.5: VA mode 3 and TS mode 4 for the Vasco da Gama Bridge.

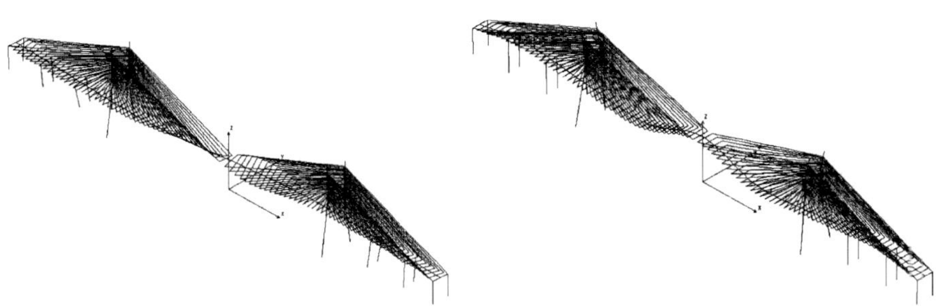

Figure 5.5.6: TA mode 5 and VS mode 6 for the Vasco da Gama Bridge.

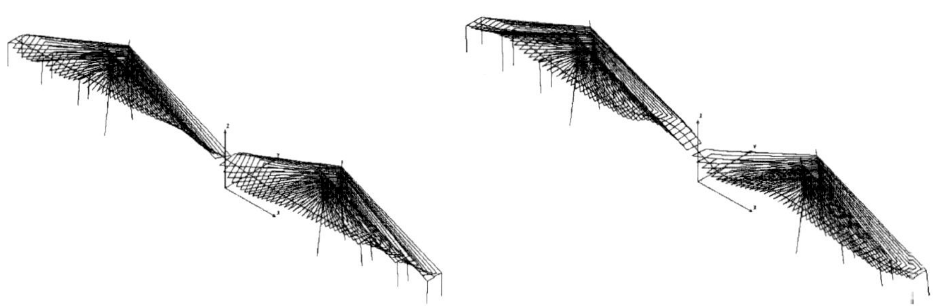

Figure 5.5.7: TA mode 7 and VA mode 8 or the Vasco da Gama Bridge.

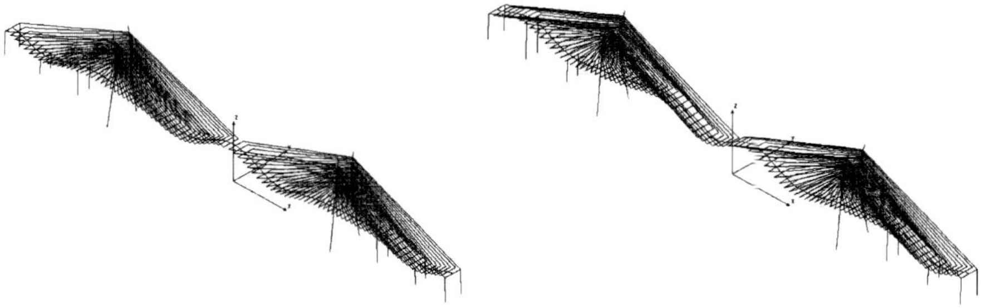

Figure 5.5.8: VS mode 9 and VA mode 10 for the Vasco da Gama Bridge.

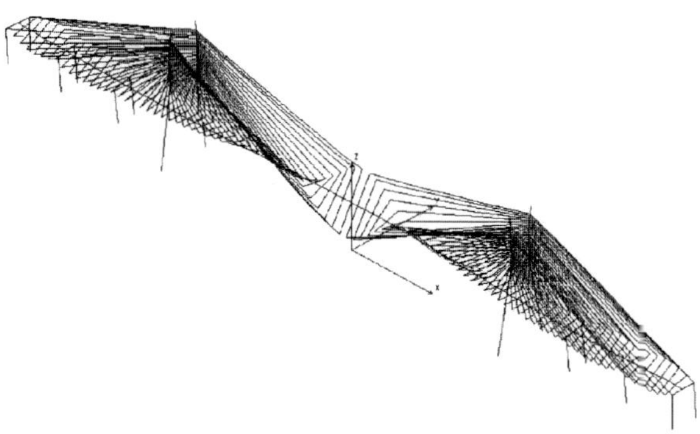

Figure 5.5.9: TS mode 11 for the Vasco da Gama Bridge.

5.5.2 Aeroelastic analysis of the Vasco da Gama Bridge

For this bridge, aeroelastic analyses were carried out using the 11 modes shown in the previous figures. Maintaining the trend set for the previous examples, all the analyses were based on the multimodal theory, applying expressions (7.1.4)–(7.1.10). Thus, the formulation remained the same even though the Vasco de Gama is a cable-stayed rather than suspension bridge.

For the flutter derivatives in this bridge's deck, the following coefficients were made available: H_1^* and A_2^* obtained by Grillaud *et al.*[12] through experimentation in the CSTB wind tunnel of Nantes, France. Thanks to these coefficients, the first aeroelastic calculation was possible, although it would have been slightly risky to test the bridge's stability by working with only two coefficients.

For further details about the other derivatives, it is possible to turn to a quasi-stationary approach for some of these. This method was tested by Singh[14] in his doctoral thesis. In this work, the derivatives obtained experimentally were compared with those that had been approximated through coefficients from aerodynamic forces (C_L, C_M, C_D) and their derivates related to the angle of attack (C_L', C_M', C_D') in the Tsurumi (cable-stayed) and Deer Isle (suspension) Bridges.

With n denoting the response frequency in Hz and B standing for deck width, the expressions related to aerodynamic coefficients and flutter derivatives are as follows:

$$\begin{aligned}
H_1^* &= \frac{1}{2\pi}\left(\frac{U}{nB}\right)C_L' \\
H_3^* &= \frac{1}{4\pi^2}\left(\frac{U}{nB}\right)^2 C_L' \\
H_5^* &= \frac{-1}{\pi}\left(\frac{U}{nB}\right)C_L \\
A_1^* &= \frac{1}{2\pi}\left(\frac{U}{nB}\right)C_M' \\
A_3^* &= \frac{1}{4\pi^2}\left(\frac{U}{nB}\right)^2 C_M' \\
A_5^* &= \frac{-1}{\pi}\left(\frac{U}{nB}\right)C_M \\
P_1^* &= \frac{-1}{\pi}\left(\frac{U}{nB}\right)C_D \\
P_2^* &= \frac{1}{2\pi}\left(\frac{U}{nB}\right)C_D' \\
P_3^* &= \frac{1}{4\pi^2}\left(\frac{U}{nB}\right)^2 C_D' \\
P_5^* &= \frac{-1}{\pi}\left(\frac{U}{nB}\right)C_D
\end{aligned} \tag{5.5.2}$$

The coefficients for aerodynamic forces in the Vasco da Gama Bridge were obtained through testing done by Mendes and Branco[13]: these are presented in Table 5.5.3.

The graphs in Figures 5.5.10–5.5.17 represent the flutter derivative values for the Vasco da Gama (VAS) Bridge along with those for a flange shape (ALA) according to Theodorsen's theory. Of the coefficients just mentioned for the Vasco de Gama, H_1^* y A_2^* were obtained through experimentation in Nantes and the rest had been calculated by the quasi-stationary approach, also referred to above. The coefficients that are absent from the graphs had not been included in the analyses.

Aeroelastic analysis of the Vasco da Gama Bridge using two vibration modes

The first aeroelastic analysis carried out on the Vasco da Gama Bridge focuses on the coupling of the two modes considered classic in the phenomenon of flutter, the first

Table 5.5.3: Coefficients for aerodynamic forces in the Vasco da Gama Bridge.

C_D	C_L	C_M	C'_D	C'_D	C'_D
0.144	−0.037	0.015	0.0	−5.85	1.00

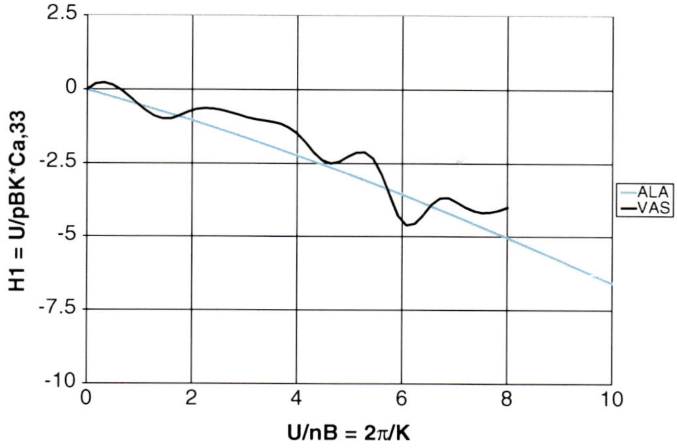

Figure 5.5.10: H_1* flutter derivatives.

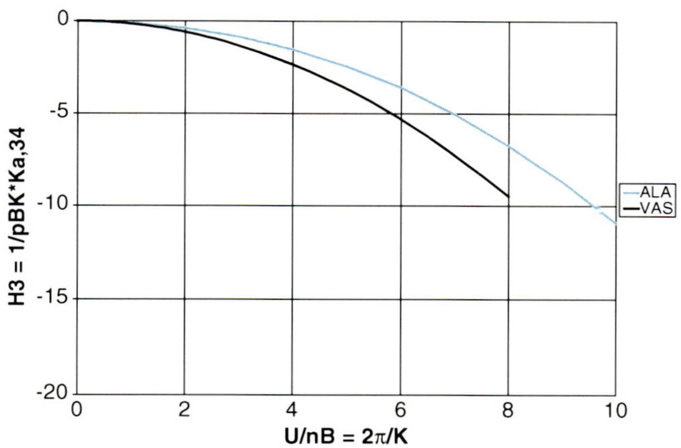

Figure 5.5.11: H_3* flutter derivatives.

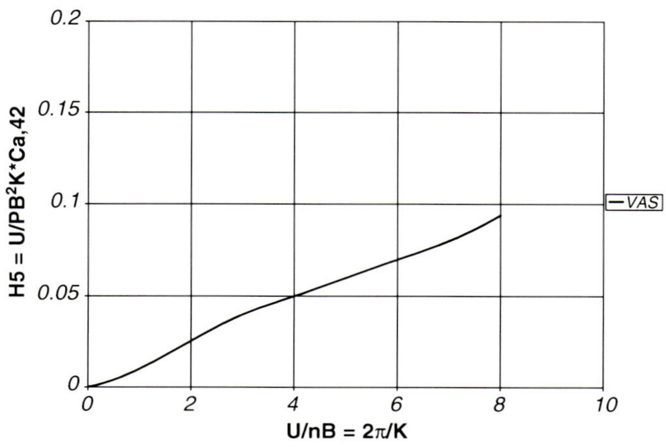

Figure 5.5.12: H_5* flutter derivatives.

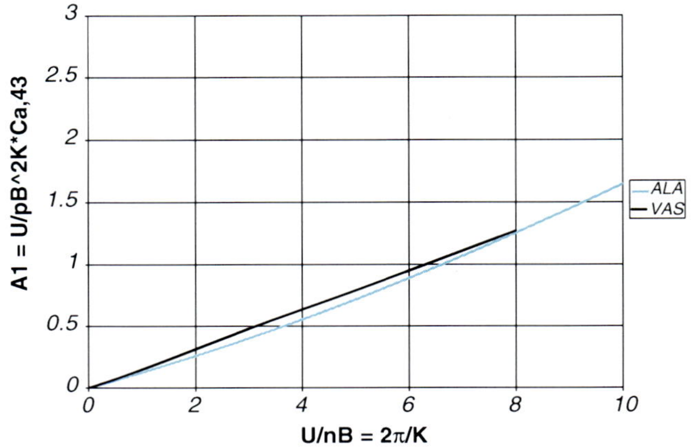

Figure 5.5.13: A_1* flutter derivatives

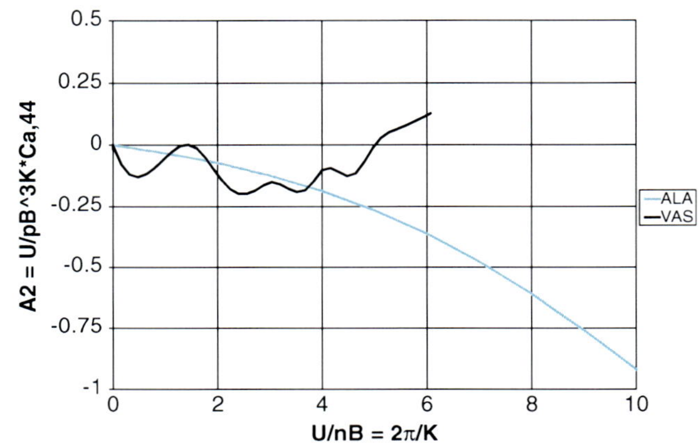

Figure 5.5.14: A_2* flutter derivatives.

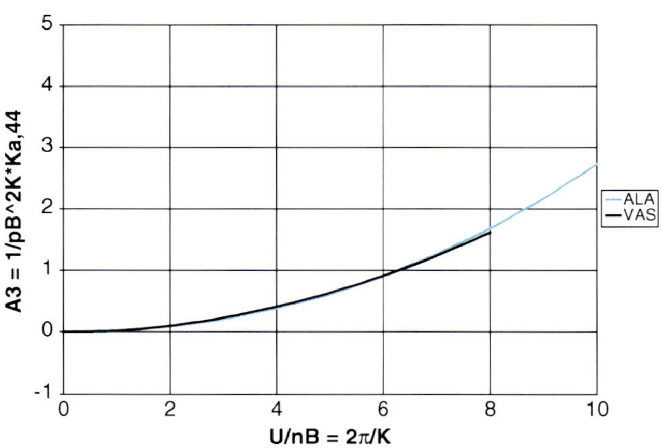

Figure 5.5.15: A_3* flutter derivatives.

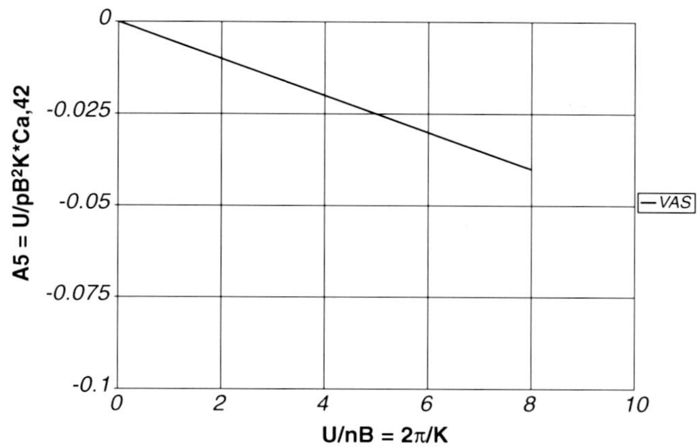

Figure 5.5.16: A_5* flutter derivatives.

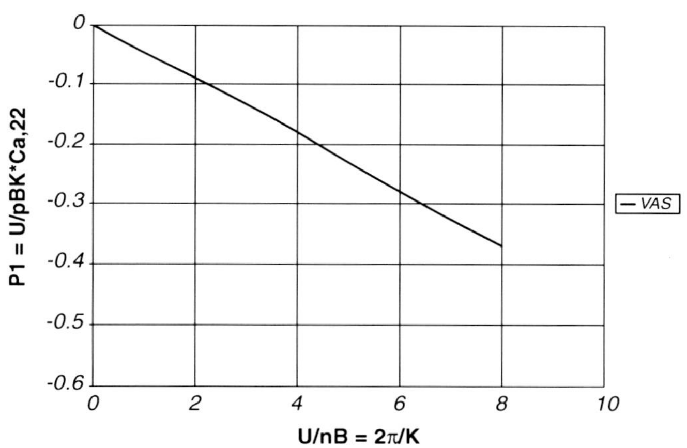

Figure 5.5.17: P_1* flutter derivatives.

symmetrical mode of vertical bending and the first symmetrical one of torsion. The graphs in Figures 5.5.18 and 5.5.19 chart the evolution of the real and imaginary parts of the eigenvalues obtained for growing wind speeds. Also shown are the values for critical wind speeds and their associated reduced frequency.

As can be gleaned from the figures, the torsional mode can be blamed for the flutter instability reached at wind speeds of $U = 80.16$ m/s, and with a reduced frequency of $K = 1.1148$, corresponding to $\omega = 2.892$ rad/s.

The oscillating shape of the curves in Figure 5.5.18, representing the damping, is caused by the similarly oscillating shape of the curves related to flutter derivatives H_1^* and A_2^*. On the other hand, Figure 5.5.19 has a representation of the frequencies, in which only a slight decrease in the torsional mode curve can be detected.

Figure 5.5.20 shows the joint evolution experienced by the real and imaginary parts of the eigenvalues. Each point on the graph is representative of an eigenvalue with a specific wind speed. Flutter arises when a point lies to the right of the vertical axis.

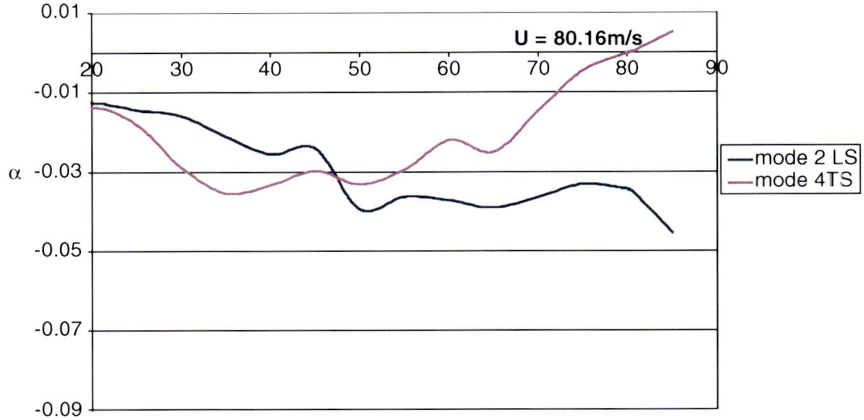

Figure 5.5.18: Evolution real part α compared with U using two modes for the Vasco da Gama Bridge.

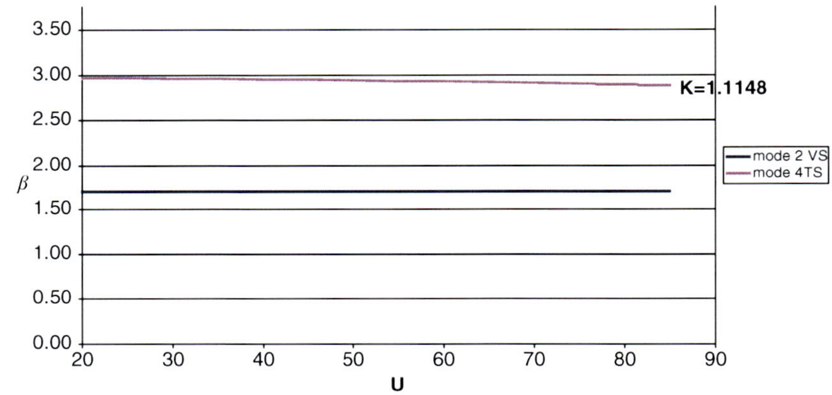

Figure 5.5.19: Evolution of the imaginary part β compared with U using two modes for the Vasco da Gama Bridge.

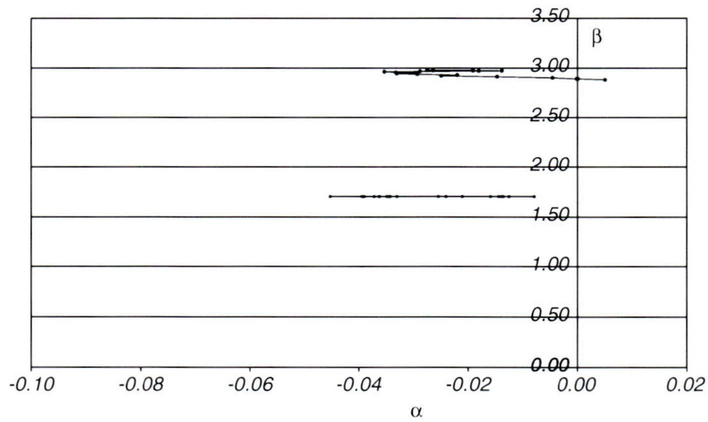

Figure 5.5.20: Evolution of β compared with α using two modes.

Another C-shaped curve is produced for the torsional mode of this bridge, revealing the trend in which the damping increases while the frequency increases.

Aeroelastic analysis of the Vasco da Gama Bridge using 11 vibration modes
Figures 5.5.21–5.5.23 have the same kind of graph used for the bimodal analysis, although this time the first eleven modes are examined.

The evolution of the damping is difficult to follow in the graph in Figure 5.5.21 due to the curves' oscillations. Nevertheless, there is the same trend seen in the other examples; the damping for the torsional modes goes up and then down; in the vertical modes, it tends to rise; and in the lateral ones, it stays more or less constant.

In the graph of Figure 5.5.22, one can easily observe a certain decrease in the frequencies for the torsional modes.

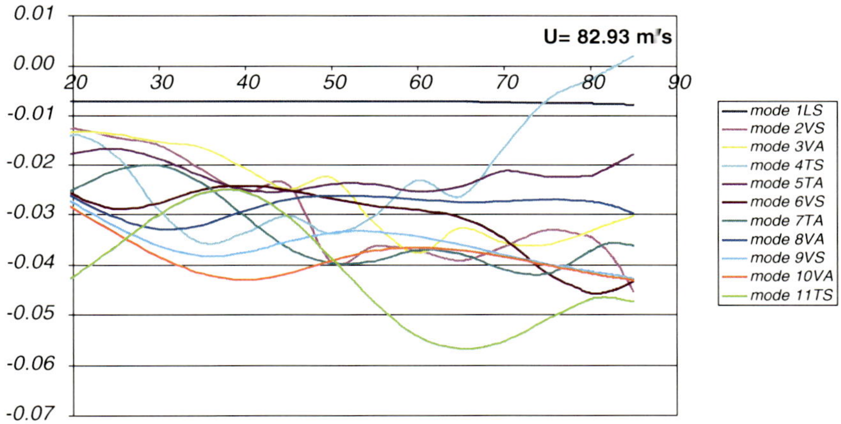

Figure 5.5.21: Evolution of real part α in contrast with U using 11 modes for the Vasco da Gama Bridge.

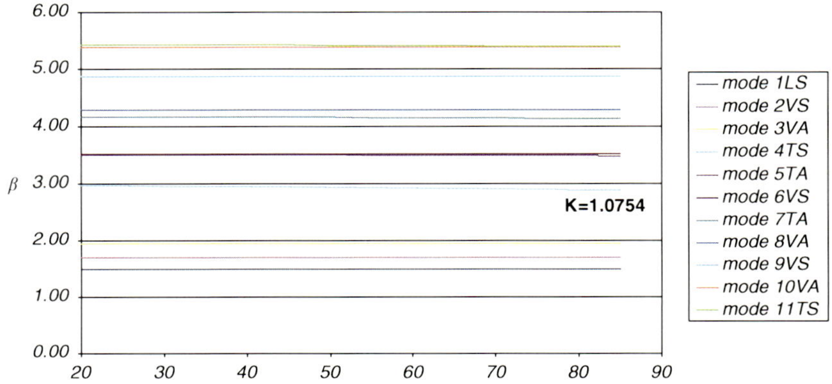

Figure 5.5.22: Evolution of imaginary part β in contrast with U using 11 modes for the Vasco da Gama Bridge.

The critical flutter speed is now U = 82.93 m/s, close to the speed obtained using two vibration modes, which is only 2.77 m/s higher. The reduced frequency is also similar at K = 1.0754. Studying Figure 5.5.21, one becomes aware that the first mode of symmetrical torsion 4 can be held responsible for flutter instability. From these results, it can be concluded that the aeroelastic analysis relying on two vibration modes provides a fairly accurate picture of the bridge's behaviour under flutter. This is not the case with suspension bridges, like the Great Belt or Akashi-Kaikyo, examined earlier. Partly this is because the vibration modes other than the two classic ones (1st VS and 1st TS) play a much more limited role in a bridge whose central span is on a more modest scale, as is the case with the Vasco da Gama. Its central span is 420 m long compared with the 1624 m of the Great Belt or the 1991 m of the Akashi-Kaikyo.

The Argand diagram in Figure 5.5.23 features yet another C-shaped curve for torsional mode 4, with little vertical variation, indicating that the response frequency changes only slightly in terms of the natural frequency for that mode.

Table 5.5.4 places the aeroelastic analyses for the Vasco da Gama Bridge alongside the analytical calculations of Semiao and Mendes[15], and the experimental results of the CSTB in Nantes obtained by Grillaud ***et al.***[12]. It is clear that the values are very close.

All of the critical speed values hover around 80 m/s. This figure indicates that for this bridge, it is possible to achieve a good approximation with a bimodal flutter analysis. In any case, given that a multimodal analysis is not much more labour intensive, it still

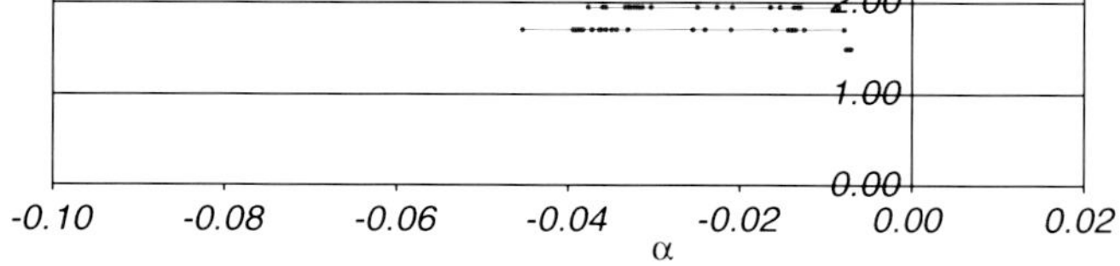

Figure 5.5.23: Evolution of b in contrast with a using 11 modes for the Vasco da Gama Bridge.

Table 5.5.4: Comparison of various aeroelastic analyses for the Vasco da Gama Bridge.

Analysis	Flutter speed Uf (m/s)
FLAS, 2 modes	80.16
FLAS, 11 modes	82.93
Semiao and Mendes	77–84
Testing at the CSTB in Nantes	79

seems recommendable to use a higher number of vibration modes when studying the aeroelastic behaviour of whatever bridge.

As was the case with other examples mentioned in this text, the exact number could be determined by the appearance of the third torsional mode, precisely the one which has the greatest impact on flutter.

5.6 References

[1] Larsen A. [1993] *Aerodynamic aspects of the final design of the 1624 m suspension bridge across the Great Belt.* Journal of Wind Engineering and Industrial Aerodynamics, Vol. 48, pp. 261–285.

[2] Cobo D. [1998] *An analysis of wind stability. Improvements to the response of suspension bridges.* Tesis doctoral de La Universitat Politécnica de Catalunya. Escola Técnica Superior D'Enginyers de Camins, Canals i Ports.

[3] Theodorsen T. [1935] *General Theory of Aerodynamic Instability and the Mechanism of Flutter.* NACA Report 741.

[4] Katsuchi T. [1997] *An Analytical Study on Flutter and Buffeting of the Akashi-Kaikyo Bridge.* Essay submitted in conformity with the requirements for Master of Science in Engineering.

[5] Miyata T., Yokoyama K., Yasuda M., Hikami Y. [1992] *Wind effects and full model wind tunnel tests.* Proceedings of the First International Symposium on *Aerodynamics of Large Bridges.* A.A. Balkema, Copenhagen, Denmark, pp. 217–236.

[6] Ammann O. [1941] *The Failure of the Tacoma Narrows Bridge.* Report to the Federal Work Agency, March 1940, pp. 37–126.

[7] Herzog M. [1990] *Die aerostatische Stabilität der Hängebrücken.* Schweizer Ingenieur und Architekt Nr.9, 1 März, pp. 223–227.

[8] Scanlan R. H., Tomko J. J. [1971] *Airfoil and bridge deck flutter derivatives.* Journal of Engineering Mechanics, Vol. 97, No. 6, pp. 1717–1737.

[9] Farquharson F. B. [1949–1954] *Aerodynamic Stability of Suspension Bridges.* Parts I–V, Bulletin No. 116, University of Washington Engineering Experimentation, Seattle.

[10] Capra A., Leveille A. [1997] *Vasco da Gama main bridge seismic design.* International Conference New Technologies in Structural Engineering. Vol. 1. Lisbon, Portugal.

[11] Machado L., Branco F. A. [1997] *The Vasco da Gama bridge project.* International Conference New Technologies in Structural Engineering. Vol. 2, Lisbon, Portugal.

[12] Grillaud G., Barré C., Flamand O. [1997] *Wind action on the Vasco Da Gama Cable stayed bridge.* International Conference on New Technologies in Structural Engineering. Vol.1, Lisbon, Portugal.

[13] Mendes P. A., Branco F. A. [1998] *Numerical wind studies for the Vasco de Gama Bridge, Portugal.* Structural Engineering International Vol. 3, No. 2, pp. 124–128.

[14] Singh L. [1997] *Experimental Determination of Aeroelastic and Aerodynamic Parameters of Long-Span Bridges.* A dissertation submitted to the Johns Hopkins University in conformity with the requirements for the degree of Doctor of Philosophy.

[15] Semiao C., Mendes P. [1998] *Avaliazao das condizoes de estabilidade aerodinâmica em pontes.* Jornadas portuguesas de engenharia de estructuras. Lisboa, Portugal.

Chapter 6

Sensitivity analysis of eigenvalue problems

6.1 Introduction

The calculation of structural sensitivity consists of changing the design variables in search of a better solution during the optimization process. Since the sensitivity analysis is often a computationally expensive process in the optimization, we need efficient algorithms to evaluate the derivatives.

An important part of engineering problems such as calculations of buckling loads or frequencies and vibration modes is associated with obtaining its eigenvalues. For this reason, the study of procedure to carry out the sensitivity analysis has great importance. Some of the recent structural dynamics textbooks such as the one by Hart and Wong[1] dedicate ample space for obtaining sensitivity for each type of dynamic problems.

However, this type of problems is more complex than the linear analysis under static loads. As a result, the method to carry out the sensitivity analysis can be very sophisticated.

The finite difference approximation is the simplest technique that allows us to calculate derivatives of the structural response with respect to the design variables. This technique is computationally expensive, but is easy to formulate and very popular. Efficiency of the analytical methods is compared with the one of the finite difference method, which unfortunately present accuracy problems.

The analytical methods of sensitivity analysis are based on the derivative of the equations that govern the structural problem with respect to the design variables. Once the solution is obtained and introduced in the derivative equations analytically, the sensitivity, which is the variation of the solution with respect to the design variables, may be obtained directly.

In the case of vibration or buckling problems, it is necessary to derive the eigenvalue problem, then to derive matrices such as stiffness, $\mathbf{K}$, masses, $\mathbf{M}$ or geometric stiffness matrix, $\mathbf{K}_G$, analytically. Depending on the design variables, x, their derivatives, $\partial \mathbf{K}/\partial x$, $\partial \mathbf{M}/\partial x$, $\partial \mathbf{K}_G/\partial x$, can give a great amount of work especially if they are design variables of a shape. The complication of deriving these matrices analytically may be resolved by deriving the same matrices for finite difference and then carrying out the rest of the procedure analytically. The methods operated in this manner are termed semi-analytical since the finite difference calculations are used in the process. This classification of the sensitivity analysis is well explained by Adelman and Haftka[2].

In this chapter, we will review the methods of sensitivity analysis in general regarding eigenvalue problems, and we will look into a specific method for bridge flutter problems in Chapter 7.

6.2 Approximation by finite difference

The simplest approximation by finite difference is the first-order forward difference. For a given function $u(x)$ with a design variable x, the approximation by finite forward difference $\Delta u/\Delta x$ to the derivative du/dx is given by:

$$\frac{\Delta u}{\Delta x} = \frac{u(x+\Delta x) - u(x)}{\Delta x} \tag{6.2.1}$$

Other common approximation is the central difference approximation

$$\frac{\Delta u}{\Delta x} = \frac{u(x+\Delta x) - u(x-\Delta x)}{2\Delta x} \tag{6.2.2}$$

We can also employ higher order approximations, but they are not practical for its high computational costs. If we need to find out the derivatives of the structural response with respect to n design variables, the approximation by forward difference requires n additional analyses, while the approximation by central difference requires $2n$ additional analyses. Higher order approximations are even more computationally expensive.

The finite difference formulation to approximate derivatives leads to the following two types of errors.

Truncation error

The truncation error $e_T(\Delta x)$ is produced for ignoring terms in the Taylor series. For example, the series of $u(x + \Delta x)$ can be written as

$$u(x+\Delta x) = u(x) + \Delta x \frac{du}{dx}(x) + \frac{(\Delta x)^2}{2}\frac{d^2u}{dx^2}(x+\varsigma\Delta x) \quad , \quad 0 \le \zeta \le 1 \tag{6.2.3}$$

From equation (6.2.3), the truncation error for the forward finite difference approximation is

$$e_T(\Delta x) = \frac{(\Delta x)}{2}\frac{d^2u}{dx^2}(x+\varsigma\Delta x), \qquad 0 \le \zeta \le 1 \tag{6.2.4}$$

In the same manner, by including one more term in the Taylor series, the truncation error for the central difference approximation is

$$e_T(\Delta x) = \frac{(\Delta x)^2}{6}\frac{d^3u}{dx^3}(x+\varsigma\Delta x), \qquad -1 \le \zeta \le 1 \tag{6.2.5}$$

Condition error

The condition error is the difference between the numerical evaluation of the function and its exact value. One of the contributors of the condition error is round-off error in calculating du/dx from the already perturbed value of u. This contribution is small for most computers except for the case that Δx is extremely small. However if is $u(x)$ obtained in a slow or badly conditioned process, the contribution of round-off to the condition error can be significant. There can be additional conditions errors if $u(x)$ is calculated by iterative methods that do not converge. If we have an absolute error limit, ε_u in calculating u, the condition error can be estimated. For example, for the finite forward difference approximation, the condition error $e_c(\Delta x)$ is estimated in a very conservative way from equation (6.2.1) as

$$e_c(\Delta x) = \frac{2}{\Delta x}\varepsilon_u \tag{6.2.6}$$

Equations (6.2.4) and (6.2.6) present us a dilemma of step size. If we take a very small step size in order to reduce the truncation error, then the condition error can be increased excessively. In some cases, we might not find the appropriate step size in order to meet the acceptable error values.

6.3 Analytical sensitivity for eigenvalue problems

The eigenvalue problems are found fundamentally in the structural stability and vibration analyses. When we consider conservative forces without an effect of damping, we have problems with real eigenvalues that represent buckling loads or vibration frequencies. In a general case with damping, the eigenvalue may be complex.

6.3.1 Sensitivity derivatives in case of vibration and buckling

Undamped vibration and buckling calculations lead to eigenvalue problems of the type

$$\mathbf{Ku} - \mu\mathbf{Mu} = \mathbf{0} \tag{6.3.1}$$

where $\mathbf{K}$ is the stiffness matrix, $\mathbf{M}$ is the mass matrix under vibration, or the geometrical stiffness matrix under buckling and $\mathbf{u}$ is the mode shape (vibration or buckling mode). μ is the square of the free vibration frequency for vibration problems, and it is the load factor for buckling problems. Both of $\mathbf{K}$ and $\mathbf{M}$ matrices are symmetric and $\mathbf{K}$ is positive definite. The mode shape is usually normalized with a positive definite matrix, $\mathbf{W}$ as

$$\mathbf{u}^T\mathbf{Wu} = 1 \tag{6.3.2}$$

where $\mathbf{W}$ is usually the mass matrix, $\mathbf{M}$ for vibration problems. Equations (6.3.1) and (6.3.2) are valid for all eigenpairs (μ_k, $\mathbf{u}^k$). Differentiating these equations and considering a generic design variable x, we obtain

$$(\mathbf{K} - \mu\mathbf{M})\frac{\partial \mathbf{u}}{\partial x} - \frac{\partial \mu}{\partial x}\mathbf{Mu} = -\left(\frac{\partial \mathbf{K}}{\partial x} - \mu\frac{\partial \mathbf{M}}{\partial x}\right)\mathbf{u} \tag{6.3.3}$$

and

$$\mathbf{u}^T\mathbf{W}\frac{\partial \mathbf{u}}{\partial x} = -\frac{1}{2}\mathbf{u}^T\frac{\partial \mathbf{W}}{\partial x}\mathbf{u} \tag{6.3.4}$$

where we have taken advantage of the symmetry of $\mathbf{W}$. Equations (6.3.3) and (6.3.4) are valid only for the cases of distinct eigenvalues (the repeated eigenvalues are generally not differentiable and only directional derivatives can be obtained according to Haug *et al.*[3]). In most cases, we are only interested in the derivatives of the eigenvalues.

These derivatives can be obtained by multiplying previously equation (6.3.3) by $\mathbf{u}^T$ in order to obtain

$$\frac{\partial \mu}{\partial x} = \frac{\mathbf{u}^T \left(\frac{\partial \mathbf{K}}{\partial x} - \mu \frac{\partial \mathbf{M}}{\partial x} \right) \mathbf{u}}{\mathbf{u}^T \mathbf{M} \mathbf{u}} \tag{6.3.5}$$

In some applications, the derivatives of the eigenvectors are required. For example, in automobile designs, the critical vibration mode should have low amplitudes at the front seats. Likewise in a bridge flutter study due to wind effect, the vibration modes and their sensitivities have great importance. For all those design problems, the derivatives of the mode shapes are required. In order to obtain the derivatives of the eigenvectors, we can use the direct approximation and combine equations (6.3.3) and (6.3.4) as

$$\begin{bmatrix} \mathbf{K} - \mu \mathbf{M} & -\mathbf{M}\mathbf{u} \\ -\mathbf{u}^T \mathbf{W} & 0 \end{bmatrix} \begin{Bmatrix} \partial \mathbf{u} / \partial x \\ \partial \mu / \partial x \end{Bmatrix} = \begin{Bmatrix} -\left(\frac{\partial \mathbf{K}}{\partial x} - \mu \frac{\partial \mathbf{M}}{\partial x} \right) \mathbf{u} \\ \frac{1}{2} \mathbf{u}^T \frac{\partial \mathbf{W}}{\partial x} \mathbf{u} \end{Bmatrix} \tag{6.3.6}$$

The system (6.3.6) can be solved to obtain the derivatives of the eigenvalues and the eigenvectors. However, we should be very cautious of the solution because $\mathbf{K} - \mu\mathrm{M}$ is singular. Cardani and Mantegazza[4] as well as Murthy and Haftka[5] propose various strategies to deal with this problem.

One of the most common solution techniques is the one of Nelson[6]. This method replaces the normalization conditions of equation (6.3.2) with a requirement that the largest component of the eigenvector is one. Designating the normalized vector in this manner $\bar{\mathbf{u}}$, and assuming that its largest component is the mth one whose value is one, equation (6.3.2) is replaced by

$$\bar{u}_m = 1 \tag{6.3.7}$$

and equation (6.3.4) is replaced by

$$\frac{\partial \bar{u}_m}{\partial x} = 0 \tag{6.3.8}$$

Equation (6.3.3) holds if $\mathbf{u}$ is replaced by $\bar{\mathbf{u}}$, but equation (6.3.8) is used to reduce its order by eliminating the mth row and the mth column. When the eigenvalue μ is found, the reduced system is no longer singular. Therefore it may be solved by standard techniques.

To calculate the derivatives of the eigenvector with the normalization of equation (6.3.2), $\mathbf{u} = u_m \bar{\mathbf{u}}$ can be employed as

$$\frac{\partial \mathbf{u}}{\partial x} = \frac{\partial u_m}{\partial x} \bar{\mathbf{u}} + u_m \frac{\partial \bar{\mathbf{u}}}{\partial x} \tag{6.3.9}$$

and du_m/dx can be obtained by substituting equation (6.3.9) into equation (6.3.4) to obtain

$$\frac{\partial u_m}{\partial x} = -u_m^2 \mathbf{u}^T \mathbf{W} \frac{\partial \mathbf{u}}{\partial x} - \frac{u_m}{2} \mathbf{u}^T \frac{\partial \mathbf{W}}{\partial x} \mathbf{u} \tag{6.3.10}$$

An adjoint or modal technique can also be used to calculate the derivatives of the eigenvectors by expressing it as a linear combination of eigenvectors. The method designates $(\mu_i, \mathbf{u}^i)$ as ith eigenpair of equation (6.3.1) assuming that

$$\frac{\partial \mathbf{u}^k}{\partial x} = \sum_{j=1}^{l} c_{kj} \mathbf{u}^j \tag{6.3.11}$$

As discussed by Rogers[7], we can demonstrate that the coefficients c_{kj} are

$$c_{kj} = \frac{\mathbf{u}^{jT}\left(\dfrac{\partial \mathbf{K}}{\partial x} - \mu_k \dfrac{\partial \mathbf{M}}{\partial x}\right)\mathbf{u}^k}{(\mu_k - \mu_i)\mathbf{u}^{jT}\mathbf{M}\mathbf{u}^j}, \quad k \neq j \tag{6.3.12}$$

Using the normalization condition of equation (6.3.7), we arrive to

$$c_{kk} = -\sum_{j \neq k} c_{kj} u_m^j \tag{6.3.13}$$

On the other hand, if we use the normalization condition of equation (6.3.2) with **W**=**M**, we obtain

$$c_{kk} = -\frac{1}{2}\left(\mathbf{u}^k\right)^T \frac{\partial \mathbf{M}}{\partial x}\mathbf{u}^k \tag{6.3.14}$$

If all the eigenvectors are included in the sum, equation (6.3.11) is exact. However, for most problems, it is not practical to calculate all the eigenvectors, so we include only the ones with large eigenvalues. Wang[8] developed a modified modal method that quickens the convergence. Instead of equation (6.3.11), we employ

$$\frac{\partial \mathbf{u}^k}{\partial x} = \mathbf{u}_s^k + \sum_{j=1}^{l} d_{kj} \mathbf{u}^j \tag{6.3.15}$$

where

$$\mathbf{u}_s^k = \mathbf{K}^{-1}\left[\frac{\partial \mu}{\partial x}\mathbf{M} - \frac{\partial \mathbf{K}}{\partial x} + \mu \frac{\partial \mathbf{M}}{\partial x}\right]\mathbf{u}^k \tag{6.3.16}$$

is a static correction term, then

$$d_{kj} = \mu_k \frac{\mathbf{u}^{jT}\left(\dfrac{\partial \mathbf{K}}{\partial x} - \mu_k \dfrac{\partial \mathbf{M}}{\partial x}\right)\mathbf{u}^k}{\mu_j(\mu_k - \mu_j)\mathbf{u}^{jT}\mathbf{M}\mathbf{u}^j}, \quad k \neq j \tag{6.3.17}$$

The coefficient d_{kk} is given by equation (6.3.14) for the normalization condition of $\mathbf{u}^T\mathbf{M}\mathbf{u} = 1$. For the normalization condition of (6.3.7)

$$d_{kk} = -u_{sm}^k - \sum_{j \neq k} d_{kj} u_m^j \tag{6.3.18}$$

Sutter[9] presents a study of convergence of the derivatives expressed as a combination of modes using the modal method and the modified modal method, and shows that the latter converges better.

When the eigenvalue μ is repeated with m different modes, there are m linearly independent vectors associated with this eigenvalue μ. Moreover, any linear combination of those eigenvectors is also an eigenvector. Therefore the selection of eigenvectors is not unique. In this case, the obtained eigenvectors are determined by the numerical analysis process used in the solution of eigenproblem. Assuming that $\mathbf{u}^1,\mathbf{u}^2,\ldots,\mathbf{u}^m$ is a set of linearly independent eigenvectors associated with μ, we can write any eigenvector associated with μ as

$$\mathbf{u} = \sum_{i=1}^{m} q_i \mathbf{u}^i = \mathbf{U}\mathbf{q} \tag{6.3.19}$$

where $\mathbf{q}$ is a vector of coefficients and $\mathbf{U}$ is a matrix with same number of m columns as $\mathbf{u}^i$. When the design variable x is changed, the eigenvalues separate and the eigenvectors become again to be unique. We can obtain the new eigenvectors by substituting equation (6.3.19) into equation (6.3.3) and premultiplying by $\mathbf{U}^T$, we obtain

$$\left(\mathbf{A} - \frac{\partial \mu}{\partial x}\mathbf{B}\right)\mathbf{q} = 0 \tag{6.3.20}$$

where

$$\mathbf{A} = \mathbf{U}^T\left(\frac{\partial \mathbf{K}}{\partial x} - \mu\frac{\partial \mathbf{M}}{\partial x}\right)\mathbf{U} \tag{6.3.21}$$

and

$$\mathbf{B} = \mathbf{U}^T\mathbf{M}\mathbf{U} \tag{6.3.22}$$

Equation (6.3.20) is an eigenvalue problem with $m \times m$ dimension for $\partial\mu/\partial x$. The m solutions correspond to the derivatives of the m eigenvalues derived from μ as x is changed. The obtained eigenvectors $\mathbf{q}$ from equation (6.3.19) provide the eigenvectors associated with perturbed eigenvalues. Ojalvo[10] proposed a generalization of Nelson's method to obtain the derivatives of the eigenvectors, and Mills-Curran[11] and Dailey[12] modified this method. Although we can calculate derivatives of any variable, these methods are valid only as directional derivatives. In these derivatives with respect to x or y, they cannot be combined linearly as

$$d\mu = \frac{\partial \mu}{\partial x}dx + \frac{\partial \mu}{\partial y}dy \tag{6.3.23}$$

which generally does not hold.

The consequences of the failure in the calculation of a derivative in an arbitrary direction from derivatives in the coordinate directions are very serious. Most optimization algorithms depend on this type of calculations to choose the direction of movement or to estimate objective functions and constraints. Therefore these algorithms may have serious difficulties in problems with repeated eigenvalues. However, the good news is as demonstrated by computational experience, very small differences in the eigenvalues will be enough to prevent such difficulties. Besides, the coincidence of eigenvalues has an unfavourable effect on the structural behaviour. For example, in vibration problems, the coincidence of the eigenvalues results in difficulties in system control. To avoid this problem in designs, we use constraints to separate the eigenvalues.

6.3.2 Sensitivity derivatives for non-Hamiltonian eigenvalue problems

When structural damping is important, or when damping comes from aerodynamic forces or control systems, the damping motion is ruled by the equation

$$\mathbf{M\ddot{u}} + \mathbf{C\dot{u}} + \mathbf{Ku} = \mathbf{0} \tag{6.3.24}$$

where $\mathbf{C}$ is the damping matrix, which is assumed to be symmetric and the dot indicates the differentiation with respect to time. The movements $\mathbf{u}$ are often formulated in the vibration mode function as

$$\mathbf{u} = \mathbf{\Phi q} \tag{6.3.25}$$

and $\mathbf{q}$ that represents the modal participation vector has the form

$$\mathbf{q} = \mathbf{w}e^{\mu t} \tag{6.3.26}$$

Equation (6.3.24) can be written as

$$\left(\mu^2\mathbf{M} + \mu\mathbf{C} + \mathbf{K}\right)\mathbf{\Phi w} = \mathbf{0} \tag{6.3.27}$$

premultiplying by Φ

$$\left(\mu^2\mathbf{M}_R + \mu\mathbf{C}_R + \mathbf{K}_R\right)\mathbf{w} = \mathbf{0} \tag{6.3.28}$$

where

$$\mathbf{M}_R = \mathbf{\Phi}^T\mathbf{M\Phi}, \mathbf{C}_R = \mathbf{\Phi}^T\mathbf{C\Phi}, \mathbf{K}_R = \mathbf{\Phi}^T\mathbf{K\Phi} \tag{6.3.29}$$

Notice that the eigenvalue m is defined similarly to the one in non-damped vibration problems where μ is the square of the vibration frequency. Here, when $\mathbf{C} = \mathbf{0}$, we get $\mu = i\omega$ where ω is the vibration frequency.

Once the eigenvector $\mathbf{w}$ is obtained, the derivative of the eigenvalue can be obtained using two methods. One is called updated-mode approximation and consists of taking the derivative of equation (6.3.28) directly. The resulting expression is

$$\frac{\partial\mu}{\partial x} = -\frac{\mu^2\mathbf{w}^T\dfrac{\partial\mathbf{M}_R}{\partial x}\mathbf{w} + \mu\mathbf{w}^T\dfrac{\partial\mathbf{C}_R}{\partial x}\mathbf{w} + \mathbf{w}^T\dfrac{\partial\mathbf{K}_R}{\partial x}\mathbf{w}}{2\mu\mathbf{w}^T\mathbf{M}_R\mathbf{w} + \mathbf{w}^T\mathbf{C}_R\mathbf{w}} \tag{6.3.30}$$

where the derivative of $\mathbf{K}_R$ is given by

$$\frac{\partial\mathbf{K}_R}{\partial x} = \mathbf{\Phi}^T\frac{\partial\mathbf{K}}{\partial x}\mathbf{\Phi} + \frac{\partial\mathbf{\Phi}^T}{\partial x}\mathbf{K\Phi} + \mathbf{\Phi}^T\mathbf{K}\frac{\partial\mathbf{\Phi}}{\partial x} \tag{6.3.31}$$

[12] Dailey R. L. [1989] *Eigenvector derivatives with repeated eigenvalues.* AIAA Journal, Vol. 27, No. 4, pp. 486–491.

[13] Bindolino G. Mantegazza P. [1987] *Aeroelastic derivatives as a sensitivity analysis of nonlinear equations.* AIAA Journal, Vol. 26, No. 8, pp. 1145–1146.

Chapter 7

Analytical sensitivity analysis of free vibration problems

7.1 Introduction

To predict how structures like cable-stayed or suspension bridges will behave under wind-induced loads, an aeroelastic calculation may be carried out. Beforehand, however, it is necessary to know, among other data, the free vibration of those structures. In other words, the first step is to calculate their natural frequencies and vibration eigenmodes.

It is essential to know frequency sensitivities and vibration eigenmodes before attempting sensitivity analyses of flutter speed for a bridge. These sensitivity analyses are obviously useful in many ways; indeed, they are suitable for any dynamic problem that is solved by modal analysis. Because they are so important, this chapter has been dedicated to them. The chapter describes in detail how to obtain the frequencies and vibration eigenmodes for a structure. It also explains how to calculate sensitivities in a completely analytical and general way using both linear theory and geometrical non-linearity, referred to here as second-order non-linear theory.

Throughout this study, linear and elastic constitutive equations have been assumed for the material. Therefore the class of non-linearity considered corresponds to the hypothesis of large displacements that is the typical behaviour of suspension and other cable-supported bridges, whose interaction with the wind is the subject of this text.

These types of structures can be analysed with the help of bar models; there is a wide variety of literature around such studies. Recent examples of aeroelastic calculation include Larsen[1], Cobo[2], Namini and Aldbrecht[3], Astiz[4], Jurado and Hernández[5] or Semiào and Mendes[6]. For this reason, the work developed within this chapter relates to three-dimensional bar structures with a constant section.

7.1.1 Matrix calculation for bar structures in linear, second-order theory

The way in which Samartín[7], Hernández[8], Livesley[9] or Saez-Benito[10] develop the matrix analysis of bar structures through the equilibrium or movement method leads to the expression

$$\mathbf{p} = \mathbf{K} \cdot \mathbf{u} \tag{7.1.1}$$

where **p** represents the load vector in nodes, while **u** is the nodal movement vector and **K** is the stiffness matrix for the structure.

In linear theory, the result for the stiffness matrix is

$$\mathbf{K}_{lin} = \mathbf{IE}\left(\mathbf{K}_E^b\right) \tag{7.1.2}$$

in which **IE** represents the operation of assembling the stiffness matrix for all the bars. $\mathbf{K}_E^b$ is thus the basic elastic stiffness matrix for each bar b, along global axes, and $\mathbf{K}_{lin}$ is the stiffness matrix for the whole structure in linear theory.

There could be a bar in three-dimensional space, like the one represented in Figure 7.1.1. Its local axes $\{x, y, z\}$ in a global, three-dimensional system are defined by axes $\{x_g, y_g, z_g\}$. Working at the elemental level of the bar and with local axes, it can be verified that

$$\mathbf{p}^b = \mathbf{K}_E^{b,l} \cdot \mathbf{u}^b \tag{7.1.3}$$

where $\mathbf{p}^b$ serves as the load vector in nodes along the bar's local axes, comprising 12 components, or three forces and three moments for each node: $\mathbf{u}^b$ is the movement vector, again along the bar's local axes, with three translations and three gyrations for each node and $\mathbf{K}_E^{b,l}$ denotes the elastic stiffness matrix along the same axes.

The elastic stiffness matrix for the bar is usually expressed subdivided into boxes as can be seen below

$$\mathbf{K}_E^{b,l} = \begin{pmatrix} \mathbf{K}_{11} & \mathbf{K}_{12} \\ \mathbf{K}_{21} & \mathbf{K}_{22} \end{pmatrix} \tag{7.1.4}$$

with $\mathbf{K}_{11}$, $\mathbf{K}_{12}$, $\mathbf{K}_{21}$, $\mathbf{K}_{22}$, defined in the following way:

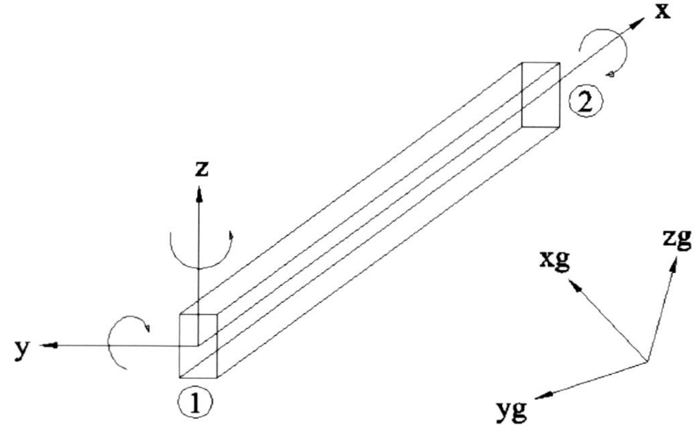

Figure 7.1.1: Bar element. Axes and positive sign criteria.

$$\mathbf{K}_{11} = \begin{pmatrix} \frac{E\cdot A}{L} & 0 & 0 & 0 & 0 & 0 \\ 0 & \frac{12\cdot E\cdot I_z}{L^3} & 0 & 0 & 0 & \frac{6\cdot E\cdot I_z}{L^2} \\ 0 & 0 & \frac{12\cdot E\cdot I_y}{L^3} & 0 & -\frac{6\cdot E\cdot I_y}{L^2} & 0 \\ 0 & 0 & 0 & \frac{G\cdot J}{L} & 0 & 0 \\ 0 & 0 & -\frac{6\cdot E\cdot I_y}{L^2} & 0 & \frac{4\cdot E\cdot I_y}{L} & 0 \\ 0 & \frac{6\cdot E\cdot I_z}{L^2} & 0 & 0 & 0 & \frac{4\cdot E\cdot I_z}{L} \end{pmatrix} \tag{7.1.5}$$

$$\mathbf{K}_{12} = \mathbf{K}_{21}{}^{T} = \begin{pmatrix} -\frac{E\cdot A}{L} & 0 & 0 & 0 & 0 & 0 \\ 0 & -\frac{12\cdot E\cdot I_z}{L^3} & 0 & 0 & 0 & -\frac{6\cdot E\cdot I_z}{L^2} \\ 0 & 0 & -\frac{12\cdot E\cdot I_y}{L^3} & 0 & \frac{6\cdot E\cdot I_y}{L^2} & 0 \\ 0 & 0 & 0 & -\frac{G\cdot J}{L} & 0 & 0 \\ 0 & 0 & \frac{6\cdot E\cdot I_y}{L^2} & 0 & \frac{2\cdot E\cdot I_y}{L} & 0 \\ 0 & -\frac{6\cdot E\cdot I_z}{L^2} & 0 & 0 & 0 & \frac{2\cdot E\cdot I_z}{L} \end{pmatrix} \tag{7.1.6}$$

$$\mathbf{K}_{22} = \begin{pmatrix} \frac{E\cdot A}{L} & 0 & 0 & 0 & 0 & 0 \\ 0 & \frac{12\cdot E\cdot I_z}{L^3} & 0 & 0 & 0 & -\frac{6\cdot E\cdot I_z}{L^2} \\ 0 & 0 & \frac{12\cdot E\cdot I_y}{L^3} & 0 & \frac{6\cdot E\cdot I_y}{L^2} & 0 \\ 0 & 0 & 0 & \frac{G\cdot J}{L} & 0 & 0 \\ 0 & 0 & \frac{6\cdot E\cdot I_y}{L^2} & 0 & \frac{4\cdot E\cdot I_y}{L} & 0 \\ 0 & -\frac{6\cdot E\cdot I_z}{L^2} & 0 & 0 & 0 & \frac{4\cdot E\cdot I_z}{L} \end{pmatrix} \tag{7.1.7}$$

The transport matrix **T** is defined along global axes as

$$\mathbf{T} = \begin{pmatrix} \cos\alpha(x_g,x) & \cos\alpha(x_g,y) & \cos\alpha(x_g,z) & 0 & 0 & 0 \\ \cos\alpha(y_g,x) & \cos\alpha(y_g,y) & \cos\alpha(y_g,z) & 0 & 0 & 0 \\ \cos\alpha(z_g,x) & \cos\alpha(z_g,y) & \cos\alpha(z_g,z) & 0 & 0 & 0 \\ 0 & 0 & 0 & \cos\alpha(x_g,x) & \cos\alpha(x_g,y) & \cos\alpha(x_g,z) \\ 0 & 0 & 0 & \cos\alpha(y_g,x) & \cos\alpha(y_g,y) & \cos\alpha(y_g,z) \\ 0 & 0 & 0 & \cos\alpha(z_g,x) & \cos\alpha(z_g,y) & \cos\alpha(z_g,z) \end{pmatrix} \tag{7.1.8}$$

$a(i,j)$ being the angle formed by the axes i and j.

Finally, the elemental elastic stiffness matrix is obtained for the bar along global axes in the following way:

$$\mathbf{K}_E^b = \begin{pmatrix} \mathbf{T}\cdot\mathbf{K}_{11}\cdot\mathbf{T}^T & \mathbf{T}\cdot\mathbf{K}_{12}\cdot\mathbf{T}^T \\ \mathbf{T}\cdot\mathbf{K}_{21}\cdot\mathbf{T}^T & \mathbf{T}\cdot\mathbf{K}_{22}\cdot\mathbf{T}^T \end{pmatrix} \tag{7.1.9}$$

leading to the definition of equation (7.1.2) and then proceeding onto the assembly of all the bars. In this way it is possible to solve the equation system given by (7.1.1) after imposing the relevant contour conditions and having obtained, in a prior step, the load vector in the nodes along global coordinates.

Second-order theory entails studying non-linear behaviour in terms of great movements and considering linear and elastic material. To carry out the matrix analysis of a bar structure using the equilibrium or movement method, as seen in Przemieniecki[11] and Hernández[8], the following expression is reached:

$$\mathbf{p} = \mathbf{K}_{nolin}\cdot\mathbf{u} \tag{7.1.10}$$

where $\mathbf{p}$ represents the nodal load vector; $\mathbf{u}$, the nodal movement vector; and $\mathbf{K}_{nolin}$, the stiffness matrix for the structure according to non-linear or second-order theory. Thus

$$\mathbf{K}_{nolin} = \mathbf{K}_{lin} + \mathbf{K}_G \tag{7.1.11}$$

in which $\mathbf{K}_G$ is the geometry stiffness matrix for the complete structure so that

$$\mathbf{K}_G = \mathbf{IE}\left(N_b\cdot\mathbf{K}_G^b\right) \tag{7.1.12}$$

Here, N_b is the axial force on the bar b and $\mathbf{K}_G^b$ is the geometry stiffness matrix in elemental terms along global axes of the bar.

Each matrix $\mathbf{K}_G^b$ is obtained from the geometry stiffness matrix along local axes of the bar $\mathbf{K}_G^{b,l}$, as well as from the transport matrix **T** defined in (7.1.8).

$$\mathbf{K}_G^{b,l} = \begin{pmatrix} \mathbf{K}_{G11} & \mathbf{K}_{G12} \\ \mathbf{K}_{G21} & \mathbf{K}_{G22} \end{pmatrix} \tag{7.1.13}$$

The result for each one of the submatrices is

$$\mathbf{K}_{G11} = \begin{pmatrix} 0 & 0 & 0 & 0 & 0 & 0 \\ 0 & \frac{6}{5\cdot L} & 0 & 0 & 0 & \frac{1}{10} \\ 0 & 0 & \frac{6}{5\cdot L} & 0 & -\frac{1}{10} & 0 \\ 0 & 0 & 0 & 0 & 0 & 0 \\ 0 & 0 & -\frac{1}{10} & 0 & \frac{2\cdot L}{15} & 0 \\ 0 & \frac{1}{10} & 0 & 0 & 0 & \frac{2\cdot L}{15} \end{pmatrix} \tag{7.1.14}$$

$$\mathbf{K}_{G12} = \mathbf{K}_{G21}^{T} = \begin{pmatrix} 0 & 0 & 0 & 0 & 0 & 0 \\ 0 & -\frac{6}{5\cdot L} & 0 & 0 & 0 & \frac{1}{10} \\ 0 & 0 & -\frac{6}{5\cdot L} & 0 & -\frac{1}{10} & 0 \\ 0 & 0 & 0 & 0 & 0 & 0 \\ 0 & 0 & \frac{1}{10} & 0 & -\frac{L}{30} & 0 \\ 0 & -\frac{1}{10} & 0 & 0 & 0 & -\frac{L}{30} \end{pmatrix} \tag{7.1.15}$$

$$\mathbf{K}_{G22} = \begin{pmatrix} 0 & 0 & 0 & 0 & 0 & 0 \\ 0 & \frac{6}{5\cdot L} & 0 & 0 & 0 & -\frac{1}{10} \\ 0 & 0 & \frac{6}{5\cdot L} & 0 & \frac{1}{10} & 0 \\ 0 & 0 & 0 & 0 & 0 & 0 \\ 0 & 0 & \frac{1}{10} & 0 & \frac{2\cdot L}{15} & 0 \\ 0 & -\frac{1}{10} & 0 & 0 & 0 & \frac{2\cdot L}{15} \end{pmatrix} \tag{7.1.16}$$

In these, L is the bar's length and, consequently, for each matrix $\mathbf{K}_G^b$ along global axes the final result is

$$\mathbf{K}_G^b = \begin{pmatrix} \mathbf{T}\cdot\mathbf{K}_{G11}\cdot\mathbf{T}^T & \mathbf{T}\cdot\mathbf{K}_{G12}\cdot\mathbf{T}^T \\ \mathbf{T}\cdot\mathbf{K}_{G21}\cdot\mathbf{T}^T & \mathbf{T}\cdot\mathbf{K}_{G22}\cdot\mathbf{T}^T \end{pmatrix} \tag{7.1.17}$$

On the other hand, the axial force N_b within each bar b is obtained in function of the nodal movements according to the expression

$$N_b = \mathbf{w}^T \cdot \mathbf{u}^b \tag{7.1.18}$$

in which $\mathbf{u}^b$ is the vector containing 12 degrees of freedom within the bar element along global coordinates (three traslation movements and three rotations in each node), and $\mathbf{w}$ is the vector linking these with the axial force. The following expression is next

$$\mathbf{w} = \begin{pmatrix} -\frac{E \cdot A}{L} \cdot \cos\alpha(x_g, x) \\ -\frac{E \cdot A}{L} \cdot \cos\alpha(y_g, x) \\ -\frac{E \cdot A}{L} \cdot \cos\alpha(z_g, x) \\ 0 \\ 0 \\ 0 \\ -\frac{E \cdot A}{L} \cdot \cos\alpha(x_g, x) \\ -\frac{E \cdot A}{L} \cdot \cos\alpha(y_g, x) \\ -\frac{E \cdot A}{L} \cdot \cos\alpha(z_g, x) \\ 0 \\ 0 \\ 0 \end{pmatrix} \tag{7.1.19}$$

defining equation (7.1.12). This leads to the assembly of all the bars and afterwards it is possible to solve the equation system given by (7.1.10). Previously, however, the boundary conditions have been imposed and the load vector for nodes along the global coordinates have been obtained.

7.1.2 Frequencies and natural vibration modes in linear and second-order theories

Obtaining natural frequencies and eigenmodes of vibration is a eigenvalue problem that appears in a varied assortment of cases in structural engineering as the identification of buckling loads, dynamic and aeroelastic analysis and so on.

A three-dimensional bar structure is used to formulate a dynamic problem based on free vibration without damping, as described in Hacar and Alarcón[12] or Chopra[13]. In this way

$$\mathbf{M} \cdot \ddot{\mathbf{u}} + \mathbf{K} \cdot \mathbf{u} = 0 \tag{7.1.20}$$

where $\mathbf{M}$ is the mass matrix for the structure, obtained by concentrating the mass of the bar elements on the nodes, including both lateral movements and rotations. $\mathbf{K}$ denotes the stiffness matrix for the structure. This expression only has to contain the degrees of freedom with associated mass. Therefore, one should begin by carrying out the static condensation, as described in Chopra[13].

Through modal analysis, this dynamic problem is formulated in the following way, using stiffness matrix in linear or non-linear theory

$$\left[\mathbf{K}_{lin} - \omega_n^2 \cdot \mathbf{M}\right] \cdot \phi_n = 0 \tag{7.1.21a}$$

$$\left[\mathbf{K}_{nonlin} - \omega_n^2 \cdot \mathbf{M}\right] \cdot \phi_n = 0 \tag{7.1.21b}$$

This is a classic example of a self-value problem, in which ω_n is the *nth* natural vibration frequency and ϕ_n is the vector that contains the *n*th vibration mode, being $n = 1,...,N$ with N the number of degrees of freedom of the structure with associated mass.

7.2 Sensitivity analysis of frequencies and vibration eigen modes in linear and second-order theories

By calculating structural sensitivity, it is possible to study the changes that take place in an effort to reach better design conditions during the optimization process. At this stage, sensitivity analysis often entails exhaustive computational work and it is therefore essential to have on hand efficient algorithms to evaluate derivatives.

In engineering, a main source of problems comes from obtaining eigenvalues, and consequently, procedures for carrying out a sensitivity analysis are worthy of careful study. Yet this concern is not adequately addressed in the body of scientific literature. The only recent text on dynamic calculation that comes to mind is by Hart and Wong[14], who touch upon the topic of sensitivity calculation within each type of dynamic problem.

There are various ways to approach sensitivity analysis in eigenvalue problems. Adelman and Haftka[15] establish a clear classification of these methods, dividing them into three basic groups:

- Approximation through finite differences
- Semi-analytical methods
- Analytical methods

Approximation through finite differences is the simplest technique; derivatives can be calculated from the response in terms of design conditions. Because it is easy to carry out formulations with it, it is very popular. However, this technique involves extensive computation. Its other drawback is its lack of precision, due to what is known as the step-size dilemma. If smaller steps are taken to avoid truncation, the likelihood of design variable errors is greater.

Analytical methods of sensitivity involve differenciating the equations that govern the structural problem in terms of design variables. Once this problem is solved and then introduced into the analytically differenciated equations, the sensitivity, that is, the variation with regard to the design variable, can be directly obtained.

The third group, semi-analytical methods, also follows an analytical formulation to solve the problem, but, at some point in the process, uses calculations through finite differences to evaluate the derivatives for some of its terms.

A good example of how this method is applied can be seen in structural vibration problems that calculate the frequencies and vibration eigenmodes for the structure. Universally renowned software codes like GENESIS[16], Nastran, Ansys or Cosmos/M rely on linear theory to find the derivatives of these frequencies and vibration modes in semi-analytical formulations. They evaluate the derivative of the stiffness matrix **K** through finite differences. When the problem has to take into account second-order effects, sensitivity analysis is hampered. These methods' weakness as far as the second-order sensitivity analysis of natural frequencies and vibration modes has yet to be resolved. The formulation presented below sidesteps this problem when dealing with size-related design conditions in three-dimensional structures made up of straight, constant-section bars.

When carrying out calculations for cable-supported bridges, such as suspension bridges, the geometrical non-linearity of cables must be taken into account due to the rigidity of axial forces they withstand. Kanok-Nukulchai *et al.*[17], Abbas and Scordelis[18], Schrefler *et al.*[19], Arzoumanidis and Bieniek[20], as well as Ge and Tanaka[21] all refer to this. Therefore, finding a second-order sensitivity analysis that can deal with this geometrical non-linearity may be considered crucial if an alternative to finite difference calculation is sought.

Imagine that the generic design variable x is taken for a three-dimensional, constant–section bar structure. Expression (7.1.20) can be used to define the free vibration problem and expression (7.1.21a) or (7.1.21b) to solve it. According to second-order theory, the vibration modes can be normalized with a matrix defined as positive, which is usually the mass matrix **M**, so that

$$\phi_n^T \cdot \mathbf{M} \cdot \phi_n = 1 \tag{7.2.1}$$

The modes have been normalized to the mass, and the variable has been verified for each of the modes. In Adelman and Haftka[15] or Haftka and Gürdal[22], there is background information about analytically differenciating this kind of eigenvalue problem

$$\frac{\partial\left(\omega_{\mathrm{n}}^2\right)}{\partial x} = \frac{\phi_n^T \cdot \left(\dfrac{\partial \mathbf{K}}{\partial \mathrm{x}} - \omega_{\mathrm{n}}^2 \cdot \dfrac{\partial \mathbf{M}}{\partial \mathrm{x}}\right) \cdot \phi_n}{\phi_n^T \cdot \mathbf{M} \cdot \phi_n} \tag{7.2.2}$$

where ω_n the *nth* natural vibration frequency; ϕ_n, the vector containing the *nth* vibration mode; **K** the stiffness matrix for the structure in second-order theory; and **M**, the mass matrix.

The evaluation of these sensitivities over the natural vibration frequencies entails calculating the derivative for the stiffness matrix with respect to generic design variable x. On the other hand, the derivatives of the vibration eigenmodes are obtained by differenciating the eigenvalue problem through the following expression

$$\left(\mathbf{K} - \omega_n^2 \cdot \mathbf{M}\right) \cdot \frac{\partial \phi_n}{\partial x} = \frac{\partial\left(\omega_n^2\right)}{\partial x} \cdot \mathbf{M} \cdot \phi_n - \left(\frac{\partial \mathbf{K}}{\partial x} - \omega_n^2 \cdot \frac{\partial \mathbf{M}}{\partial x}\right) \cdot \phi_n \tag{7.2.3}$$

This provides an equation system for each frequency and eigenmode; within this system, the sensitivities for these vibration eigenmodes are unknown.

Special care must be taken when solving this last equation system owing to its singularity, a point reinforced by, for instance, Jurado[23]. The most commonly used technique for getting around this problem, and the one implemented here, can be attributed to Nelson[24]. Instead of following the normalization conditions for equation (7.2.1), he opts for the condition in which the greatest component of the eigenvector be equal to one. Denominating the normalized eigenvector $\bar{\phi}$, and assuming that its greatest component is the *m*th and equal to one, equation (7.2.1) can be replaced by

$$\bar{\phi}^m = 1 \tag{7.2.4}$$

and therefore it can be assumed that

$$\frac{\partial \bar{\phi}^m}{\partial x} = 0 \tag{7.2.5}$$

Equation (7.2.3) is valid if ϕ is replaced by $\bar{\phi}$. However, equation (7.2.5) is used to reduce it an order eliminating row and column m. In this way, the reduced system is no longer singular. It can be solved for each one of the N pairs (ω^2, $\bar{\phi}$) through standard techniques.

To calculate the derivative of the eigenvector through the original normalization of equation (7.2.1), the procedure would be as follows. The equation is derived and the results of re-ordering the terms are

$$\phi^T \cdot \mathbf{M} \cdot \frac{\partial \phi}{\partial x} = -\frac{1}{2} \cdot \phi^T \cdot \frac{\partial \mathbf{M}}{\partial x} \cdot \phi \tag{7.2.6}$$

Remembering that $\phi = \phi^m . \bar{\phi}$, after deriving, this is the outcome

$$\frac{\partial \phi}{\partial x} = \frac{\partial \phi^m}{\partial x} \cdot \bar{\phi} + \phi^m \cdot \frac{\partial \bar{\phi}}{\partial x} \tag{7.2.7}$$

Then $\frac{\partial \phi^m}{\partial x}$ can be got by substituting equation (7.2.7) in equation (7.2.6), so that

$$\frac{\partial \phi^m}{\partial x} = -\left(\phi^m\right)^2 \cdot \phi^T \cdot \mathbf{M} \cdot \frac{\partial \bar{\phi}}{\partial x} - \frac{\phi^m}{2} \cdot \phi^T \cdot \frac{\partial \mathbf{M}}{\partial x} \cdot \phi \tag{7.2.8}$$

and the former equation (7.2.7) remains completely defined and the derivative of each vibration eigenvector is obtained by normalizing it to the mass.

There are alternatives to Nelson[24], whose method has been described above. Rogers[25] or Wang[26], for example, use adjuncts or modal techniques, while Ojalvo[27], Mills-Curran[28] or Dailey[29] propose methods based on a generalization of Nelson. Once again, the evaluation of these sensitivities over the vibration eigenmodes entails calculating the derivative of the stiffness matrix with respect to generic design variable x.

7.2.1 Sensitivity analysis in linear theory

Consider a three-dimensional bar structure with a constant section in which vector $\mathbf{x}$ is the design variable for the bar. Therefore, $\mathbf{x} = (I_x, I_y, I_z)$, with $I_x = J$, I_y, I_z being the inertia moments for the bars with respect to the three local axes: that is, two under bending and one under torsion. Deriving equation (7.1.2) related to design variable x, the derivative of the structure's stiffness matrix is obtained through linear theory in this way

$$\frac{\partial \mathbf{K}_{lin}}{\partial x} = \mathbf{IE}\left(\frac{\partial \mathbf{K}_E^b}{\partial x}\right) \tag{7.2.9}$$

Once the bars contained by the design variable have been assembled, the derivatives of the elemental stiffness matrices for the bar, which in function of expressions (7.1.4)–(7.1.7), are arrived at through each and every one of its components. These may be null. Alternatively, they may not be dependent on that variable, or they may depend on it in linear terms and can be directly obtained by dividing by that variable.

This is one way of solving the sensitivity analysis for the natural frequencies and vibration eigenmodes through linear theory with equations (7.2.2) and (7.2.3).

7.2.2 Sensitivity analysis in second-order theory

Here, the non-linear geometry of the structure is taken into account. The three-dimensional bar structure with a constant section makes a re-appearance and once again vector $\boldsymbol{x}$ is taken from the design variables. Similar to the case mentioned above, $\mathbf{x} = (I_x, I_y, I_z)$. Thus $I_x=J$, I_y, I_z are the inertia moments of the bar along its three local axes, with two inertia moments under bending and one under torsion. Deriving equation (7.1.11) with respect to design variable x_i, as described in Mosquera and Hernández[30], the derivative of the stiffness matrix for the structure is thus obtained through non-linear theory

$$\frac{\partial \mathbf{K}_{nolin}}{\partial x} = \frac{\partial \mathbf{K}_{lin}}{\partial x} + \frac{\partial \mathbf{K}_G}{\partial x} \tag{7.2.10}$$

The derivative of the linear stiffness matrix for the structure $\mathbf{K}_{lin}$ has already been obtained in the section above. However, the one for its geometry stiffness matrix $\mathbf{K}_G$ is got by means of equation (7.1.12)

$$\frac{\partial \mathbf{K}_G}{\partial x} = \frac{\partial}{\partial x}\left(\mathbf{IE}\left(N_b \cdot \mathbf{K}_G^b\right)\right) = \mathbf{IE}\left(\mathbf{K}_G^b \cdot \frac{\partial N_b}{\partial x}\right) \tag{7.2.11}$$

It is important to note that, within equation (7.2.11), the derivatives of the bar's elemental geometry matrices defined in equations (7.1.13)–(7.1.16) $\mathbf{K}_G^b$, are null. They have the geometrical parameters for the bars, but not their mechanical ones. The axial force N_b acting on each bar can be reached through equation (7.1.18). Consequently, the previous equation (7.2.11) can be developed like this

$$\frac{\partial \mathbf{K}_G}{\partial x} = \mathbf{IE}\left(\mathbf{K}_G^b \cdot \mathbf{w}^T \cdot \frac{\partial \mathbf{u}^b}{\partial x}\right) + \mathbf{IE}\left(\mathbf{K}_G^b \cdot \frac{\partial \mathbf{w}^T}{\partial x} \cdot \mathbf{u}^b\right) \tag{7.2.12}$$

What remains unknown in the non-linear problem is the derivative of vector $\mathbf{u}^b$- that is, the vector of nodal movements – with respect to the design variable. This vector can be obtained through a static problem in second-order theory defined through equation (7.1.10)

$$\mathbf{p} = \left(\mathbf{K}_{lin} + \mathbf{K}_G\right) \cdot \mathbf{u} \tag{7.2.13}$$

where **p** represents the nodal load vector and **u**, the one for the nodal movement. Load vector **p** will comprise the loads considered for the structure load state to calculate the frequencies and eigenmodes, which generally tend to be its own weight. This will be obtained by concentrating the bar's weight along its nodes. Deriving this expression with respect to design variable x_i, the result is

$$\frac{\partial \mathbf{p}}{\partial x} = \frac{\partial}{\partial x}\left(\mathbf{K}_{lin} \cdot \mathbf{u}\right) + \frac{\partial}{\partial x}\left(\mathbf{K}_G \cdot \mathbf{u}\right) \tag{7.2.14}$$

With equation (7.1.12) the result of adding the second member of this equation is

$$\frac{\partial}{\partial x}\left(\mathbf{K}_G \cdot \mathbf{u}\right) = \frac{\partial}{\partial x}\left[\mathbf{IE}\left(N_b \cdot \mathbf{K}_G^b\right) \cdot \mathbf{u}\right] \tag{7.2.15}$$

Out of convenience, this expression can be modified in this way

$$\frac{\partial}{\partial x}\left(\mathbf{K}_G \cdot \mathbf{u}\right) = \frac{\partial}{\partial x}\left[\mathbf{IE}\left(\mathbf{K}_G^b \cdot \mathbf{u}^b \cdot N_b\right)\right] \tag{7.2.16}$$

After the derivatives have been carried out, the following is reached:

$$\frac{\partial}{\partial x}\left(\mathbf{K}_G \cdot \mathbf{u}\right) = \mathbf{IE}\left(\frac{\partial \mathbf{K}_G^b}{\partial x} \cdot \mathbf{u}^b \cdot N_b + \mathbf{K}_G^b \cdot \frac{\partial \mathbf{u}^b}{\partial x} \cdot N_b + \mathbf{K}_G^b \cdot \mathbf{u}^b \cdot \frac{\partial N_b}{\partial x}\right) \tag{7.2.17}$$

By substituting N_b within equation (7.1.18), one gets

$$\frac{\partial}{\partial x}\left(\mathbf{K}_G \cdot \mathbf{u}\right) = \mathbf{K}_G \cdot \frac{\partial \mathbf{u}}{\partial x} + \mathbf{IE}\left(\mathbf{K}_G^b \cdot \mathbf{u}^b \cdot \mathbf{w}^T\right) \cdot \frac{\partial \mathbf{u}}{\partial x} + \mathbf{IE}\left(\mathbf{K}_G^b \cdot \mathbf{u}^b \cdot \frac{\partial \mathbf{w}^T}{\partial x} \cdot \mathbf{u}^b\right) \tag{7.2.18}$$

This expression is introduced into equation (7.2.14), and this expression is reached

$$\frac{\partial \mathbf{p}}{\partial x} = \frac{\partial \mathbf{K}_{lin}}{\partial x} \cdot \mathbf{u} + \mathbf{K}_{lin} \cdot \frac{\partial \mathbf{u}}{\partial x} + \mathbf{K}_G \cdot \frac{\partial \mathbf{u}}{\partial x} + \mathbf{IE}\left(\mathbf{K}_G^b \cdot \mathbf{u}^b \cdot \mathbf{w}^T\right) \cdot \frac{\partial \mathbf{u}}{\partial x} + \mathbf{IE}\left(\mathbf{K}_G^b \cdot \mathbf{u}^b \cdot \frac{\partial \mathbf{w}^T}{\partial x} \cdot \mathbf{u}^b\right) \tag{7.2.19}$$

By reordering and regrouping the terms, the above may be formulated differently

$$\left[\mathbf{K}_{lin} + \mathbf{K}_G + \mathbf{IE}\left(\mathbf{K}_G^b \cdot \mathbf{u} \cdot \mathbf{w}^T\right)\right] \cdot \frac{\partial \mathbf{u}}{\partial x} = \frac{\partial \mathbf{p}}{\partial x} - \frac{\partial \mathbf{K}_{lin}}{\partial x} \cdot \mathbf{u} - \mathbf{IE}\left(\mathbf{K}_G^b \cdot \mathbf{u}^b \cdot \frac{\partial \mathbf{w}^T}{\partial x} \cdot \mathbf{u}^b\right) \tag{7.2.20}$$

This step produces a linear system of equations whose unknown factors are the derivatives of the nodal movements crucial for solving equation (7.2.12). It is interesting how, with the chosen design variables, the term $\partial \mathbf{w}^T/\partial x$ is null because it does not depend on those design variables according to (7.1.19), and the term $\partial \mathbf{p}/\partial x$ is also null. The final task is to introduce these terms into equation (7.2.10). Taking non-linearity into account and employing second-order theory, it is possible to obtain the stiffness matrix derivative

$$\frac{\partial \mathbf{K}_{nolin}}{\partial x} = \mathbf{IE}\left(\frac{\partial \mathbf{K}_E^b}{\partial x}\right) + \mathbf{IE}\left(\mathbf{K}_G^b \cdot \mathbf{w}^T \cdot \frac{\partial \mathbf{u}^b}{\partial x}\right) + \mathbf{IE}\left(\mathbf{K}_G^b \cdot \frac{\partial \mathbf{w}^T}{\partial x} \cdot \mathbf{u}^b\right) \qquad (7.2.21)$$

In this way, the derivatives for the natural frequencies and vibration eigenmodes derived from equations (7.2.2) and (7.2.3) can also be obtained analytically if these kinds of effects are considered.

7.3 Description of the *"ADISNOL3D"* code

A programme has been designed to help with implementing all of the analytical formulations described above. It also facilitates the task of carrying out, through linear, second-order theory, the calculations necessary to obtain the natural frequencies and vibration eigenmodes for a three-dimensional bar structure with constant sections. This code for doing calculations on three-dimensional structures is called *ADISNOL3D* (Figure 7.3.1). In Spanish, this acronym stands for Dynamic Analysis through Linear and Non-linear Sensitivity in Three Dimensions.

ADISNOL3D is a complete code for static and dynamic or free-reign calculations through linear and second-order, that is, non-linear, geometrical theory using three-dimensional bar structures with constant sections as subjects. At the same time, the code has a module related to the sensitivity analysis for dynamic calculation problems involving natural frequencies and vibration eigenmodes in linear, second-order theory.

Entirely developed through the language FORTRAN 90, *ADISNOL3D* has various modules shown in Figures 7.3.2 and 7.3.3, whose basic features are:

Graphical pre-processing:
Generating the geometry for the elements in the bar structure is achieved through the graphical processor GEOSTAR, the same pre-processing module found in the

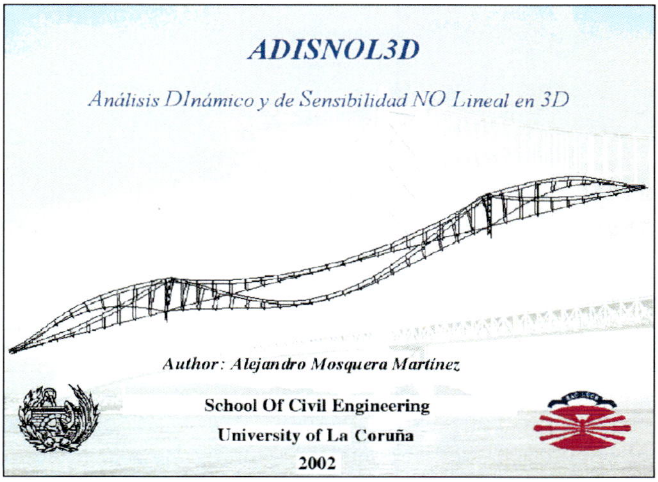

Figure 7.3.1: Opening screen for *ADISNOL3D*.

commercially available COSMOS/M[31] finite element package. Bar orientation is also defined, along with section type and material and contour conditions.

ADISNOL pre-processing:
This is a form of pre-processing complementary to the one mentioned above. It serves to define the properties of the materials, types and features of the different sections and design variables, all of which are necessary for the sensitivity analysis.

Static calculation module:
Help is provided to solve the static and linear structural problem defined by equations (7.1.1)–(7.1.9), as well as the second-order theory static structural one, taking into account the non-linear geometry defined by equations (7.1.10)–(7.1.19).

To solve the diverse equation systems, a factorization developed by Cholesky has been implemented, following an algorithm for solving systems similar to SPOSL, as shown in Dongarra and Moler[32].

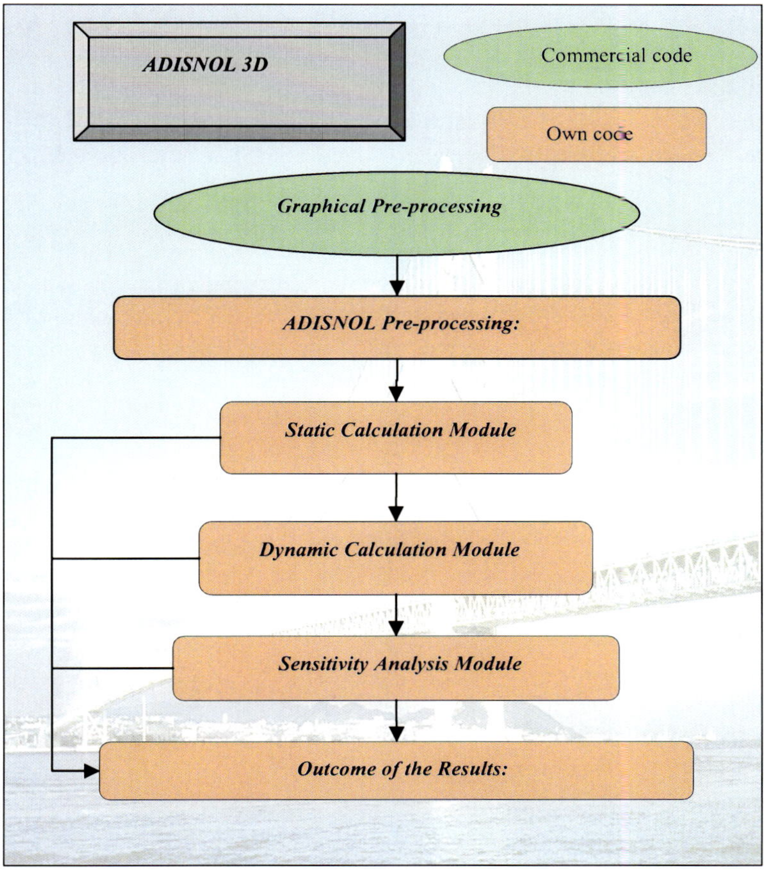

Figure 7.3.2: Provides a flow chart explaining how this code works and which modules it contains.

The convergence of the non-linear problem has been imposed on the movements and axial forces. It is possible to obtain both the nodal movements and all the forces at the ends of the bar in each iteration.

Dynamic calculation module:
The module solves the dynamic problem related to free vibration defined by equations (7.1.20) and (7.1.21) the obtain the structure's natural frequencies and vibration eigenmodes in both linear and second-order theory. To this end, a static condensation of the dynamic problem has been implemented, and it is therefore operated with the degrees of freedom associated with mass.

To deal with self-value problems, there is a solution system based on the factorization developed by Cholesky, as seen in Martin and Wilkingson[33], following a matrix algorithm along the lines of QR, as found in Smith *et al.*[34] or Hanson *et al.*[35].

It is possible to obtain the natural frequencies and all the components of the vibration eigenvalues, normalized to the mass according to (7.2.1), with or without associated mass, by de-condensing the dynamic problem, as described in Chopra[13].

Sensitivity analysis module:
This module is for obtaining the sensitivities for the natural frequencies and vibration eigenmodes in both linear and second-order theory, with design variables associated with the mechanical parameters of the bars, implementing the analytical formulation presented in equations (7.2.2) and (7.2.3). To solve the equation systems given in (7.2.3), an $\mathbf{LDU}^T$ factorization system has been implemented using the SSISL solution algorithm, described in Dongarra and Moler[32].

To calculate the derivative for the structure's rigidity matrix using the analytical formulations from equation (7.2.1.1) for linear theory, and (7.2.10)–(7.2.21) for its non-linear geometrical counterpart, this module relies on an **LU** type of factorization and a SGESL algorithm to solve the system in (7.2.20), see Dongarra and Moler[32].

The natural frequency derivatives and vectors are obtained through the derivatives for all the components of each vibration eigenmode calculated.

Outcome of the results:
The outcome of the results is carried out with various files. There are different options as far as their visualization: nodal movements, bar stresses, natural vibration frequencies, vibration eigenmodes and the sensitivities of frequencies and eigenmodes, all of which can be through linear or second-order theory. The results of the sensitivity analysis are directly compatible with the FLAS code (whose initials stand for, in Spanish, Flutter and Analysis of Sensitivities); for more details, see Jurado[23]. The following chapter will deal with this calculation code.

To provide an overview of all these features, flow charts follow:

Currently on the market one can find powerful and feature-loaded finite element packages for bar structures. However, the main reason for creating a structural calculation code like *ADISNOL3D* is that implementing the analytical formulation described for a sensitivity analysis of natural frequencies and vibration eigenmodes entails having unlimited access to the entire code during the solution process. A case in point is how it would be necessary to have the stiffness matrix for both the assembled structure and its elemental,

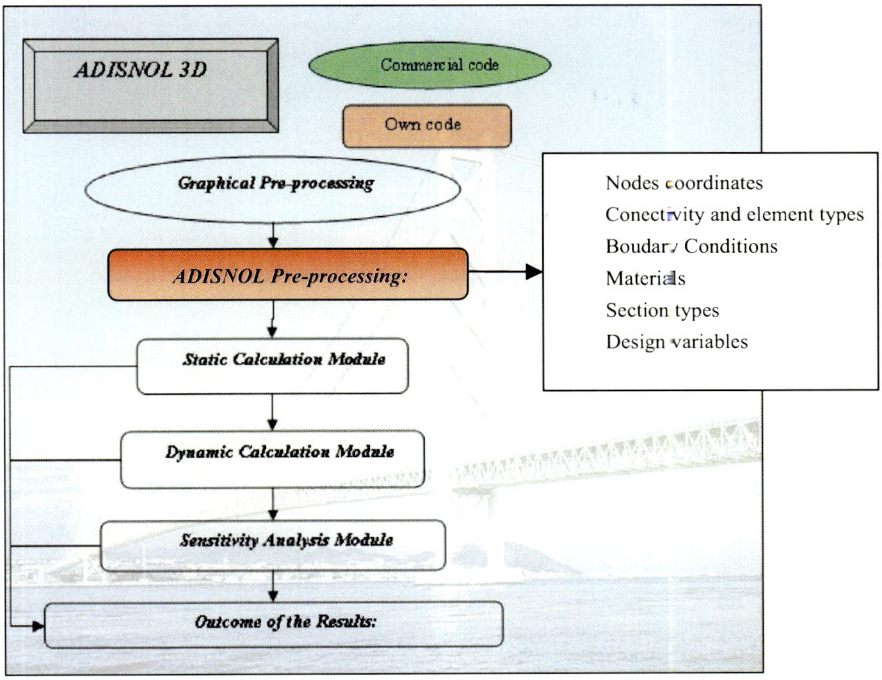

Figure 7.3.3: Flow chart for the modules of the *ADISNOL3D* code.

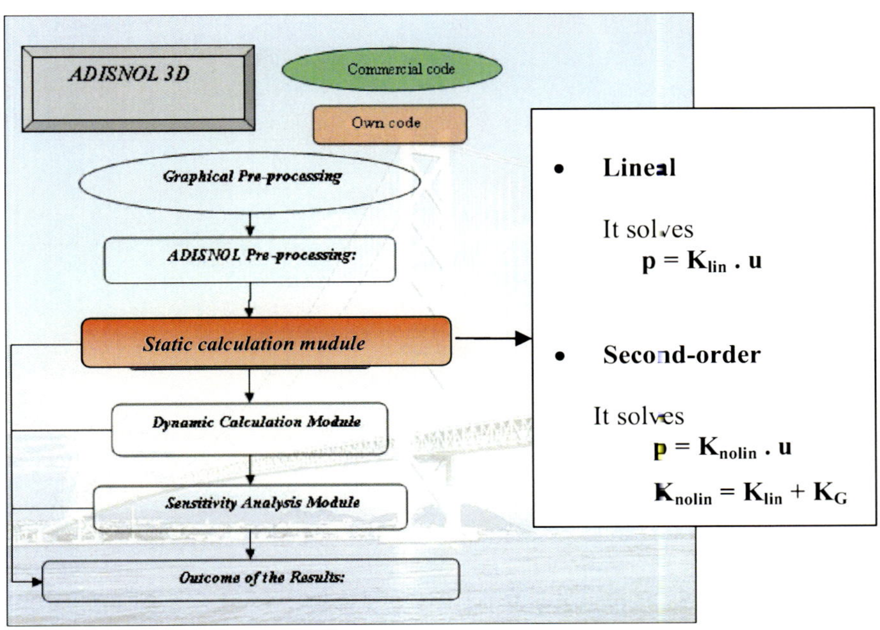

Figure 7.3.3: *Continued.*

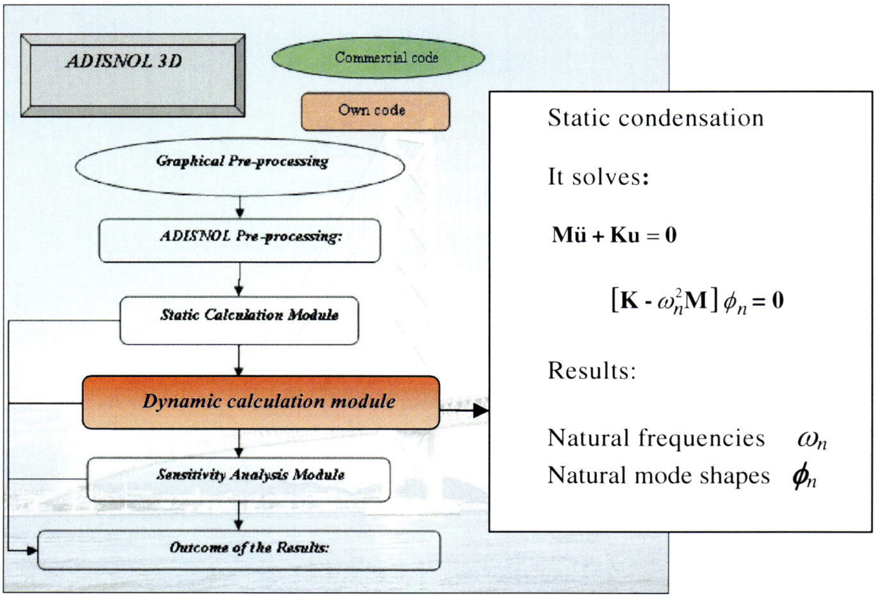

Figure 7.3.3: *Continued.*

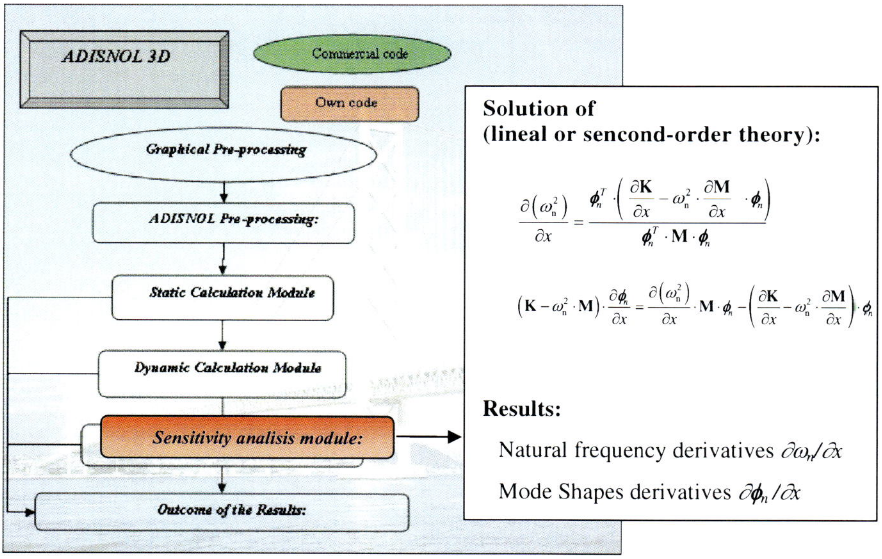

Figure 7.3.3: *Continued.*

unassembled state to calculate the derivative in terms of the chosen design variable and along the chosen bars. Most commercial codes fail to provide this kind of information; it is impossible to work with these matrices because they are not made available to the user. Even when that matrix can be obtained, the difficulty lies in identifying within it the

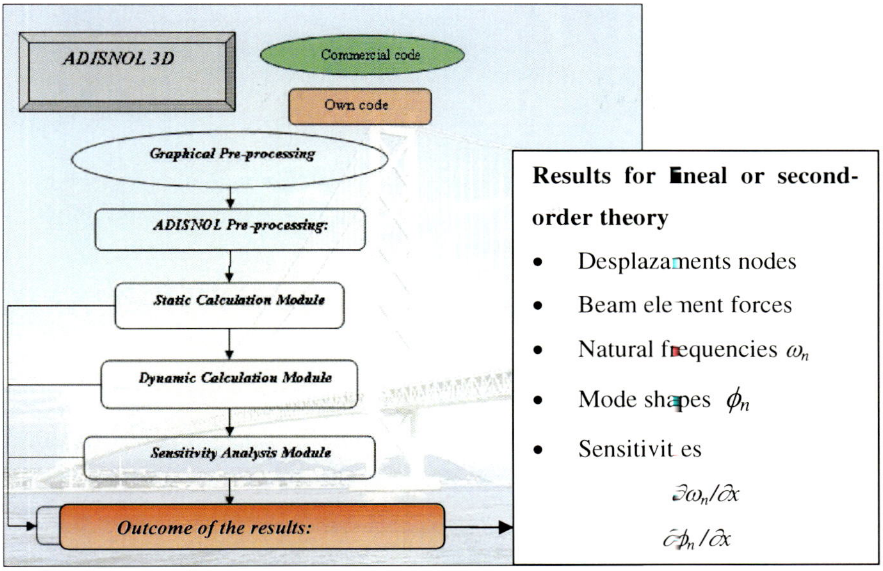

Figure 7.3.3: *Continued.*

terms for the bars. The matrix is provided once the corresponding re-numeration routine has been executed. This is only one small example. After observing how the formulation is implemented, one will also reach the conclusion that a special code had to be developed from the beginning.

7.4 Practical examples with *ADISNOL3D*

The objective of these examples will differ. The following will serve as a summary:

- Verifying that the geometric non-linearity has been analysed correctly when this is done through static loads.
- Verifying that the natural frequencies and vibration modes have been accurately obtained through second-order theory with geometric non-linearity in a real structure that is sufficiently complex and analysed.
- Obtaining the analytical sensitivity analyses through second-order theory for the natural frequencies and vibration eigenmodes and comparing these with the results of a finite differential calculation.

7.4.1 Example 1: main cable of the Golden Gate Bridge

The starting point here is the static calculation with non-linear geometry for one of the main cables of the Golden Gate Bridge, the subject of various other studies, such as Abbas and Scordelis[18], as well as Cobo[2], whose approaches differ Table 7.4.1 shows the characteristics of this example, while Figure 7.4.1 has an *ADISNOL3D*-produced image of the model with 84 bars, identical to one used by the authors mentioned above.

Serving as a reference point is the configuration of a hanger in which the cable is withstanding its own weight and that of the deck, modelled as specific loads along the cable's nodes. Its sag has an initial value of f = 143.9 m. Table 7.4.2 compares the results obtained from the sag and the maximum tension in the cable with loads varying for each case.

The results are satisfactory, demonstrating that the formulation implemented within *ADISNOL3D* produces results that basically coincide with those presented in technical literature.

7.4.2 Example 2: suspension bridge over the Great Belt

In this second example, the frequencies and vibration eigenmodes are calculated for the Great Belt Bridge in Denmark. This structure is described in the previous chapter, comparing the results obtained through the *ADISNOL3D* code with those presented by other authors. Here, the sensitivity analysis of those same frequencies and modes will be carried out in terms of numerous design variables and the results will be contrasted with those obtained through differential calculation.

7.4.2.1 Caracteristics of the Great Belt suspension bridge

14 km in length, the Great Belt of Denmark lies between the islands of Funen and Zealand. On June 14 1998, traffic was opened along this new communication link across the strait; the most impressive element of this link was the suspension bridge. It is best known as the Great Belt Bridge, and can be seen in Figures 7.4.2 and 7.4.3. The bridge's design was left to the hands of the consultancy firm COWI. It is characterized by the unbroken line of the deck beneath its towers, the great relationship

Table 7.4.1: Characteristics of the model for one of the main cables in the Golden Gate.

Total span length	1280.2 m
Distance between the vertical hangers	15.2 m
Cable area	0.537 m^2
Cable weight	48.75 kN/m
Young's modulus	200000 MPa
Deck weight	118.35 kN/m
Overload	29.2 kN/m

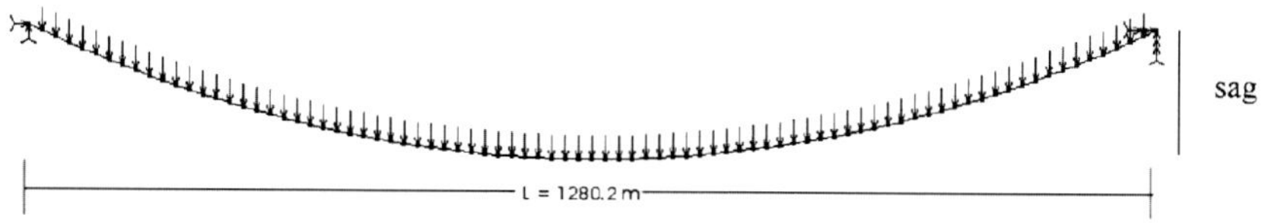

Figure 7.4.1: Bar model of the cable from the Golden Gate.

Table 7.4.2: Comparison of results from the static calculation of the main cable in the Golden Gate.

	Abbas and Scordelis[18]		Cobo[2] 1998		ADISNOL 3D	
	Displace-ment (m)	σ_{max} (MPa)	Displace-ment (m)	σ_{max} (MPa)	Displace-ment (m)	σ_{max} (MPa)
Cable weight + Deck weight (Reference)	143.9	483	143.9	486	143.9	474
- Deck weight	139.9	145	140.0	146	140.0	144
+ Overload	144.8	565	144.8	567	144.8	553
+ 7.8 × Overload	150.6	1100	150.4	1104	150.4	1070

Figure 7.4.2: Great Belt Bridge.

Table 7.4.3: Geometrical and mechanical properties of the Great Belt.

Main span length (m)	1624
Length of side spans (m)	535
Sag of main cables (m)	180
Distance between main cables (m)	27
Deck width (m)	31
Deck depth (m)	4.4
Section of each main cable (m^2)	0.44
Mass of each main cable (t/m)	3.74
Inertia moment under vertical bending for deck I_y (m^4)	4.0
Inertia moment under lateral bending for deck I_z (m^4)	100.0
Inertia moment under torsion for deck J (m^4)	7.6
Deck mass (T/m)	15.26
Inertia polar moment for deck (Tm^2/m)	1106.8

between its cable sag and the length of its main span (1/9), the studied geometry of the deck's box section and the sensitivity shown towards the structure's environmental impact. Noteworthy among the aerodynamic studies of this bridge are the ones done by Larsen[1] in 1993, prior to the definitive design of the deck. Table 7.4.3 provides the main geometrical and mechanical properties of the bridge, obtained by Larsen[1], Cobo[2] and Jurado and Hernández[5].

7.4.2.2 Free vibration analysis of the Great Belt Bridge

A three-dimensional model was made with 286 bar elements and 208 nodes, a detailed graphical representation of which can be found in Figure 7.4.4. This was done to model the bridge and carry out the calculation for the natural frequencies and mode shapes through the dynamic analysis code *ADISNOL3D*. This calculation involved second-order theory, taking into account the non-linear geometry of the structure's behaviour, as described above.

The mass of all the beam elements was considered, for the deck as well as the main cables and hangers, taking into account their own weight. Also defined was a series of fictitious transversal bars within the deck; these had a high stiffness and were weightless to make the deck movements compatible with the lower ends of the vertical hangers.

Attention must be drawn to how the numeration of the deck nodes goes from 1 to 21; in this way, it is easier to understand the earlier graphical representation of the vibration modes and their sensitivities.

Table 7.4.4 shows the results, in rad/s, for the natural vibration frequencies obtained alongside those published by other authors, such as Larsen[1] and Cobo[2]. Where necessary, the codes they used are provided. The data only include those frequencies associated with the modes that present movements along the deck, given that these alone are necessary for calculating flutter and therefore the sensitivity analysis. However, the data

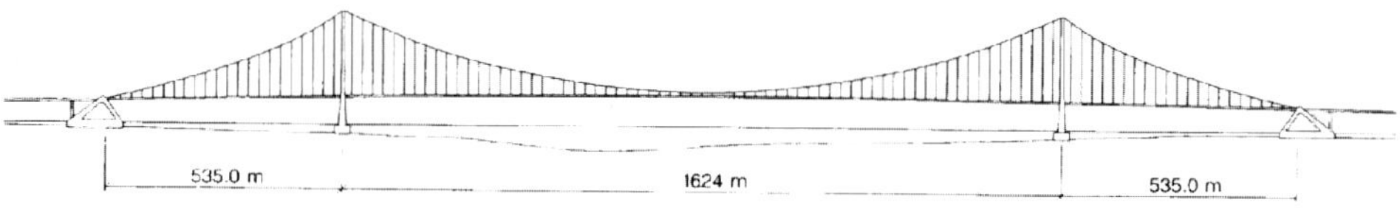

Figure 7.4.3: Elevation of the Great Belt.

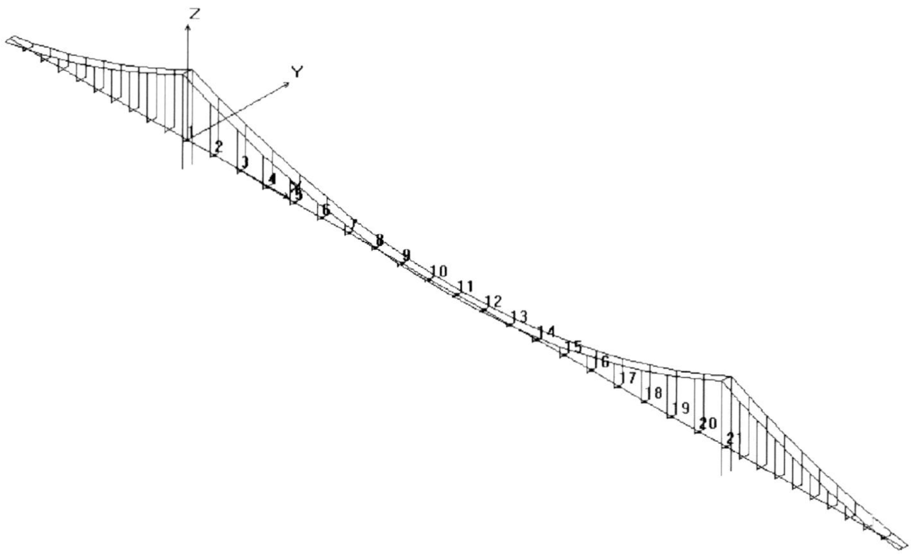

Figure 7.4.4: Beam model of the Great Belt.

Table 7.4.4: Comparison of the natural frequencies (rad/s) for the Great Belt.

Mode type	*ADISNOL3D*	Larsen[1]	Cobo[2] (NOLIN3D)
(1) LS	0.326	0.326	0.328
(2) VS	0.606	0.628	0.622
(3) VA	0.693	0.723	0.712
(4) LA	0.781	0.773	0.765
(5) VS	0.813	0.848	0.832
(6) VA	1.104	–	1.123
(11) VA	1.284	–	1.189
(12) VS	1.360	–	1.351
(13) LS	1.382	1.175	1.251
(18) VS	1.559	–	1.527
(19) VA	1.747	–	–
(21) VA	1.821	–	–
(22) TS	1.855	1.747	1.730
(27) VS	2.115	–	–
(28) LTA	2.207	–	–
(29) LS	2.434	–	–
(30) TA	2.461	2.406	2.216
(37) TS	2.816	3.154	–

fail to indicate, for instance, the modes in which the cables are deformed. Here, the terms explaining what each type of vibration mode relates to are as follows:

V: vertical vibration mode. *L*: lateral vibration mode. *T*: torsional vibration mode.
S: symmetrical vibration mode. *A*: asymmetrical vibration mode.

By analyzing Table 7.4.4, it is possible to discern how well the values for the frequencies obtained in this study coincide with their counterparts found in other research.

Figures 7.4.5–7.4.22 list the results of the vibration modes associated with previously mentioned natural frequencies. From these, the first 18 vibration modes of the deck were chosen as these would be included in the sensitivity analysis for the aeroelastic response of the bridge appearing in later chapters.

7.4.2.3 Free vibration sensitivity analysis of the suspension bridge over the Great Belt

Once the natural frequencies and vibration modes for the Great Belt Bridge were known, the next step was to carry out a sensitivity analysis of those frequencies and modes by using the sensitivity analysis module implemented in *ADISNOL3D*.

The design variables that were derived included the one related to the mechanical characteristics of the deck section: the two inertia bending moments – both lateral (I_z) and vertical (I_y) – and the inertia moment under torsion (J). This choice was due to the fact that earlier aeroelastic sensitivity analyses had taken these characteristics as variables.

As already mentioned, there is currently no commercially available package for sensitivity analyses and optimization that could carry out this kind of sensitivity analysis

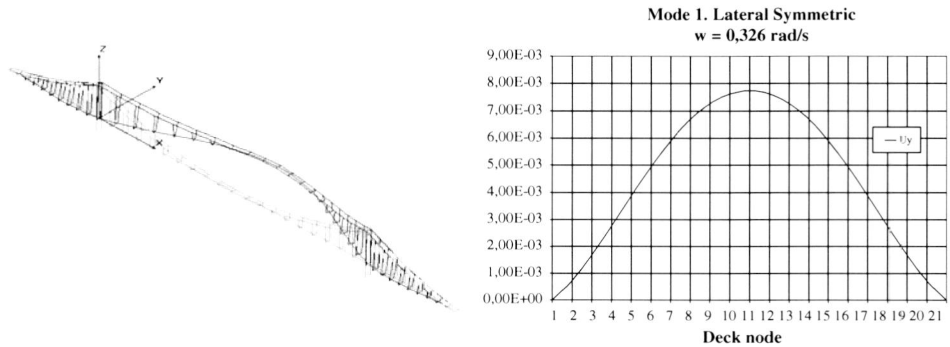

Figure 7.4.5: Mode (1) LS of the Great Belt.

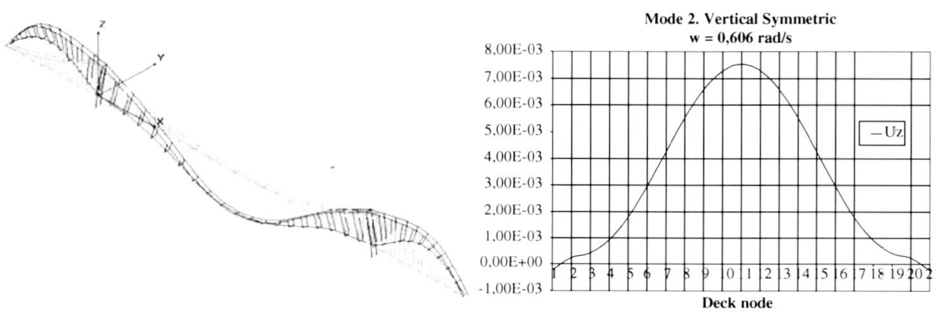

Figure 7.4.6: Mode(2) VS of the Great Belt.

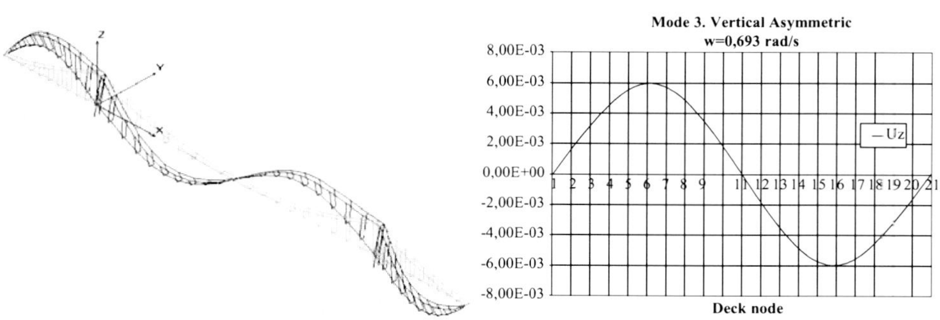

Figure 7.4.7: Mode (3) VA of the Great Belt.

through second-order theory, taking geometric non-linearity into account. Thus, the only way to make a comparison is with finite differential calculations, which, unfortunately, may be inaccurate. In any case, it is worthwhile contrasting these results, to verify good behaviour within the formulation provided and to evaluate in an efficient way the accuracy of those finite differential calculations.

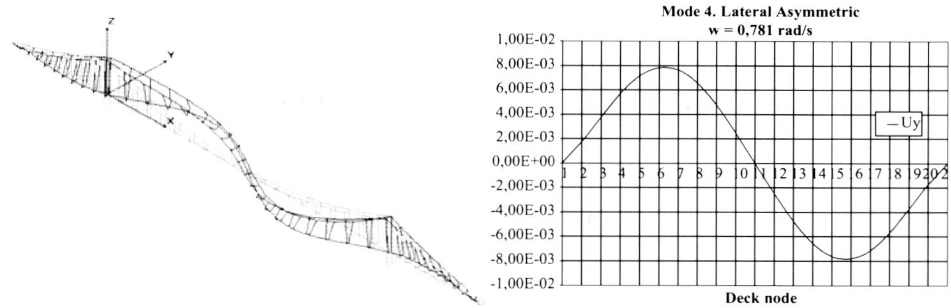

Figure 7.4.8: Mode (4) LA of the Great Belt.

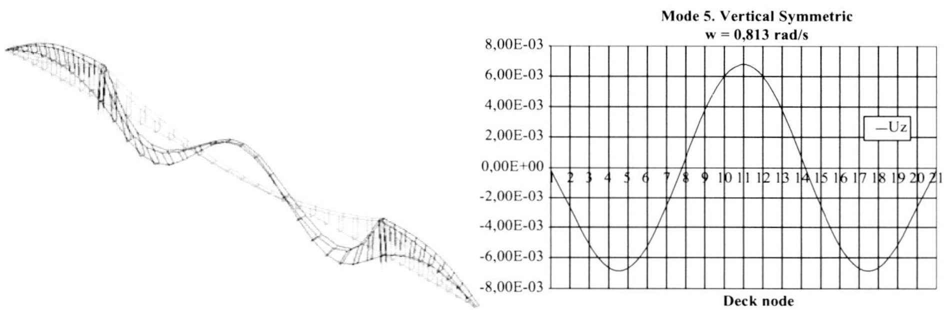

Figure 7.4.9: Mode (5) VS of the Great Belt.

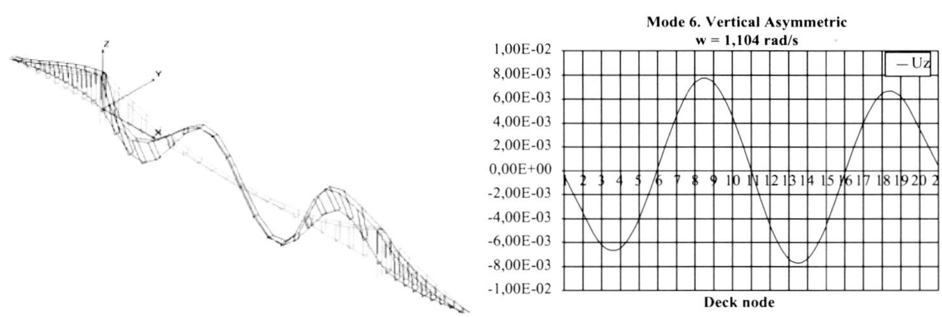

Figure 7.4.10: Mode (6) VA of the Great Belt.

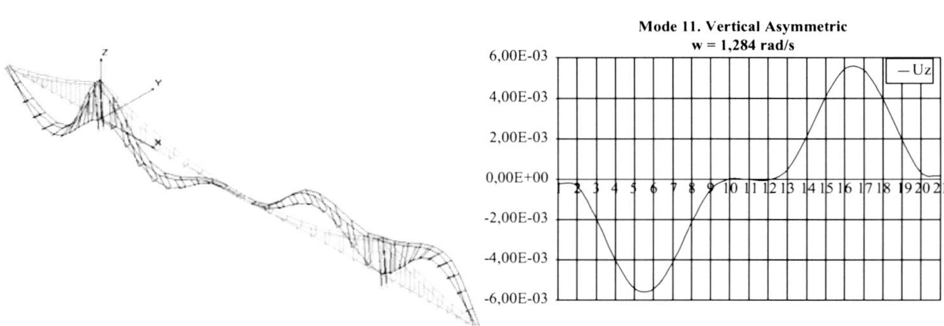

Figure 7.4.11: Mode (11) VA of the Great Belt.

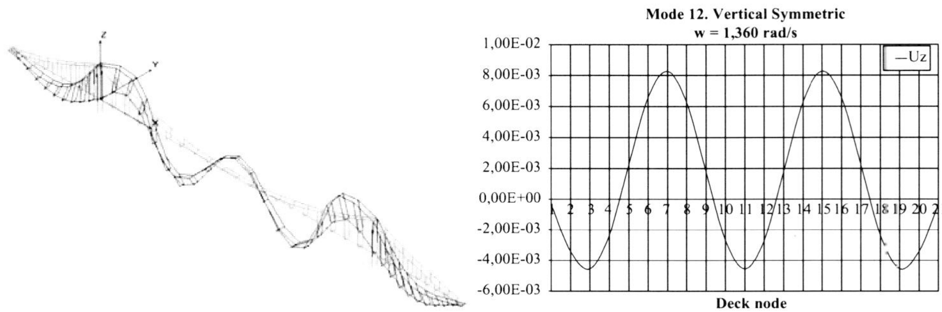

Figure 7.4.12: Mode (12) VS of the Great Belt.

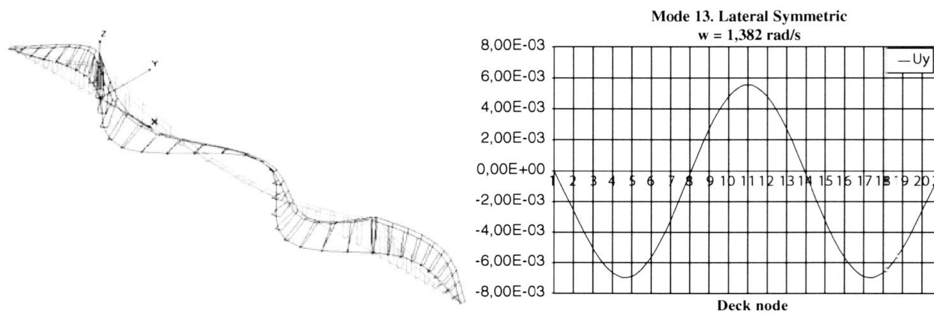

Figure 7.4.13: Mode (13) LS of the Great Belt.

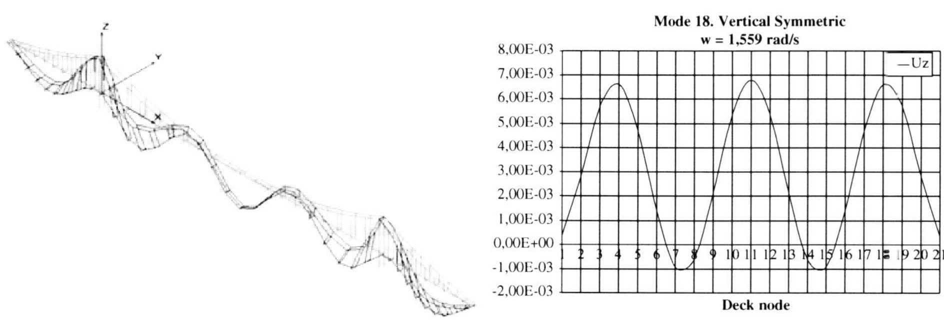

Figure 7.4.14: Mode (18) VS of the Great Belt.

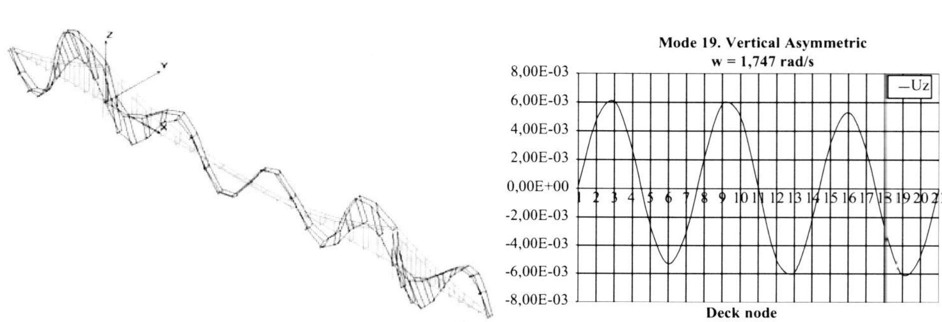

Figure 7.4.15: Mode (19) VA of the Great Belt.

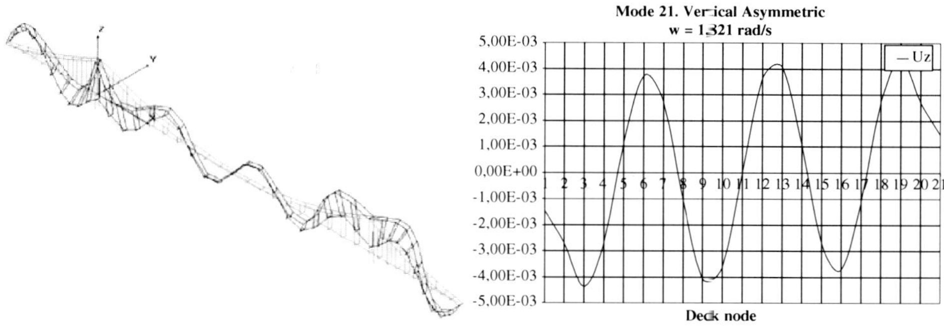

Figure 7.4.16: Mode (21) VA of the Great Belt.

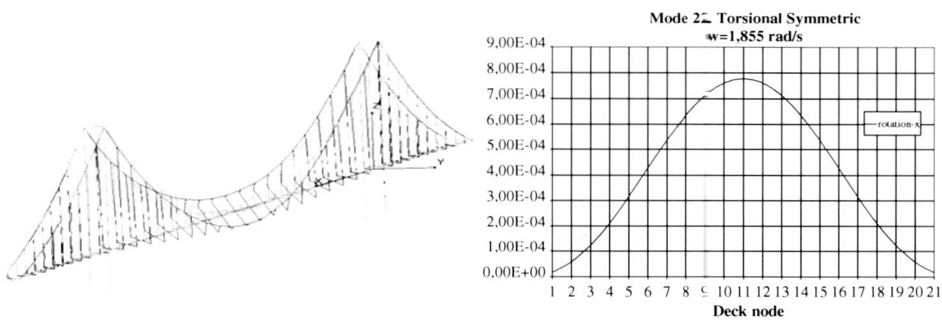

Figure 7.4.17: Mode (22) TS of the Great Belt.

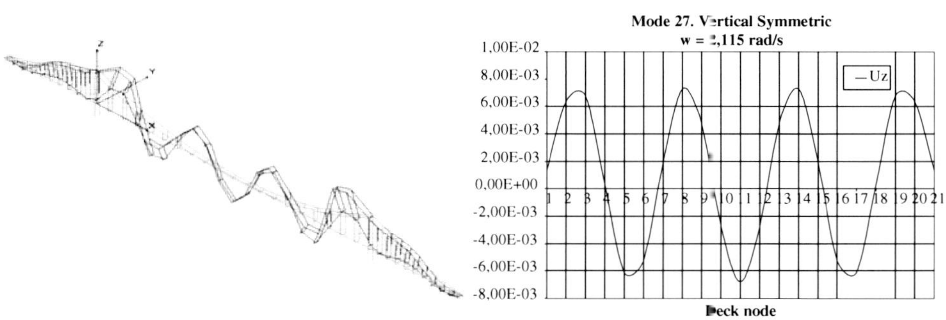

Figure 7.4.18: Mode (27) VS of the Great Belt.

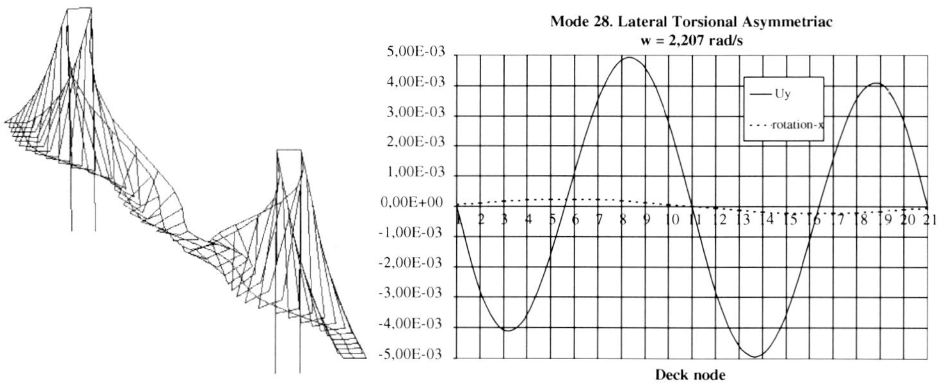

Figure 7.4.19: Mode (28) LTA of the Great Belt.

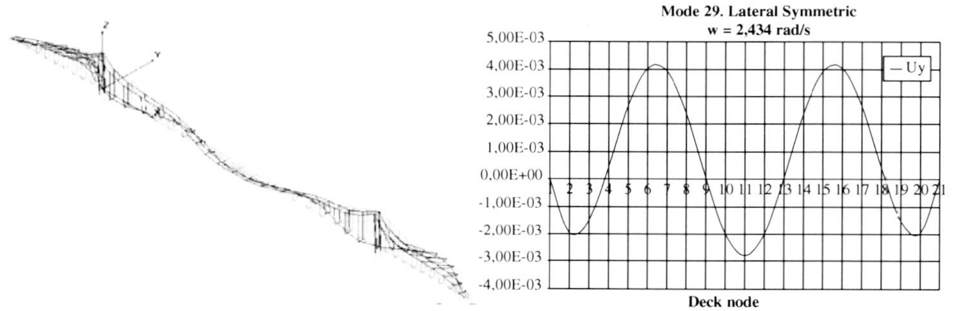

Figure 7.4.20: Mode (29) LS of the Great Belt.

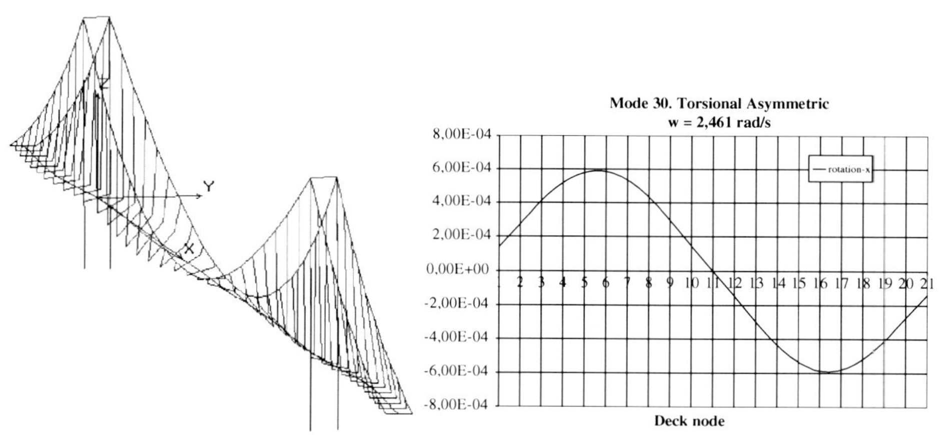

Figure 7.4.21: Mode (30) TA of the Great Belt.

As a starting point there are the results obtained for the natural frequency sensitivities. Table 7.4.5 provides an overview of these sensitivities in terms of the inertia under vertical bending for the deck (I_y). These derivatives were evaluated through finite differentials centred between 1%, 2% and 5%.

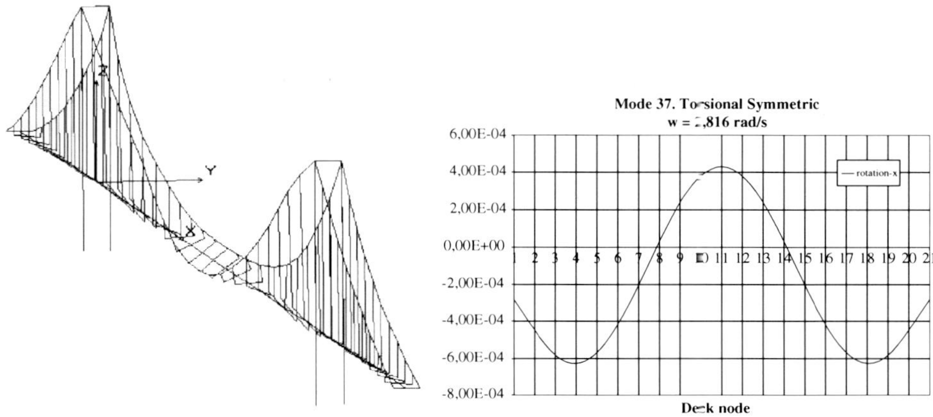

Figure 7.4.22: Mode (37) TS of the Great Belt.

Table 7.4.5: Comparison of the results for the sensitivity analysis, in terms of the inertia under vertical bending for the deck I_y, and the natural frequencies of the Great Belt.

Mode	ω_n rad/s	Frequency sensitivities in terms of I_y			
		Finite difference 5%	Finite difference2%	Finite difference1%	*ADISNOL3D*
(1) LS	0.326	-6.000E-05	-6.100E-05	-6.647E-05	-6.071e-5
(2) VS	0.606	4.780E-03	4.810E-03	4.802E-03	4.810e-3
(3) VA	0.693	3.810E-03	3.818E-03	3.824E-03	3.819e-3
(4) LA	0.781	-8.500E-05	-8.453E-05	-8.299E-05	-8.325e-5
(5) VS	0.813	6.790E-03	6.797E-03	6.794E-03	6.797e-3
(6) VA	1.104	1.645E-02	1.650E-02	1.648E-02	1.650e-2
(11) VA	1.284	8.600E-03	8.650E-03	8.576E-03	8.650e-3
(12) VS	1.360	2.070E-02	2.092E-02	2.075E-02	2.091e-2
(13) LS	1.382	-2.000E-04	-1.919E-04	-2.009E-04	-1.879e-4
(18) VS	1.559	1.635E-02	1.629E-02	1.620E-02	1.629e-2
(19) VA	1.747	4.670E-02	4.683E-02	4.671E-02	4.683e-2
(21) VA	1.821	5.495E-02	5.533E-02	5.520E-02	5.533e-2
(22) TS	1.855	-1.000E-04	-7.201E-05	-1.439E-04	-7.014e-5
(27) VS	2.115	7.515E-02	7.561E-02	7.541E-02	7.561e-2
(28) LTA	2.207	-3.000E-04	-3.169E-04	-2.654E-04	-2.931e-4
(29) LS	2.434	-1.500E-04	-1.262E-04	-2.044E-04	-1.235e-4
(30) TA	2.461	-2.500E-04	-2.173E-04	-2.916E-04	-2.052e-4
(37) TS	2.816	-3.500E-04	-3.466E-04	-2.815E-04	-3.485e-4

Various conclusions can be reached by perusing Table 7.4.5. One observation is how well the sensitivities obtained analytically and those taken from finite differentials coincide. Indeed, the results are identical in some cases. Another point to notice is that, due to information previously made available in a qualitative manner, the number of steps taken in the finite calculation decreases. Nevertheless, the solution is not necessarily improved. The most striking differences between the values are produced in those frequencies whose

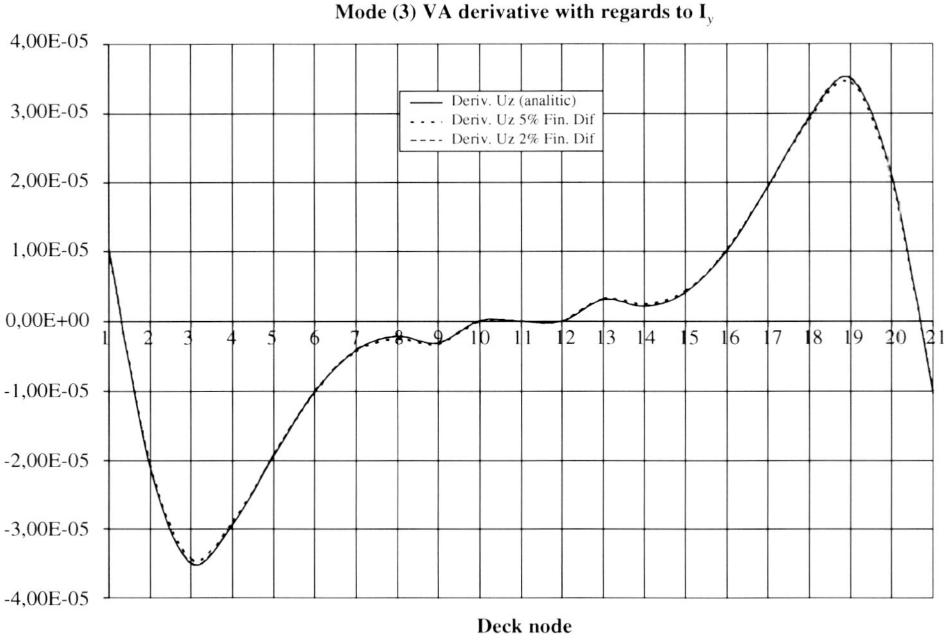

Figure 7.4.24: Derivative of vibration mode (3) VA related to I_y of the deck.

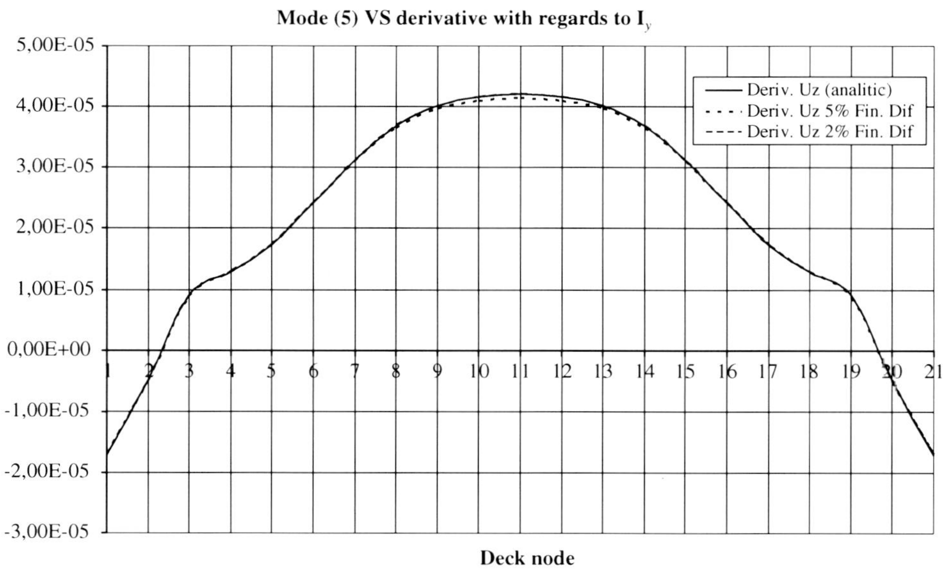

Figure 7.4.25: Derivative of vibration mode (5) VS related to I_y of the deck.

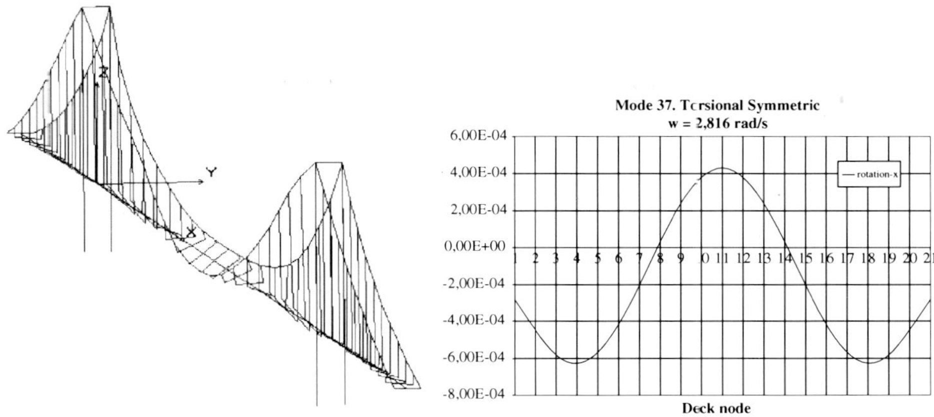

Figure 7.4.22: Mode (37) TS of the Great Belt.

Table 7.4.5: Comparison of the results for the sensitivity analysis, in terms of the inertia under vertical bending for the deck I_y, and the natural frequencies of the Great Belt.

Mode	ω_n rad/s	Frequency sensitivities in terms of I_y			
		Finite difference 5%	Finite difference2%	Finite difference1%	*ADISNOL3D*
(1) LS	0.326	-6.000E-05	-6.100E-05	-6.647E-05	-6.071e-5
(2) VS	0.606	4.780E-03	4.810E-03	4.802E-03	4.810e-3
(3) VA	0.693	3.810E-03	3.818E-03	3.824E-03	3.819e-3
(4) LA	0.781	-8.500E-05	-8.453E-05	-8.299E-05	-8.325e-5
(5) VS	0.813	6.790E-03	6.797E-03	6.794E-03	6.797e-3
(6) VA	1.104	1.645E-02	1.650E-02	1.648E-02	1.650e-2
(11) VA	1.284	8.600E-03	8.650E-03	8.576E-03	8.650e-3
(12) VS	1.360	2.070E-02	2.092E-02	2.075E-02	2.091e-2
(13) LS	1.382	-2.000E-04	-1.919E-04	-2.009E-04	-1.879e-4
(18) VS	1.559	1.635E-02	1.629E-02	1.620E-02	1.629e-2
(19) VA	1.747	4.670E-02	4.683E-02	4.671E-02	4.683e-2
(21) VA	1.821	5.495E-02	5.533E-02	5.520E-02	5.533e-2
(22) TS	1.855	-1.000E-04	-7.201E-05	-1.439E-04	-7.014e-5
(27) VS	2.115	7.515E-02	7.561E-02	7.541E-02	7.561e-2
(28) LTA	2.207	-3.000E-04	-3.169E-04	-2.654E-04	-2.931e-4
(29) LS	2.434	-1.500E-04	-1.262E-04	-2.044E-04	-1.235e-4
(30) TA	2.461	-2.500E-04	-2.173E-04	-2.916E-04	-2.052e-4
(37) TS	2.816	-3.500E-04	-3.466E-04	-2.815E-04	-3.485e-4

Various conclusions can be reached by perusing Table 7.4.5. One observation is how well the sensitivities obtained analytically and those taken from finite differentials coincide. Indeed, the results are identical in some cases. Another point to notice is that, due to information previously made available in a qualitative manner, the number of steps taken in the finite calculation decreases. Nevertheless, the solution is not necessarily improved. The most striking differences between the values are produced in those frequencies whose mode presents no vertical movement. At the same time, they are two or three orders of magnitude lower than those which do present vertical movement and are therefore seen as more affected by the variation in the deck's inertia under vertical bending.

Table 7.4.6 presents the results obtained for the natural frequency sensitivities in terms of the deck's inertia under lateral bending (I_z). Once again, to evaluate these derivatives, finite differentials between 1%, 2% and 5% were used.

The conclusions reached on examining Table 7.4.6 are similar to what has been seen earlier. Perhaps it is worth mentioning that, for the modes, such as the vertical ones, in which inertia under lateral bending has no influence on the results from the finite differentials given that the step size is reduced, numerical errors make the results stray from what may be considered correct.

The results obtained for the natural frequencies with respect to the deck's inertia under torsion (J) are found in Table 7.4.7. Just as before, differentials between 1%, 2% and 5% were chosen to evaluate these derivatives.

The outcome is also similar. Here, attention should be drawn to how the best results obtained from finite differentials are produced through a step size of 2% for the deck's inertia under torsion. However, as for the derivatives with relevant values, specifically those for the torsional modes, the results from the three approaches are good. These results may have been predicted through intuition, but they would not have been known *a priori*.

Figures 7.4.23–7.4.41 provide a graphical representation of the sensitivity analysis carried out on the vibration eigenmodes for the Great Belt. As was the case with the natural frequency sensitivities, the analytical results obtained *ADISNOL3D* are shown, along with the ones achieved from differential calculation with different step sizes.

Table 7.4.6: Comparison of the results for the sensitivity analysis, with respect to the deck's inertia under lateral bending I_z, of the natural frequencies for the Great Belt.

Mode	ω_n (rad/s)	Frequency sensitivity in terms of *Iz*			
		Finite difference5%	Finite difference2%	Finite difference1%	*ADISNOL3D*
(1) LS	0.326	6.492E-04	6.532E-04	6.521E-04	6.351e-4
(2) VS	0.606	0.000E+00	-4.874E-12	-8.102E-08	0.0
(3) VA	0.693	0.000E+00	1.255E-10	2.886E-07	0.0
(4) LA	0.781	2.376E-03	2.412E-03	2.404E-03	2.411e-3
(5) VS	0.813	0.000E+00	-2.565E-11	-3.985E-08	0.0
(6) VA	1.104	0.000E+00	8.566E-11	-3.210E-07	0.0
(11) VA	1.284	0.000E+00	1.109E-09	-2.639E-06	0.0
(12) VS	1.360	0.000E+00	7.100E-13	-4.899E-06	0.0
(13) LS	1.382	2.946E-03	4.131E-03	3.922E-03	4.142e-3
(18) VS	1.559	0.000E+00	-9.350E-12	-4.471E-06	0.0
(19) VA	1.747	0.000E+00	1.914E-10	-3.707E-06	0.0
(21) VA	1.821	0.000E+00	8.866E-11	-2.203E-06	0.0
(22) TS	1.855	3.800E-05	3.675E-05	3.411E-05	3.674e-5
(27) VS	2.115	0.000E+00	-4.800E-12	-4.569E-06	0.0
(28) LTA	2.207	6.200E-03	6.212E-03	6.219E-03	6.213e-3
(29) LS	2.434	8.536E-03	1.083E-02	1.078E-02	1.084e-2
(30) TA	2.461	4.120E-03	2.841E-03	2.833E-03	2.839e-3
(37) TS	2.816	1.200E-05	1.172E-05	1.406E-05	1.171e-5

Table 7.4.7: Comparison of the results for the sensitivity analysis, with respect to the deck's torsional inertia J, of the natural vibration frequencies for the Great Belt.

Mode	ω_n (rad/s)	Frequency sensitivities with respect to J			
		Finite difference 5%	Finite difference2%	Finite difference1%	*ADISNOL3D*
(1) LS	0.326	2.632E-06	3.445E-06	3.987E-07	3.444e-6
(2) VS	0.606	0.000E+00	-9.543E-11	-1.066E-06	0.0
(3) VA	0.693	0.000E+00	4.018E-10	3.796E-06	0.0
(4) LA	0.781	9.158E-04	9.441E-04	9.389E-04	9.440e-4
(5) VS	0.813	0.000E+00	-9.428E-11	-5.240E-07	0.0
(6) VA	1.104	0.000E+00	2.998E-10	-4.225E-06	0.0
(11) VA	1.284	0.000E+00	3.369E-09	-3.474E-05	0.0
(12) VS	1.360	0.000E+00	-1.474E-11	-6.446E-05	0.0
(13) LS	1.382	2.632E-05	2.665E-05	2.166E-05	2.665e-5
(18) VS	1.559	0.000E+00	-3.270E-11	-5.883E-05	0.0
(19) VA	1.747	0.000E+00	6.223E-10	-4.877E-05	0.0
(21) VA	1.821	0.000E+00	3.033E-10	-2.899E-05	0.0
(22) TS	1.855	3.663E-02	3.737E-02	3.718E-02	3.737e-2
(27) VS	2.115	0.000E+00	-1.803E-11	-6.012E-05	0.0
(28) LTA	2.207	9.132E-03	9.998E-03	9.834E-03	9.994e-3
(29) LS	2.434	1.053E-04	1.211E-04	7.860E-05	1.210e-4
(30) TA	2.461	5.321E-02	5.363E-02	5.352E-02	5.363e-2
(37) TS	2.816	9.153E-02	9.380E-02	9.338E-02	9.380e-2

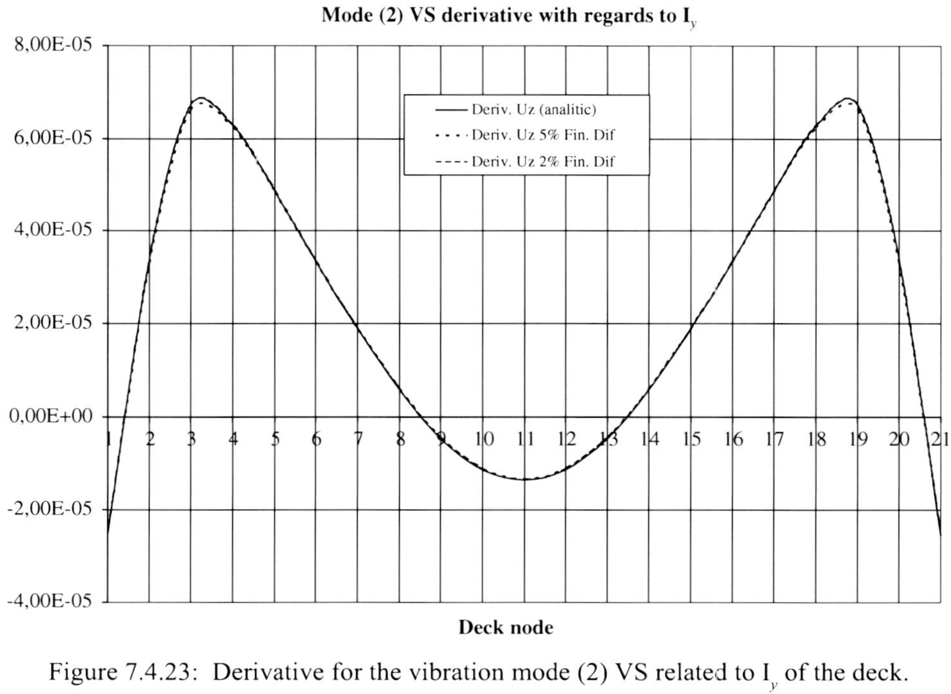

Figure 7.4.23: Derivative for the vibration mode (2) VS related to I_y of the deck.

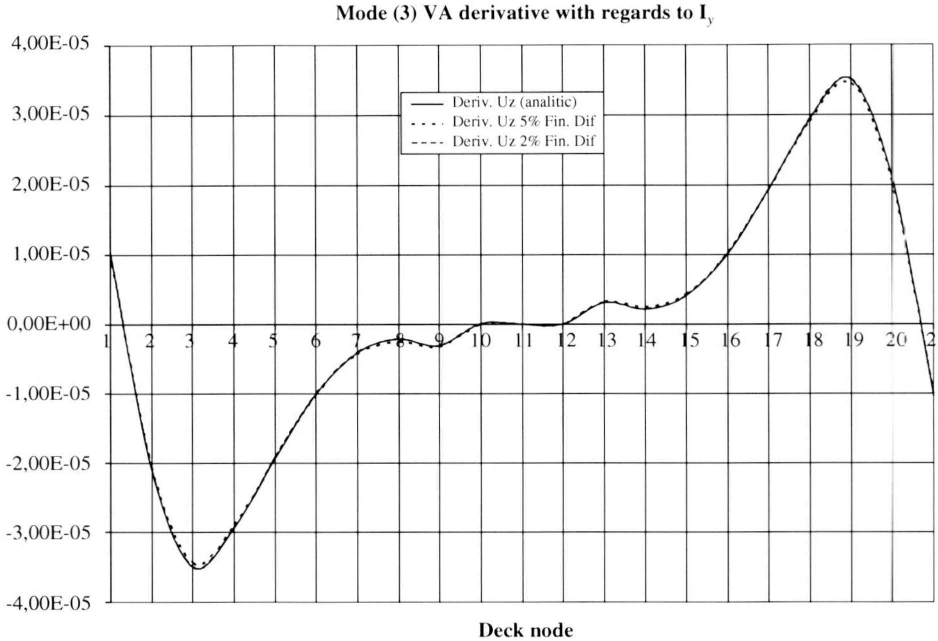

Figure 7.4.24: Derivative of vibration mode (3) VA related to I_y of the deck.

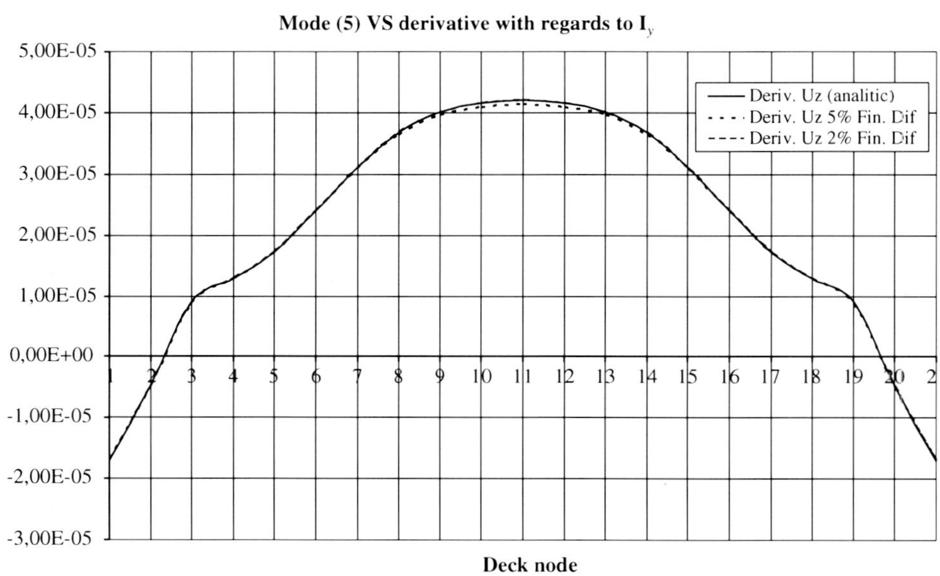

Figure 7.4.25: Derivative of vibration mode (5) VS related to I_y of the deck.

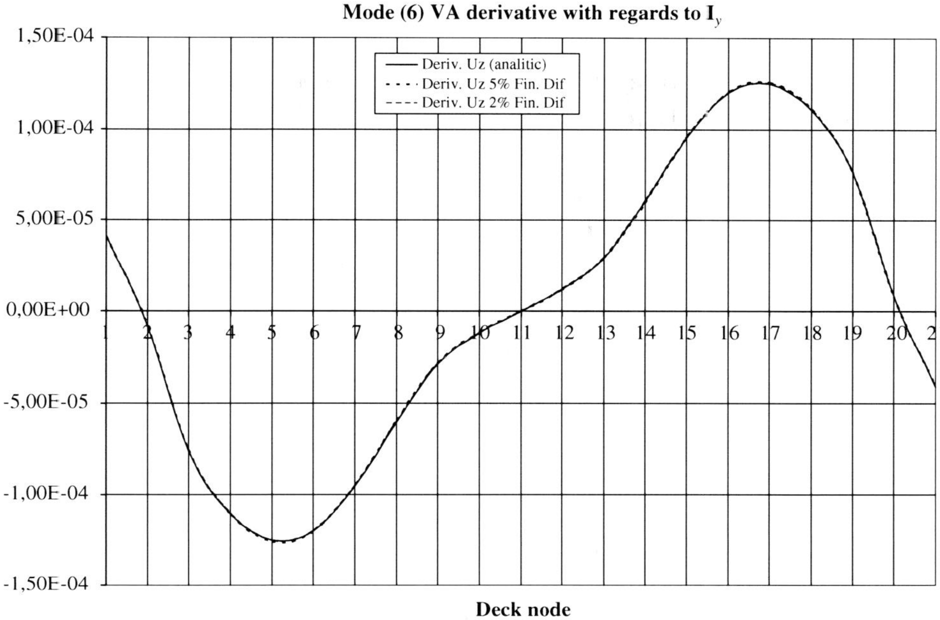

Figure 7.4.26: Derivative of the vibration mode (6) VA related to I_y of the deck.

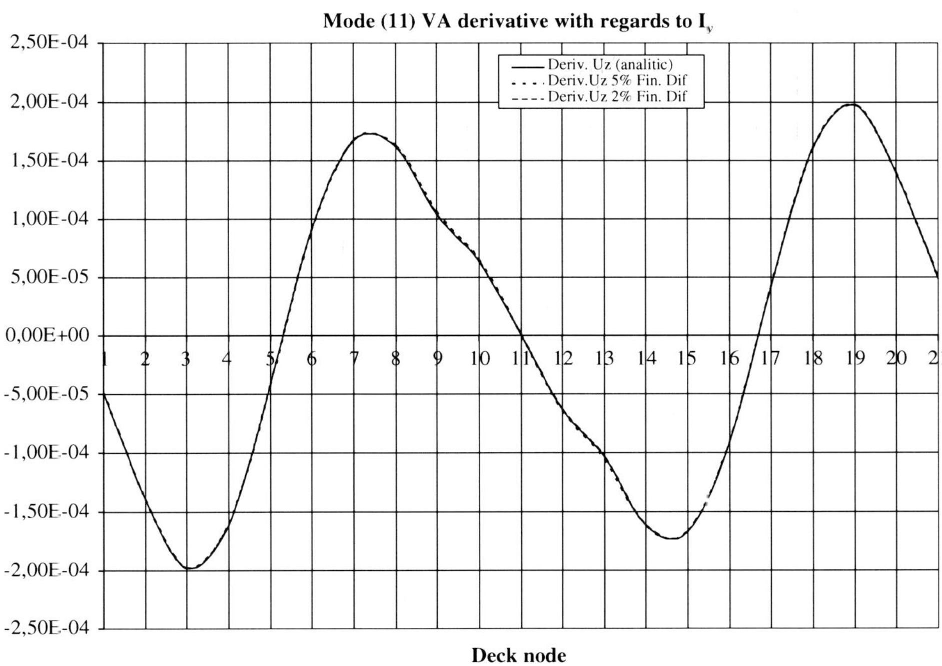

Figure 7.4.27: Derivative of the vibration mode (11) VA related to I_y of the deck.

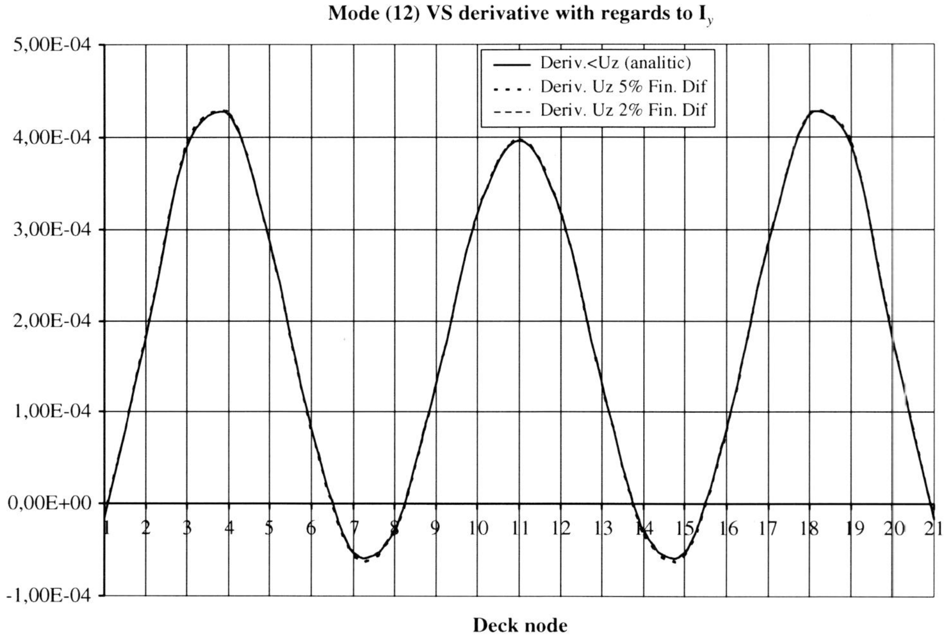

Figure 7.4.28: Derivative of the vibration mode (12) VS related to I_y of the deck.

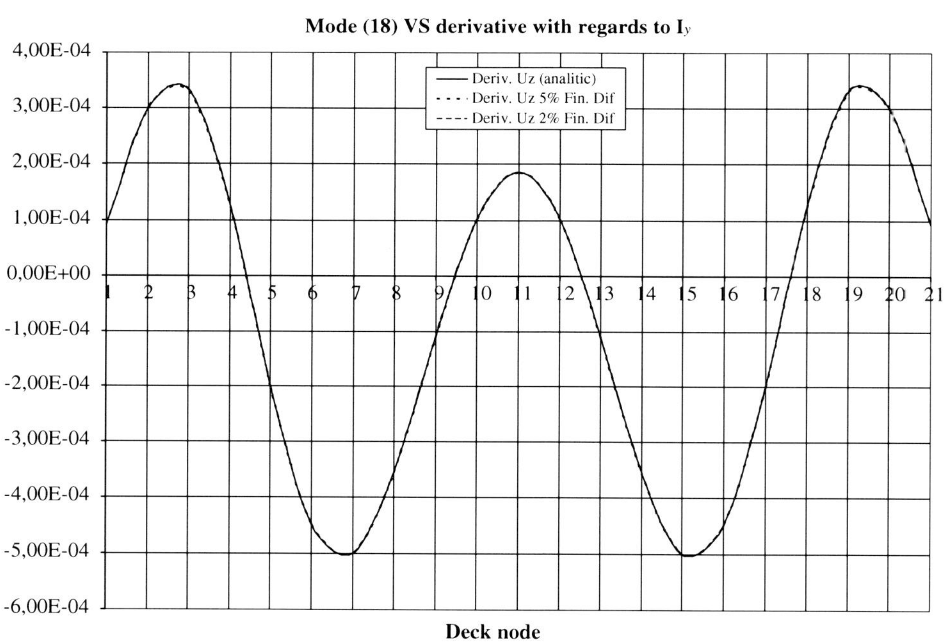

Figure 7.4.29: Derivative of the vibration mode (18) VS related to I_y of the deck.

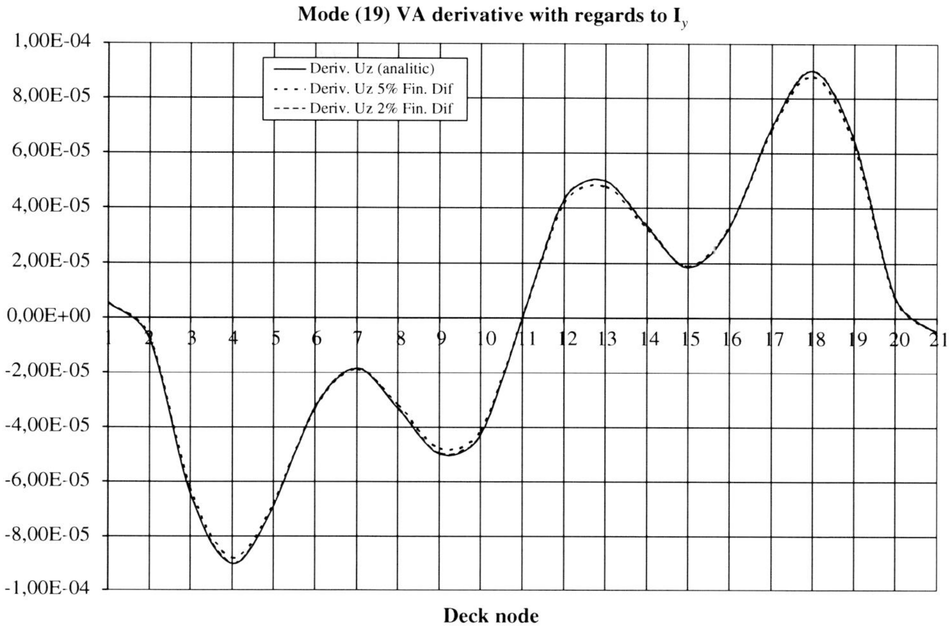

Figure 7.4.30: Derivative of the vibration mode (19) VA related to I_y of the deck.

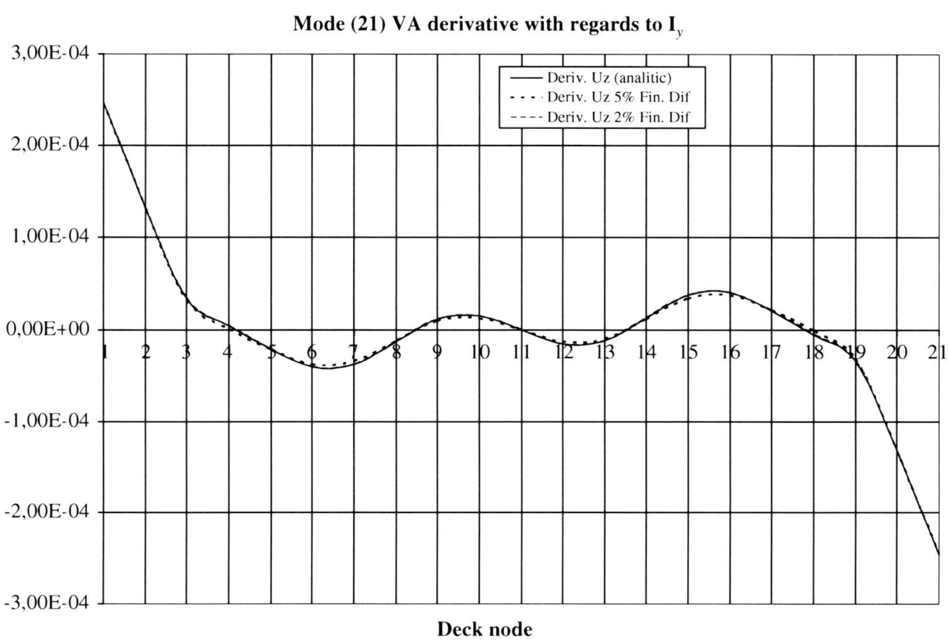

Figure 7.4.31: Derivative of the vibration mode (21) VA with respect to I_y of the deck.

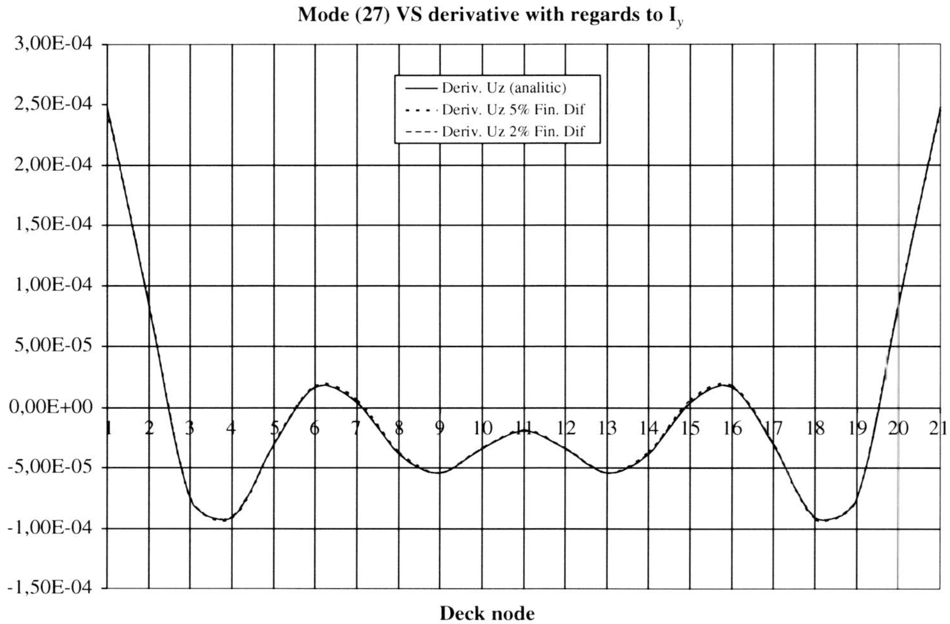

Figure 7.4.32: Derivative of the vibration mode (27) VS related to I_y of the deck.

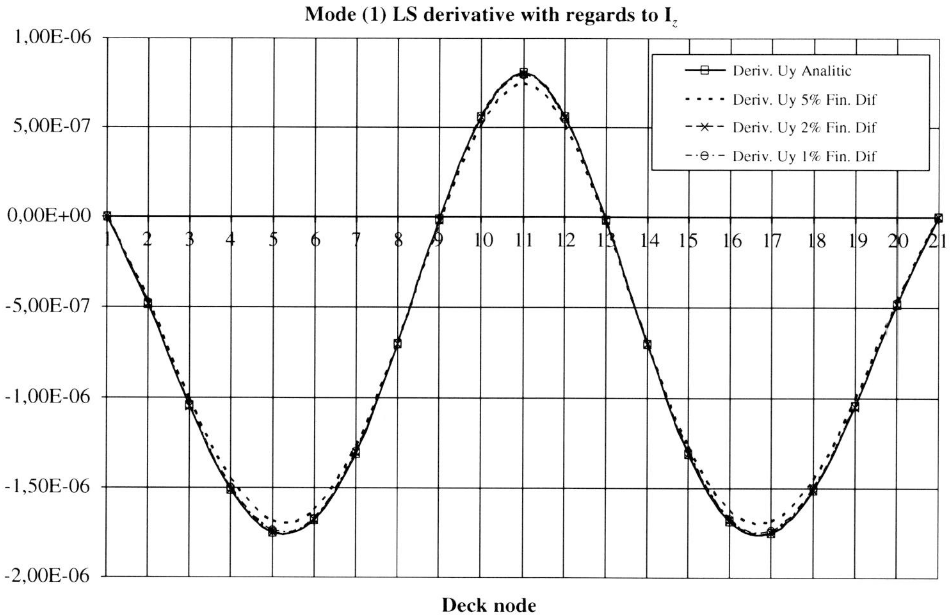

Figure 7.4.33: Derivative of the vibration mode (1) LS related to I_z of the deck.

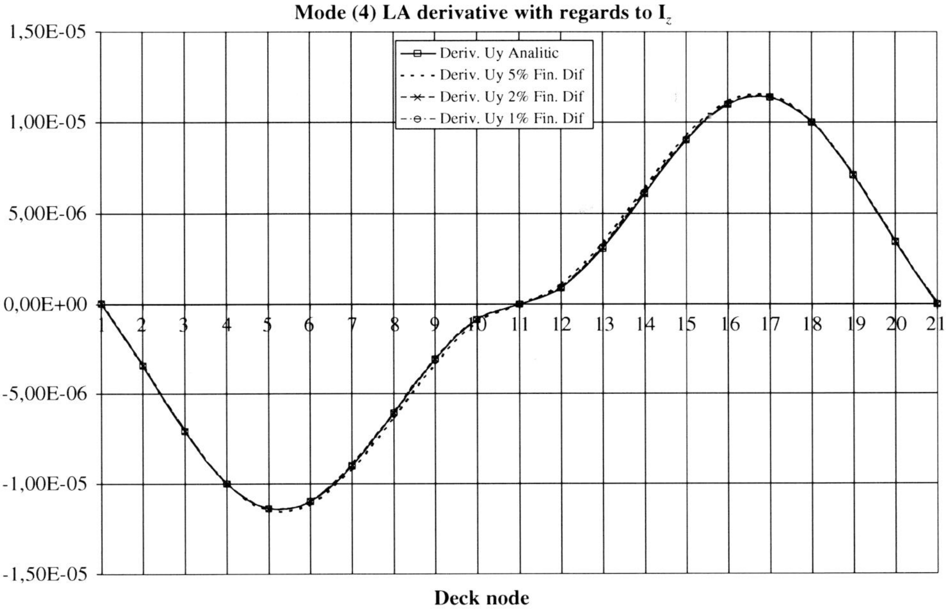

Figure 7.4.34: Derivative of the vibration mode (4) LA related to I_z of the deck.

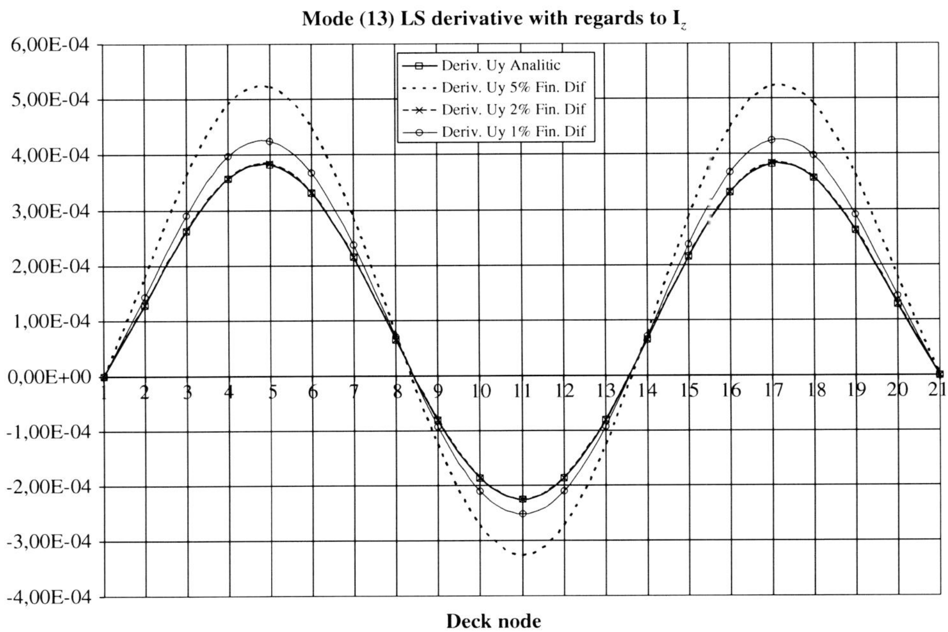

Figure 7.4.35: Derivative of the vibration mode (13) LS related to I_z of the deck.

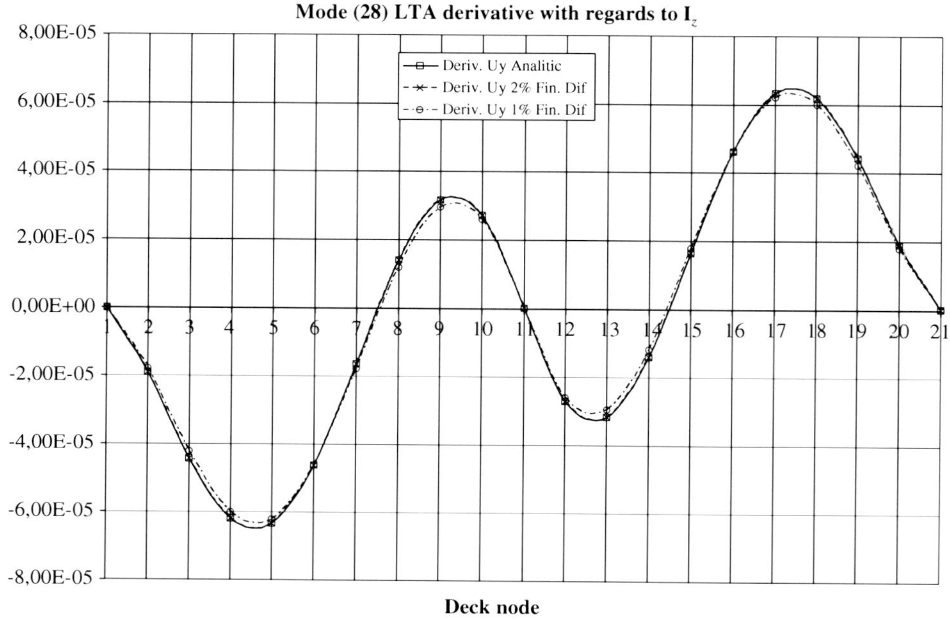

Figure 7.4.36: Derivative of the vibration mode (28) LTA related to I_z of the deck.

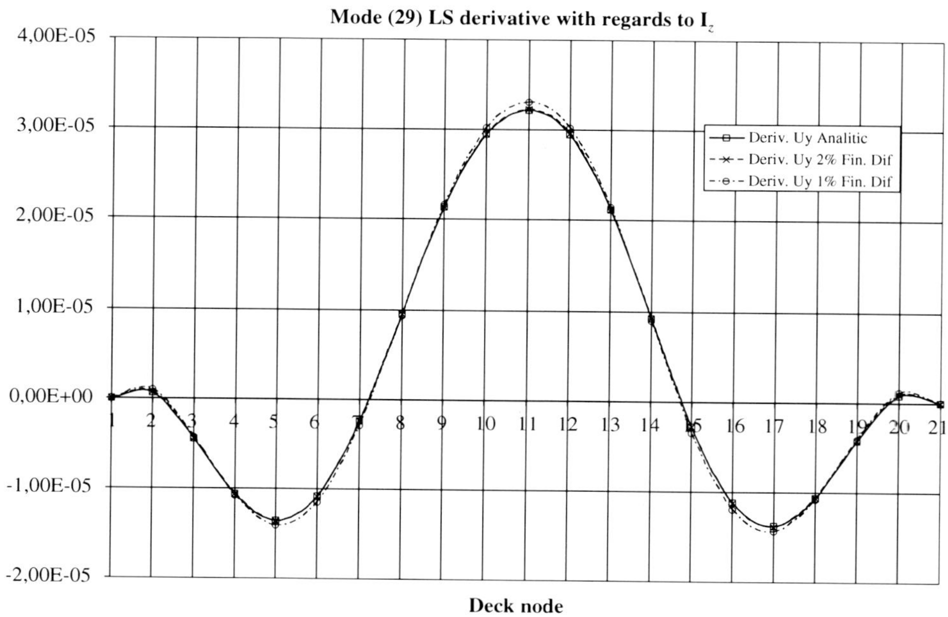

Figure 7.4.37: Derivative of the vibration mode (29) LS related to I_z of the deck.

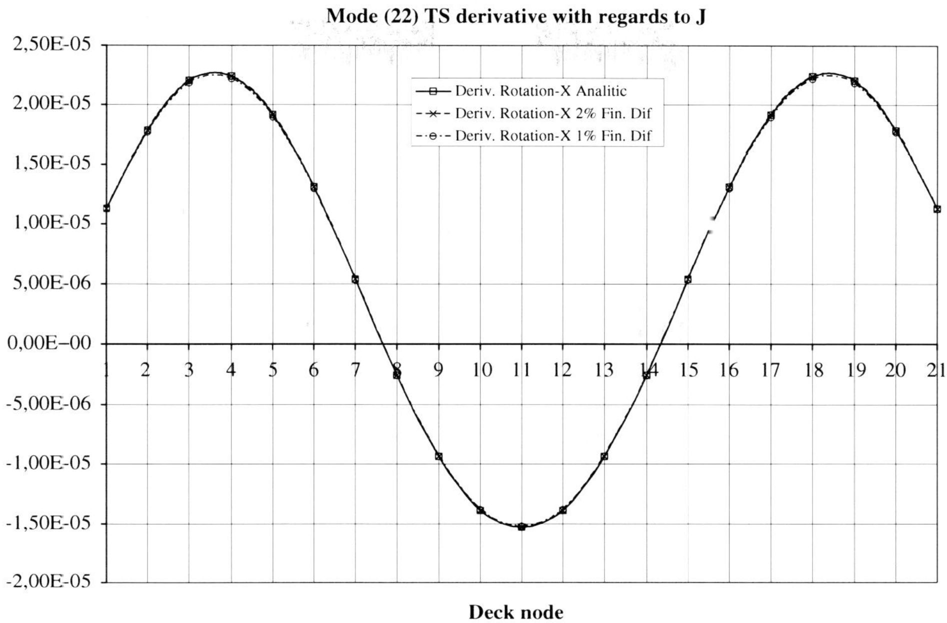

Figure 7.4.38: Derivative of the vibration mode (22) TS related to *J* of the deck.

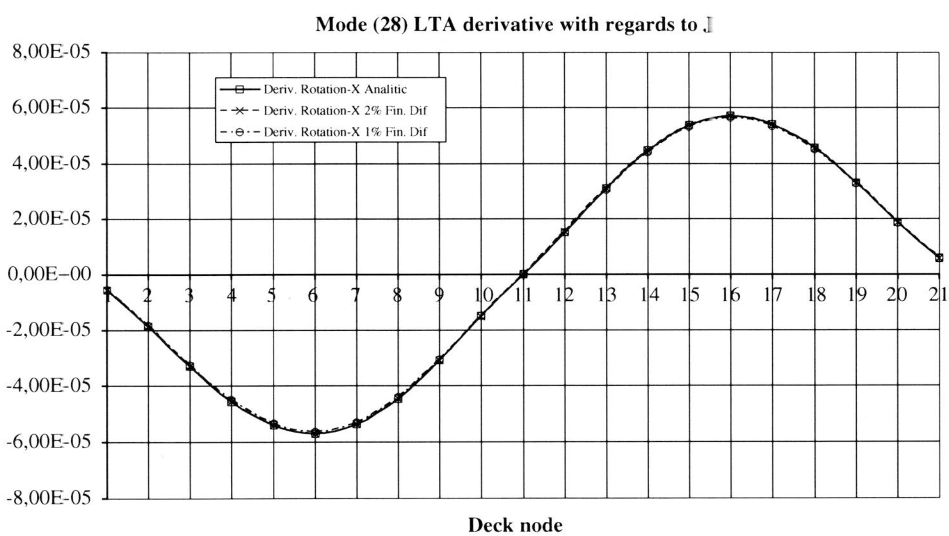

Figure 7.4.39: Derivative of the vibration mode (28) LTA related to *J* of the deck.

The methodology underpinning these results is as follows: Only the sensitivities or derivatives are chosen from those modes that are deemed significant: that is, those in which the design variable or derivation has an influence. For example, a derivative of a lateral vibration mode with respect to the deck's inertia under vertical bending that was null would not be selected.

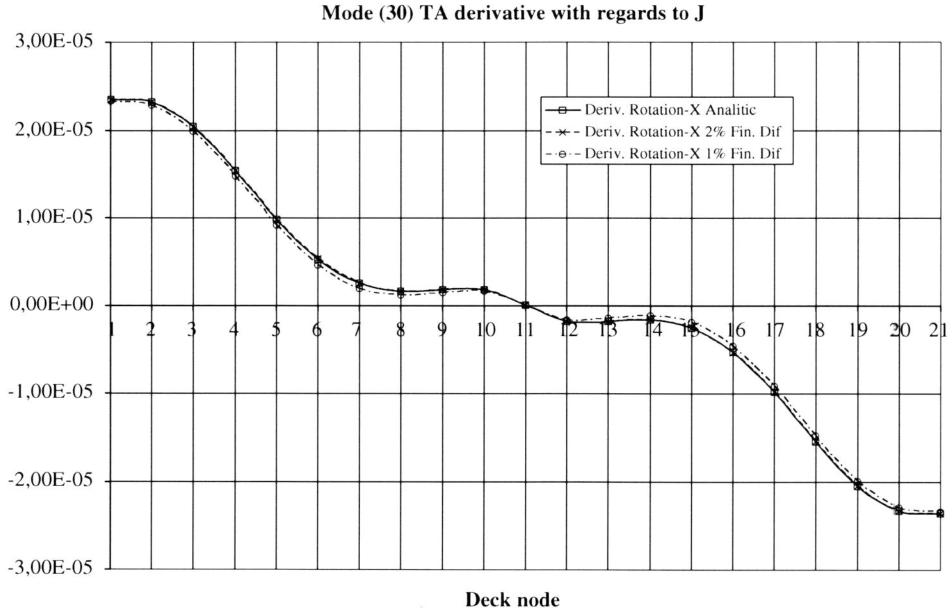

Figure 7.4.40: Derivative of the vibration mode (30) TA related to J of the deck.

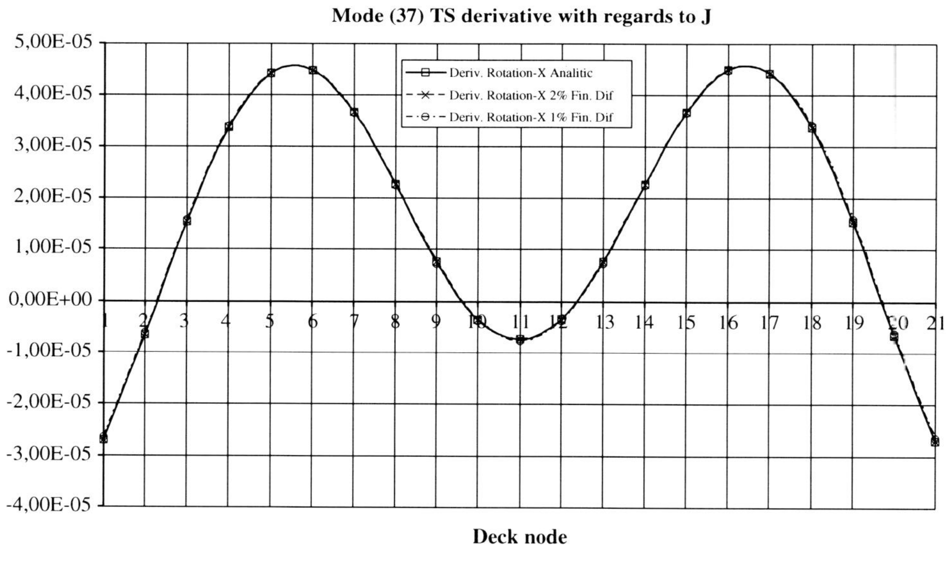

Figure 7.4.41: Derivative of the vibration mode (37) TS related to J of the deck.

Among the derivatives or sensitivities of those modes for which results are included, only the components of a significant derivative are shown. A case in point is when the components for the sensitivities of a mode's vertical movement are shown in the derivative for the vertical vibration mode. The other sensibilities are either null or are lower by several orders of magnitude.

Only the components of the sensitivities or mode derivatives along the main span are shown, as these are the largest and will have the greatest influence on earlier aeroelastic sensitivity analyses.

From the results taken from the sensitivity analysis of the vibration eigenmodes found in Figures 7.4.23–7.4.41, the following conclusions may be drawn:

- On the whole, the results obtained through the proposed analytical formulation and through finite differentials coincide well, thus validating the formulation implemented in *ADISNOL3D* for the second-order sensitivity analysis of vibration eigenmodes.
- The results from the sensitivity analysis through finite differentials are generally good. A greater stability and accuracy in the results can be observed than is the case with the natural frequency sensitivities.
- Once again, it is possible to observe how the results are not necessarily better when the finite differential step size is reduced. As seen in Figure 7.4.34, an increase in the variable of 2% brings better results than perturbations of 5 or 1%.
- By way of general comments about the example provided above, it is important to note that the sensitivity analyses done through finite differentials provide good results. However, the time needed to do the calculations given the high number of analyses required, the laborious task of post-processing and the uncertainty inherent in this method are all factors that make the analytical method proposed above more attractive as an efficient tool.

7.5 References

[1] Larsen A. [1993] *Aerodynamic aspects of the final design of the 1624 m suspension bridge across the Great Belt*. Journal of Wind Engineering and Industrial Aerodynamics. Vol. 48, No. 1–2, pp. 261–285.

[2] Cobo D. [1998] *An analysis of wind stability. Improvements to the response of suspension bridges*. Doctoral thesis from the Universitat Politécnica de Catalunya. Escola Técnica Superior D'Enginyers de Camins, Canals i Ports. (Director: Prof. Ángel Aparicio)

[3] Namini A., Aldbrecht P. [1992] *Finite Element-Based Flutter Analysis of Cable Suspended Bridges*. Journal of Structural Engineering. Vol. 118, No. 6, pp.1509–1526.

[4] Astiz M. A. [1999] *Flutter stability of very long suspension bridges*. Journal of Bridge Engineering, Vol. 3, No. 3, pp.132–139.

[5] Jurado J. A., Hernández S. [1999] *Design of ultra-long span bridges with aeroelastic constraints. Optimization in Industry II*, Junio, Banff, (Alberta) Canada.

[6] Semiào C., Mendes P. [1998] *Avaliazao das condizoes de estabilidade aerodinâmica em pontes*. Jornadas portuguesas de engenharia de estructuras. Lisboa, Portugal.

[7] Samartín A., González de Cangas J.R. [2001] *Cálculo matricial de estructuras*. Colegio de Ingenieros de Caminos, Canales y Puertos. Servicio de publicaciones – Colección Escuelas. Madrid.

[8] Hernández S. [1997] *Análisis Lineal y No Lineal de Estructuras de Barras*. Publication service of the Universidade da Coruña.

[9] Livesley R. K. [1970] *Métodos Matriciales para Cálculo de Estructuras*. Editorial Blume.

[10] Saez-Benito J. [1975] *Cálculo Matricial de Estructuras Formadas por Piezas Prismáticas*. Fondo Editorial de Inenieria Naval.

[11] Przemieniecki J.S. [1968] *Theory of Matrix Structural Analysis*. Dover, New York.

[12] Hacar M. A., Alarcón E. [1971] *Elementos de dinámica aplicada a las estructuras*. Publication service, E.T.S.I.I. Madrid.

[13] Chopra A. K. [1995] *Dynamics of Structures*. Prentice-Hall, Englewood Cliffs, NJ.

[14] Hart G. C., Wong K. [2000] *Structural Dynamics for Structural Engineers*. John Wiley & Sons, New York.

[15] Adelman H. M., Haftka R. T. [1992] *Sensitivity analysis of discrete systems. Structural Optimization Status and Promise*. Vol. 150 Progress in Aeronautics and Astronautics AIAA, Boulder, CO, 291–315.

[16] GENESIS [1997] *Manuals of GENESIS Version 6.0*. VMA Engineering, Colorado Spring.

[17] Kanok-Nukulchai W., Yiu P. K., Brotton D. M. [1992] *Mathematical modelling of cable-stayed bridges*. Structural Engineering International, IABSE, Vol. 2, No. 2, pp. 108–113.

[18] Abbas S., Scordelis A. [1994] *Non-linear Analysis of Cable-stayed Bridges*. International Conference A.I.P.C. – F.I.P. on Cable-Stayed and Suspension Bridges, Deauville, Vol. 2, pp.195–210.

[19] Schrefler B. A., Odorizzi S., Wood R. D. [1983] *A total Lagrangian geometricalally non-linear analysis of combined beam and cable structures*. Computers and Structures, Vol. 17, No 1, pp.115–127.

[20] Arzoumanidis S. G., Bieniek M. P. [1985] *Finite element analysis of suspension bridges*. Computers and Structures, Vol. 21, No. 6, pp. 1237–1253.

[21] Ge Y. J., Tanaka H. [2000] *Aerodynamic stability of long-span suspension bridges under erection*. Journal of Structural Engineering. Vol. 126, No. 12, 2000, pp.1404–1412.

[22] Haftka R. T., Gürdal Z. [1992] *Elements of Structural Optimization*. Kluwer, Dordrecht.

[23] Jurado J. A. [2001] *Análisis Aeroelástico y de Sensibilidad del Fenómeno de Flameo en Puentes Soportados por Cables*. Doctoral Thesis from the Universidade da Coruña. Escuela Técnica Superior de Ingenieros de Caminos, Canales y Puertos.

[24] Nelson R. B. [1976] *Simplified calculation of eigenvector derivatives*. AIAA Journal, Vol. 14, pp. 1201–1205.

[25] Rogers L. C. [1970] *Derivatives of eigenvalues and eigenvectors*. AIAA Journal, Vol. 8, No. 5, pp. 943–944.

[26] Wang B. P. [1991] *Improve Aproximate methods for computing eigenvector derivatives in structural dynamics*. AIAA Journal, Vol. 29, No. 6, pp. 1018–1120.

[27] Ojalvo I. U. [1987] *Efficient computation of mode-shape derivatives for large dynamic systems*. AIAA Journal, Vol. 25, No. 10, pp. 1386–1390.

[28] Mills-Curran W. C. [1988] *Calculation of eigenvector derivatives for structures with repeat eigenvalues*. AIAA Journal, Vol. 26, No. 7, pp. 567–871.

[29] Dailey R. L. [1989] *Eigenvector derivatives with repeated eigenvalues*. AIAA Journal, Vol. 27, No. 4, pp. 486–491.

[30] Mosquera, A., Hernández, S. [2002] *Linear and Non Linear Sensitivity Analysis of Eigenvalue Problems*. 9th AIAA/ISSMO Symposium on Multidisciplinary Analysis and Optimization. Atlanta, Georgia, USA. 4–6 September 2002

[31] COSMOS/M [1997] *User's Manual, 2.0 version*. Structural Research & Analysis Corp. Santa Monica, CA.

[32] Dongarra, J. J., Moler C. B. [1977] *EISPACK – A package for solving matrix eigenvalue problems*. Argonne National Laboratory, Argonne, IL.

[33] Martin, R. S., Wilkinson J. W. [1968] *Reduction of the symmetric eigenproblem Ax = lBx and related problems to standard form*. Numerische Mathematik, Vol. 11, pp. 99–119.

[34] Smith, B. T., Boyle J. M., Dongarra J. J., Garbow B. S., Ikebe Y., Klema V. C., Moler C. B. [1976] *Matrix Eigensystem Routines – EISPACK Guide*. Springer, New York.

[35] Hanson, Richard J., Lehoucq R., Stolle J., Belmonte A. [1990] *Improved performance of certain matrix eigenvalue computations for the IMSL/MATH Library*. IMSL Technical Report 9007, IMSL, Houston.

Chapter 8

Sensitivity analysis of flutter response for cable-supported bridges

8.1 Introduction

Stable behaviour in long-span bridges under dynamic wind loads is obviously essential; no phenomenon related to instability, such as flutter, can appear in the final design. An important condition, therefore, is that the average critical wind speed acting on the bridge, at which point flutter is produced, exceeds a design value predetermined by the corresponding extramal analysis.

Through the sensitivity analysis of flutter, it is possible to obtain the variation in the critical wind speed and know if, on changing the design parameters, this speed rises above the one established as the allowable minimum.

This chapter will first refer back to the eigenvalue problems underpinning the calculation of critical flutter speed in a bridge. Thereafter, the sensitivity analysis associated with this condition will be developed. This will be followed by a detailed explanation of how to numerically solve the eigenvalue problems that arise. Lastly, there will be an overview of the computer code designed to apply the method aimed at long-span bridges; this code will be used in examples from the book.

8.2 Obtaining flutter speed

Once the integrated formulation developed in Chapter 3 for analyzing bridge flutter is worked through, a non-linear eigenvalue problem can be solved. This is formulated by means of

$$(\mathbf{A}-\mu\mathbf{I})\mathbf{w}_{\mu}e^{\mu t}=\mathbf{0} \tag{8.2.1}$$

In this expression, the matrix **A** is constructed from the rigidity and damping matrices for the structure, modified by the aeroelastic forces of the wind and reduced through modal analysis. Thus, working with vibration modes m, the result for **A** is an asymmetrical matrix whose dimensions are $2m \times 2m$. Generally, complex eigenvalues are obtained along these lines

$$\mu_j=\alpha_j+i\beta_j \qquad j=1,2,\ldots \quad j\le 2m \tag{8.2.2}$$

The real parts of these eigenvalues are related to the damping ξ_a for the response so that

$$\alpha_j = -\xi_{ja}\omega_j = -\xi_{ja}\sqrt{\alpha_j^2 + \beta_j^2} \tag{8.2.3}$$

and the imaginary part is the response frequency for the system

$$\beta_j = \omega_{aj} \tag{8.2.4}$$

As already explained in Chapter 3, the problem must be solved in an iterative way given that the matrix **A** depends on the aeroelastic matrices $\mathbf{K}_a$ and $\mathbf{C}_a$, which in turn rely on frequency $K = B\omega_a/U$. This frequency is unknown until the eigenvalue problem has been solved. Once the $2m$ eigenvalue problem verifying the equation below has been solved

$$\left[\mathbf{A}(\beta_j)-(\alpha_j+i\beta_j)\mathbf{I}\right]\mathbf{w}_\mu=\mathbf{0} \tag{8.2.5}$$

successive calculations are done. U is increased until, as the problem is solved, the real parts of the eigenvalues go from negative to positive. That is to say, the damping goes from positive to negative. The critical value U_f with which the phenomenon of flutter begins to appear, is reached through the solution in which an eigenvalue has a real part that is null

$$\alpha_j = 0 \tag{8.2.6}$$

Figures 8.2.1 and 8.2.2 show how the real and imaginary parts of the eigenvalues evolve for an example in which two vibration modes are used. By glancing at the graphical representation in Figure 8.2.1, it is possible to see how one of the real parts is annulled for a value of almost 90 m/s, which will be flutter speed U_f.

8.3 Sensitivity analysis of the flutter parameters in a bridge

Studying a bridge's stability under flutter leads to the solution of the eigenvector problem proposed in equation (8.2.1). When the real and imaginary part of the eigenvalue is decomposed $\mu_j = \alpha_j + i\beta_j$ the outcome is

$$\left[\mathbf{A}-(\alpha_j+i\beta_j)\mathbf{I}\right]\mathbf{w}_\mu=\mathbf{0} \tag{8.3.1}$$

The matrix **A** is constructed using the matrices for the structure's mass, damping, and rigidity as a basis. This matrix therefore depends on the others' structural properties, the geometry of the deck section, the area of the section, and inertias under bending or torsion. In other words, **A** is a function of possible design variables within a structural optimization problem. In the formation of **A,** the damping and aeroelastic rigidity matrices also appear, and these depend on the wind incidence speed U and the reduced frequency of the response K. By calling the design variable vector **x**, these dependencies can be described as

$$\mathbf{A}(\mathbf{x},U,K) \tag{8.3.2}$$

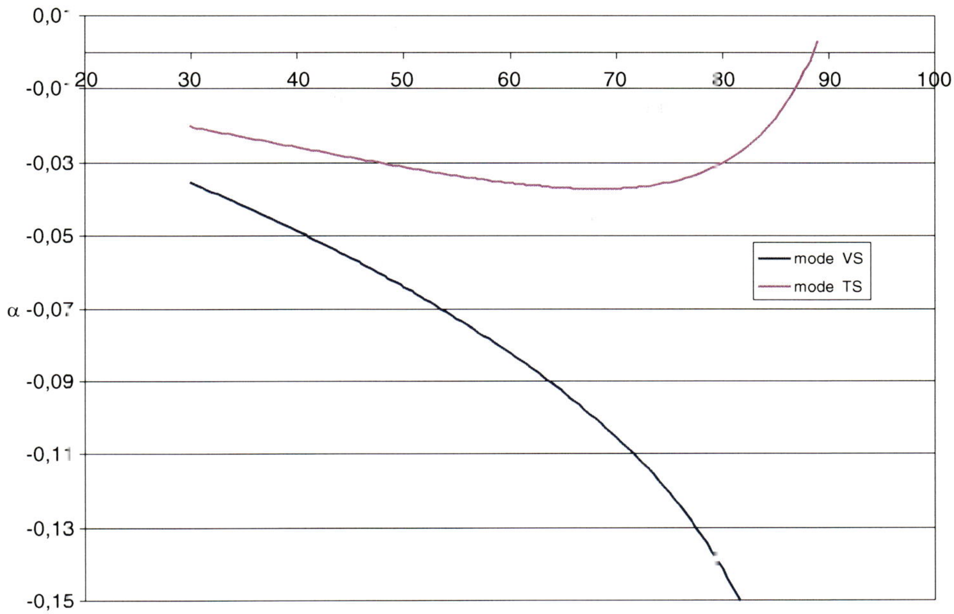

Figure 8.2.1: Evolution of the real part α versus U.

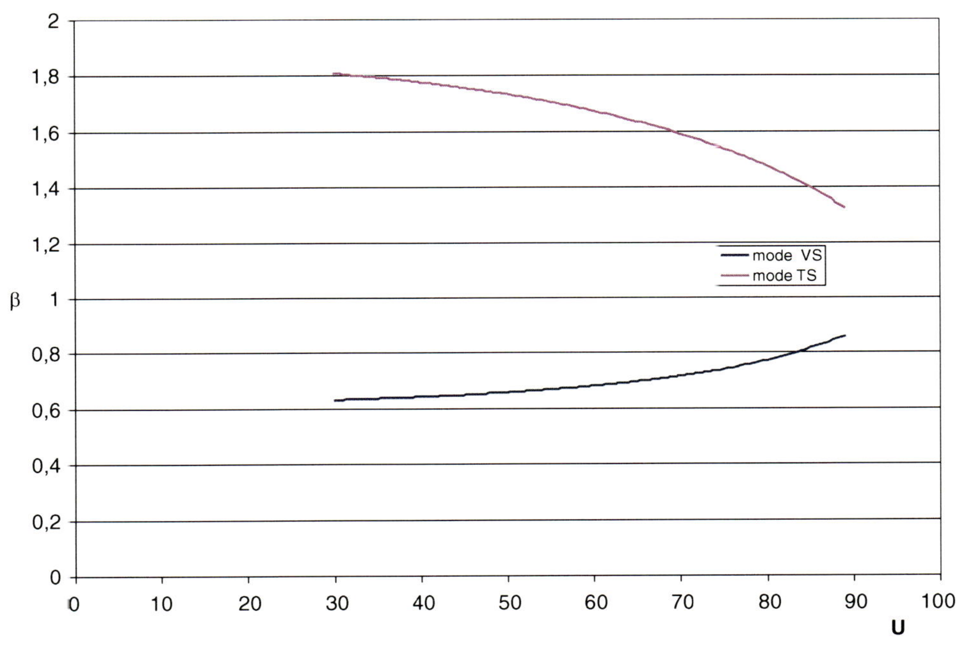

Figure 8.2.2: Evolution of the imaginary part β versus U.

In the critical situation, when flutter instability starts to appear, the real part of some of the eigenvalues is null. In this state, the wind incidence speed U_f and the reduced frequency K_f are related through

$$K_f = \frac{B\beta_j}{U_f} \tag{8.3.3}$$

β_j being the imaginary part of the eigenvalue whose real part is null ($\alpha_j = 0$). The equation that is being fulfilled in the situation of initial flutter is, as a result

$$\mathbf{A}\left(\mathbf{x}, U_f, K_f\right)\mathbf{w}_\mu + \frac{K_f U_f}{B} i\mathbf{I}\mathbf{w}_\mu = \mathbf{0} \tag{8.3.4}$$

Differentiating the state equation (8.3.4) (Jurado[1,2]), it can be written

$$d\left[\mathbf{A}\left(\mathbf{x}, U_f, K_f\right)\mathbf{w}_\mu + \frac{K_f U_f}{B} i\mathbf{I}\mathbf{w}_\mu\right] = \mathbf{0} \tag{8.3.5}$$

and for each design variable x, it become the following condition:

$$\begin{gathered}\mathbf{v}_\mu^T\left(\frac{\partial \mathbf{A}}{\partial x} + \frac{\partial \mathbf{A}}{\partial U_f}\frac{dU_f}{dx} + \frac{\partial \mathbf{A}}{\partial K_f}\frac{dK_f}{dx}\right)\mathbf{w}_\mu + \mathbf{v}_\mu^T\left(\mathbf{A} + \frac{K_f U_f}{B} i\mathbf{I}\right)\frac{\partial \mathbf{w}_\mu}{\partial x} + \\ + \mathbf{v}_\mu^T\left[\frac{i}{B}\left(\frac{\partial K_f}{\partial x}U_f + K_f\frac{\partial U_f}{\partial x}\right)\mathbf{I}\right]\mathbf{w}_\mu = \mathbf{0}\end{gathered} \tag{8.3.6}$$

The way (8.3.4) in which an eigenvector for the right $\mathbf{w}_\mu$ is also valid for the left. The expression is

$$\mathbf{v}_\mu^T\mathbf{A} + \frac{K_f U_f}{B} i\mathbf{v}_\mu^T\mathbf{I} = \mathbf{0} \tag{8.3.7}$$

Multiplying (8.3.6) by the eigenvector for the left $\mathbf{v}_\mu^T$, the result is

$$\begin{gathered}\mathbf{v}_\mu^T\left(\frac{\partial \mathbf{A}}{\partial x} + \frac{\partial \mathbf{A}}{\partial U_f}\frac{dU_f}{dx} + \frac{\partial \mathbf{A}}{\partial K_f}\frac{dK_f}{dx}\right)\mathbf{w}_\mu + \mathbf{v}_\mu^T\left(\mathbf{A} + \frac{K_f U_f}{B} i\mathbf{I}\right)\frac{\partial \mathbf{w}_\mu}{\partial x} + \\ + \mathbf{v}_\mu^T\left[\frac{i}{B}\left(\frac{\partial K_f}{\partial x}U_f + K_f\frac{\partial U_f}{\partial x}\right)\mathbf{I}\right]\mathbf{w}_\mu = \mathbf{0}\end{gathered} \tag{8.3.8}$$

developing this expression

$$\begin{gathered}\mathbf{v}_\mu^T\frac{\partial \mathbf{A}}{\partial x}\mathbf{w}_\mu + \mathbf{v}_\mu^T\frac{\partial \mathbf{A}}{\partial U_f}\mathbf{w}_\mu\frac{dU_{cr}}{dx} + \mathbf{v}_\mu^T\frac{\partial \mathbf{A}}{\partial K_f}\mathbf{w}_\mu\frac{\partial K_f}{\partial x} + \\ + \frac{i}{B}\left(\frac{\partial K_f}{\partial x}U_f + K_f\frac{\partial U_f}{\partial x}\right)\mathbf{v}_\mu^T\mathbf{I}\mathbf{w}_\mu = 0\end{gathered} \tag{8.3.9}$$

and defining the following complex numbers:

$$h_{Ax} = \mathbf{v}_{\mu}^{T} \frac{\partial \mathbf{A}}{\partial x} \mathbf{w}_{\mu}, \qquad h_{AU} = \mathbf{v}_{\mu}^{T} \frac{\partial \mathbf{A}}{\partial U_f} \mathbf{w}_{\mu}, \qquad h_{AK} = \mathbf{v}_{\mu}^{T} \frac{\partial \mathbf{A}}{\partial K_f} \mathbf{w}_{\mu} \tag{8.3.10}$$

(8.3.9) the outcome is

$$h_{Ax} + \frac{\partial U_f}{\partial x}\left[h_{AU} + \frac{iK_f}{B}\mathbf{v}_{\mu}^{T}\mathbf{I}\mathbf{w}_{\mu}\right] + \frac{\partial K_f}{\partial x}\left[h_{AK} + \frac{iU_f}{B}\mathbf{v}_{\mu}^{T}\mathbf{I}\mathbf{w}_{\mu}\right] = 0 \tag{8.3.11}$$

If two more complex numbers are defined

$$\begin{aligned} g_U &= h_{AU} + \frac{iK_f}{B}\mathbf{v}_{\mu}^{T}\mathbf{I}\mathbf{w}_{\mu} \\ g_K &= h_{AK} + \frac{iU_f}{B}\mathbf{v}_{\mu}^{T}\mathbf{I}\mathbf{w}_{\mu} \end{aligned} \tag{8.3.12}$$

equation (8.3.11) can be written in a more condensed form

$$g_U \frac{\partial U_f}{\partial x} + g_K \frac{\partial K_f}{\partial x} = -h_{Ax} \tag{8.3.13}$$

whereby $\partial U_f / \partial x$ and $\partial K_f / \partial x$ are real numbers. Multiplying equation (8.3.13) by the complex conjugate of g_U, which can be denoted as $\bar{g}_U$, this is obtained

$$\left|g_U\right|^2 \frac{\partial U_f}{\partial x} + \bar{g}_U g_K \frac{\partial K_f}{\partial x} = -\bar{g}_U h_{Ax} \tag{8.3.14}$$

and equalizing the imaginary parts of this equation, it is possible to find the value of $\partial K_f/\partial x$

$$\frac{\partial K_f}{\partial x} = \frac{-\mathrm{Im}\left(\bar{g}_U h_{Ax}\right)}{\mathrm{Im}\left(\bar{g}_U g_K\right)} \tag{8.3.15}$$

Similarly, multiplying equation (8.3.13) by $\bar{g}_K$, the complex conjugate of g_K, the result is the derivative for the critical speed in terms of the design variable under consideration

$$\frac{\partial U_f}{\partial x} = \frac{-\mathrm{Im}\left(\bar{g}_K h_{Ax}\right)}{\mathrm{Im}\left(\bar{g}_K g_U\right)} \tag{8.3.16}$$

Expressions (8.3.15) and (8.3.16) determine the variation, with respect to design variable x, of the two parameters defining the phenomenon of flutter instability in the bridge: that is, the average wind incidence speed over the bridge deck U_f and the reduced response frequency K_f.

In the following sections, the design variables are chosen and the analytical calculation for derivatives $\partial A/\partial x$, $\partial A/\partial U_f$ and $\partial A/\partial K_f$ is described in detail. A numerical method for solving the eigenvalue problem is explained; this problem is present when the flutter speed is obtained and when the sensitivity analysis is done.

8.3.1 Design variables x

There is a multitude of design variables that come into play when a structure as complex as a cable-supported bridge, either a suspension or cable-stayed one, is being defined. Affecting the bridge's dynamic behaviour and, as a result, its flutter speed is a vast array of mechanical properties; geometrical shapes as well as dimensions; and unions in any element within the bridge: the deck, towers, cables, supports or foundation. This work aims to establish the bases for carrying out the optimization of long-span bridges under an aeroelastic regime. To do so, the starting point is to verify how to undertake an efficient sensitivity analysis in terms of the flutter speed U_f. Three variables have been chosen when applying the method developed here:

- The inertia moment under vertical bending for the deck, or $x = I_y$.
- The inertia moment under lateral bending for the deck, or $x = I_z$.
- The inertia moment under torsion for the deck, or $x = J$.

These mechanical parameters are essential if the structure is to have the necessary rigidity and be able to withstand both the permanent loads and estimated overloads. For this reason, these values are subject to variation within the design process. Moreover, they are properties related to the deck, an element of the bridge in which the phenomenon of flutter-induced aeroelastic instability is found.

8.3.2 Calculating $\partial A/\partial x$

The matrix **A**, with a dimension of $2m$, is constructed from the reduced damping and rigidity matrices, taking the form

$$\mathbf{A} = \begin{pmatrix} -\mathbf{C}_R & -\mathbf{K}_R \\ \mathbf{I} & \mathbf{0} \end{pmatrix} \tag{8.3.17}$$

Deriving this in terms of x the following is reached:

$$\frac{\partial \mathbf{A}}{\partial x} = \begin{pmatrix} -\dfrac{\partial \mathbf{C}_R}{\partial x} & -\dfrac{\partial \mathbf{K}_R}{\partial x} \\ \mathbf{0} & \mathbf{0} \end{pmatrix} \tag{8.3.18}$$

To calculate the derivatives of the matrices $\mathbf{C}_R$ and $\mathbf{K}_R$, it is necessary to remember how they are constructed

$$\begin{aligned} \mathbf{C}_R &= \mathbf{C}_{e,R} - \mathbf{C}_{a,R} = \mathbf{\Phi}^T\mathbf{C}\mathbf{\Phi} - \mathbf{\Phi}^T\mathbf{C}_a\mathbf{\Phi} \\ \mathbf{K}_R &= \mathbf{K}_{e,R} - \mathbf{K}_{a,R} = \mathbf{\Phi}^T\mathbf{K}\mathbf{\Phi} - \mathbf{\Phi}^T\mathbf{K}_a\mathbf{\Phi} \end{aligned} \tag{8.3.19}$$

where **C** and **K** are, respectively, the damping and rigidity matrices of the structure; $\mathbf{C}_a$ and $\mathbf{K}_a$, the aeroelastic matrices; $\mathbf{C}_{e,R}$ and $\mathbf{K}_{e,R}$, the reduced matrices of **C** and **K**; and $\mathbf{C}_{a,R}$ and $\mathbf{K}_{a,R}$, the reduced aeroelastic matrices. Matrices $\mathbf{C}_{e,R}$ and $\mathbf{K}_{e,R}$ are diagonal, with a dimension of m (number of modes); there is an option to write its terms according to natural frequencies ω_i and the damping ξ_i for each mode.

$$\begin{aligned} C_{e,R,_{ii}} &= 2\xi_i \omega_i \\ K_{e,R,_{ii}} &= \omega_i^2 \end{aligned} \tag{8.3.20}$$

The structural damping for each mode is normally considered constant. In cable-supported bridges, it usually takes values between 1% and 5% for logarithmical decreases $\delta = 2\pi\xi / \sqrt{1-\xi^2}$. Therefore, $\partial\xi_i / \partial x$ is zero and does not have any influence over the derivative in expressions (8.3.20)

$$\begin{aligned} \partial C_{e,R,_{ii}} \big/ \partial x &= 2\xi_i \frac{\partial \omega_i}{\partial x} \\ \partial K_{e,R,_{ii}} \big/ \partial x &= 2\omega_i \frac{\partial \omega_i}{\partial x} \end{aligned} \tag{8.3.21}$$

Deriving the terms corresponding with the aeroelastic matrices, the following expressions are obtained:

$$\begin{aligned} \frac{\partial(\mathbf{C}_{a,R})}{\partial x} &= \mathbf{\Phi}^T \frac{\partial \mathbf{C}_a}{\partial x}\mathbf{\Phi} + \frac{\partial \mathbf{\Phi}^T}{\partial x}\mathbf{C}_a\mathbf{\Phi} + \mathbf{\Phi}^T\mathbf{C}_a\frac{\partial \mathbf{\Phi}}{\partial x} \\ \frac{\partial(\mathbf{K}_{a,R})}{\partial x} &= \mathbf{\Phi}^T \frac{\partial \mathbf{K}_a}{\partial x}\mathbf{\Phi} + \frac{\partial \mathbf{\Phi}^T}{\partial x}\mathbf{K}_a\mathbf{\Phi} + \mathbf{\Phi}^T\mathbf{K}_a\frac{\partial \mathbf{\Phi}}{\partial x} \end{aligned} \tag{8.3.22}$$

The design variables under consideration (I_y, I_z, J) do not explicitly intervene in the formation of the aeroelastic matrices $\mathbf{C}_a$ and $\mathbf{K}_a$. Thus, the first sums of (8.3.22) are null. The difficulty in attempting to calculate $\partial\mathbf{A}/\partial x$ stems from obtaining the derivative for the natural vibration frequencies $\partial\omega_i/\partial x$, appearing in (8.3.21), and for vibration modes $\partial\mathbf{\Phi}/\partial x$, intervening in (8.3.22).

The sensitivity analysis of a dynamic problem in linear theory have been skilfully handled by Haug *et al.*[5], among others. For instance, as already seen in Chapter 7, the Nelson[3] method makes it possible to obtain the sensitivity of each mode and its associated frequency simultaneously by solving the equation system

$$\begin{bmatrix} \mathbf{K} - \omega_i^2\mathbf{M} & -\mathbf{M}\phi \\ -\phi^T\mathbf{M} & 0 \end{bmatrix} \begin{Bmatrix} d\phi/dx \\ d\omega_i^2/dx \end{Bmatrix} = \begin{Bmatrix} -\left(\dfrac{d\mathbf{K}}{dx} - \omega_i^2\dfrac{d\mathbf{M}}{dx}\right)\phi \\ \dfrac{1}{2}\phi^T\dfrac{d\mathbf{M}}{dx}\phi \end{Bmatrix} \tag{8.3.23}$$

Structural optimization programs, like GENESIS[4], are commercially available. These provide the sensitivities for the frequencies and vibration modes. However, they do so by solving the problem in a linear fashion, rather than through second-order theory, which is the technique used for these kinds of bridges. To carry out an analytical calculation for the sensitivity within the proposed dynamic problem – that is, when obtaining frequencies and modes – this equation is used

$$\left[\mathbf{K} + \mathbf{K}_G - \omega^2\mathbf{M}\right]\phi = \mathbf{0} \tag{8.3.24}$$

This problem with the geometrical stiffness matrix $\mathbf{K}_G$ had previously been unpublished but the solution can be found in Chapter 7.

Once the sensitivities $\partial\omega_i/\partial x$ y $\partial\phi/\partial x$ have been obtained analytically, they are substituted in expressions (8.3.21) and (8.3.22) to calculate $\partial\mathbf{A}/\partial x$.

8.3.3 Calculating $\partial A/\partial U_f$

Developing this derivative from (8.3.17), one has

$$\frac{\partial\mathbf{A}}{\partial U_f}=\begin{pmatrix}-\dfrac{\partial\mathbf{C}_R}{\partial U_f} & -\dfrac{\partial\mathbf{K}_R}{\partial U_f}\\ \mathbf{0} & \mathbf{0}\end{pmatrix} \tag{8.3.25}$$

Again, it is decomposed

$$\begin{aligned}\mathbf{C}_R&=\mathbf{C}_{e,R}-\mathbf{C}_{a,R}=\mathbf{\Phi}^T\mathbf{C}\mathbf{\Phi}-\mathbf{\Phi}^T\mathbf{C}_a\mathbf{\Phi}\\ \mathbf{K}_R&=\mathbf{K}_{e,R}-\mathbf{K}_{a,R}=\mathbf{\Phi}^T\mathbf{K}\mathbf{\Phi}-\mathbf{\Phi}^T\mathbf{K}_a\mathbf{\Phi}\end{aligned} \tag{8.3.26}$$

Here the terms for the reduced damping and structural rigidity matrices

$$\begin{aligned}C_{e,R,_{ii}}&=2\xi_i\omega_i\\ K_{e,R,_{ii}}&=\omega_i^2\end{aligned} \tag{8.3.27}$$

do not explicitly depend on flutter speed U_f. Nevertheless, they are in the aeroelastic matrices $\mathbf{K}_a$ and $\mathbf{C}_a$. Chapter 4 has already dealt with how these matrices are built by assembling matrices $\mathbf{K}_{ai}$ and $\mathbf{C}_{ai}$ in each node of the deck, which are like this

$$\mathbf{K}_{ai}=\frac{l_e}{2}\begin{pmatrix}0&0&0&0&0&0\\ 0&pK^2P_4^*&pK^2P_6^*&pBK^2P_3^*&0&0\\ 0&pK^2H_6^*&pK^2H_4^*&pBK^2H_3^*&0&0\\ 0&pBK^2A_6^*&pBK^2A_4^*&pB^2K^2A_3^*&0&0\\ 0&0&0&0&0&0\\ 0&0&0&0&0&0\end{pmatrix} \tag{8.3.28}$$

$$\mathbf{C}_{ai}=\frac{l_e}{2}\begin{pmatrix}0&0&0&0&0&0\\ 0&pBKP_1^*/U&pBKP_5^*/U&pB^2KP_2^*/U&0&0\\ 0&pBKH_5^*/U&pBKH_1^*/U&pB^2KH_2^*/U&0&0\\ 0&pB^2KA_5^*/U&pB^2KA_1^*/U&pB^3KA_2^*/U&0&0\\ 0&0&0&0&0&0\\ 0&0&0&0&0&0\end{pmatrix} \tag{8.3.29}$$

where $p=0.5\rho U^2$ is the dynamic wind pressure. With (8.3.28) and (8.3.29), it is easy to calculate the derivatives for the aeroelastic matrices in terms of U_f, leaving

$$\frac{\partial \mathbf{K}_a}{\partial U_f} = \frac{2\mathbf{K}_a}{U_f}$$
$$\frac{\partial \mathbf{C}_a}{\partial U_f} = \frac{\mathbf{C}_a}{U_f} \qquad (8.3.30)$$

As the vibration modes do not rely on U_f, the derivatives for the reduced matrices $\mathbf{K}_R$ and $\mathbf{C}_R$ become

$$\frac{\partial \mathbf{K}_R}{\partial U_f} = -\mathbf{\Phi}^T \frac{2\mathbf{K}_a}{U_f}\mathbf{\Phi}$$
$$\frac{\partial \mathbf{C}_R}{\partial U_f} = -\mathbf{\Phi}^T \frac{\mathbf{C}_a}{U_f}\mathbf{\Phi} \qquad (8.3.31)$$

It is enough to substitute (8.3.31) in (8.3.25) to calculate $\partial \mathbf{A}/\partial U_f$.

8.3.4 Calculating $\partial A/\partial K_f$

Operating in a similar way to what has been done above, these results are reached

$$\frac{\partial \mathbf{A}}{\partial K_f} = \begin{pmatrix} -\dfrac{\partial \mathbf{C}_R}{\partial K_f} & -\dfrac{\partial \mathbf{K}_R}{\partial K_f} \\ \mathbf{0} & \mathbf{0} \end{pmatrix} = \begin{pmatrix} \mathbf{\Phi}^T \dfrac{\partial \mathbf{C}_a}{\partial K_f}\mathbf{\Phi} & \mathbf{\Phi}^T \dfrac{\partial \mathbf{C}_a}{\partial K_f}\mathbf{\Phi} \\ \mathbf{0} & \mathbf{0} \end{pmatrix} \qquad (8.3.32)$$

where it is understood that the following derivatives are null:

$$\frac{\partial \omega_i}{\partial K_f} = 0; \quad \frac{\partial \phi}{\partial K_f} = 0 \qquad (8.3.33)$$

Looking at expressions (8.3.28) and (8.3.29), it is noticeable that the aeroelastic matrices explicitly depend on the reduced frequency K. Moreover, all the flutter derivatives (H_i^*, A_i^*, P_i^*) are also functions of K. After the expressions related to K are derived, these matrices appear

$$\frac{\partial \mathbf{K}_{ai}}{\partial K} = \frac{l_e}{2}\begin{pmatrix} 0 & 0 & 0 & 0 & 0 & 0 \\ 0 & 2pKP_4^* + pK^2\dfrac{\partial P_4^*}{\partial K} & 2pKP_6^* + pK^2\dfrac{\partial P_6^*}{\partial K} & 2pBKP_3^* + pBK^2\dfrac{\partial P_3^*}{\partial K} & 0 & 0 \\ 0 & 2pKH_6^* + pK^2\dfrac{\partial H_6^*}{\partial K} & 2pKH_4^* + pK^2\dfrac{\partial H_4^*}{\partial K} & 2pBKH_3^* + pBK^2\dfrac{\partial H_3^*}{\partial K} & 0 & 0 \\ 0 & 2pBKA_6^* + pBK^2\dfrac{\partial A_6^*}{\partial K} & 2pBKA_4^* + pBK^2\dfrac{\partial A_4^*}{\partial K} & 2pBKA_3^* + pBK^2\dfrac{\partial A_3^*}{\partial K} & 0 & 0 \\ 0 & 0 & 0 & 0 & 0 & 0 \\ 0 & 0 & 0 & 0 & 0 & 0 \end{pmatrix}$$

(8.3.34)

isolated along the diagonal or are the eigenvalues of a 2 × 2 sub-matrix over the diagonal. In this method, the following relations are achieved:

$$\begin{aligned} \mathbf{Q}_s\left(\mathbf{A}_s - k_s\mathbf{I}\right) &= \mathbf{R}_s \\ \mathbf{A}_{s+1} &= \mathbf{R}_s\mathbf{Q}_s^T + k_s\mathbf{I} = \mathbf{Q}_s\mathbf{A}_s\mathbf{Q}_s^T \end{aligned} \tag{8.4.5}$$

where **Q** is orthogonal and **R** is an upper triangular. The algorithm will be done in two-by-two steps and it is particularly advisable to take what are two complex conjugate numbers for what are known as cut values, or k_s and k_{s+1}. It must be remembered that eigenvalues for an asymmetrical matrix may be complex. The following matrix is obtained:

$$\mathbf{A}_{s+2} = \mathbf{Q}_{s+1}\mathbf{Q}_s\mathbf{A}_s\mathbf{Q}_s^T\mathbf{Q}_{s+1}^T \tag{8.4.6}$$

Moreover, these relations should be achieved:

$$\mathbf{A}_s - k_s\mathbf{I} = \mathbf{Q}_s^T\mathbf{R}_s \tag{8.4.7}$$

$$\mathbf{A}_{s+1} = \mathbf{Q}_s\mathbf{A}_s\mathbf{Q}_s^T \tag{8.4.8}$$

$$\mathbf{A}_{s+1} - k_{s+1}\mathbf{I} = \mathbf{Q}_{s+1}^T\mathbf{R}_{s+1} \tag{8.4.9}$$

Using equation (8.4.8), it is possible to write equation (8.4.9) in this way

$$\mathbf{A}_s - k_{s+1}\mathbf{I} = \mathbf{Q}_s^T\mathbf{Q}_{s+1}^T\mathbf{R}_{s+1}\mathbf{Q}_s \tag{8.4.10}$$

If one defines

$$\mathbf{M} = \left(\mathbf{A}_s - k_{s+1}\mathbf{I}\right)\left(\mathbf{A}_s - k_s\mathbf{I}\right) \tag{8.4.11}$$

equations (8.4.7) and (8.4.10) give

$$\mathbf{R} = \mathbf{QM} \tag{8.4.12}$$

where

$$\begin{aligned} \mathbf{Q} &= \mathbf{Q}_{s+1}\mathbf{Q}_s \\ \mathbf{R} &= \mathbf{R}_{s+1}\mathbf{R}_s \end{aligned} \tag{8.4.13}$$

Equation (8.4.6) can be written (8.4.13) as

$$\mathbf{A}_s\mathbf{Q}^T = \mathbf{Q}^T\mathbf{A}_{s+2} \tag{8.4.14}$$

To calculate the orthogonal matrix **Q** that makes the real matrix **M** triangular, a sequence of transformations are carried out through Householder $\mathbf{P}_1$ matrices (acting on the first column), $\mathbf{P}_2$ (acting on the second column), ... , $\mathbf{P}_{n-1}$. In this way,

$\mathbf{Q} = \mathbf{P}_{n-1} \ldots \mathbf{P}_2\mathbf{P}_1$, and the first row of **Q** is also the first one of $\mathbf{P}_1$, given that $\mathbf{P}_i$ is an identity matrix of $(i-1) \times (i-1)$ in its upper, left-hand corner.

$$\frac{\partial \mathbf{K}_a}{\partial U_f} = \frac{2\mathbf{K}_a}{U_f}$$
$$\frac{\partial \mathbf{C}_a}{\partial U_f} = \frac{\mathbf{C}_a}{U_f} \tag{8.3.30}$$

As the vibration modes do not rely on U_f, the derivatives for the reduced matrices $\mathbf{K}_R$ and $\mathbf{C}_R$ become

$$\frac{\partial \mathbf{K}_R}{\partial U_f} = -\mathbf{\Phi}^T \frac{2\mathbf{K}_a}{U_f}\mathbf{\Phi}$$
$$\frac{\partial \mathbf{C}_R}{\partial U_f} = -\mathbf{\Phi}^T \frac{\mathbf{C}_a}{U_f}\mathbf{\Phi} \tag{8.3.31}$$

It is enough to substitute (8.3.31) in (8.3.25) to calculate $\partial \mathbf{A}/\partial U_f$.

8.3.4 Calculating $\partial A/\partial K_f$

Operating in a similar way to what has been done above, these results are reached

$$\frac{\partial \mathbf{A}}{\partial K_f} = \begin{pmatrix} -\dfrac{\partial \mathbf{C}_R}{\partial K_f} & -\dfrac{\partial \mathbf{K}_R}{\partial K_f} \\ \mathbf{0} & \mathbf{0} \end{pmatrix} = \begin{pmatrix} \mathbf{\Phi}^T \dfrac{\partial \mathbf{C}_a}{\partial K_f}\mathbf{\Phi} & \mathbf{\Phi}^T \dfrac{\partial \mathbf{C}_a}{\partial K_f}\mathbf{\Phi} \\ \mathbf{0} & \mathbf{0} \end{pmatrix} \tag{8.3.32}$$

where it is understood that the following derivatives are null:

$$\frac{\partial \omega_i}{\partial K_f} = 0; \quad \frac{\partial \phi}{\partial K_f} = 0 \tag{8.3.33}$$

Looking at expressions (8.3.28) and (8.3.29), it is noticeable that the aeroelastic matrices explicitly depend on the reduced frequency K. Moreover, all the flutter derivatives (H_i^*, A_i^*, P_i^*) are also functions of K. After the expressions related to K are derived, these matrices appear

$$\frac{\partial \mathbf{K}_{ai}}{\partial K} = \frac{l_e}{2}\begin{pmatrix} 0 & 0 & 0 & 0 & 0 & 0 \\ 0 & 2pKP_4^* + pK^2\dfrac{\partial P_4^*}{\partial K} & 2pKP_6^* + pK^2\dfrac{\partial P_6^*}{\partial K} & 2pBKP_3^* + pBK^2\dfrac{\partial P_3^*}{\partial K} & 0 & 0 \\ 0 & 2pKH_6^* + pK^2\dfrac{\partial H_6^*}{\partial K} & 2pKH_4^* + pK^2\dfrac{\partial H_4^*}{\partial K} & 2pBKH_3^* + pBK^2\dfrac{\partial H_3^*}{\partial K} & 0 & 0 \\ 0 & 2pBKA_6^* + pBK^2\dfrac{\partial A_6^*}{\partial K} & 2pBKA_4^* + pBK^2\dfrac{\partial A_4^*}{\partial K} & 2pBKA_3^* + pBK^2\dfrac{\partial A_3^*}{\partial K} & 0 & 0 \\ 0 & 0 & 0 & 0 & 0 & 0 \\ 0 & 0 & 0 & 0 & 0 & 0 \end{pmatrix} \tag{8.3.34}$$

$$\frac{\partial \mathbf{C}_{ai}}{\partial K}=\frac{l_e}{2}\begin{pmatrix} 0 & 0 & 0 & 0 & 0 & 0 \\ 0 & \left(pBP_1^* + pBK\frac{\partial P_1^*}{\partial K}\right)\Big/U & \left(pBP_5^* + pBK\frac{\partial P_5^*}{\partial K}\right)\Big/U & \left(pB^2P_2^* + pB^2K\frac{\partial P_2^*}{\partial K}\right)\Big/U & 0 & 0 \\ 0 & \left(pBH_5^* + pBK\frac{\partial H_5^*}{\partial K}\right)\Big/U & \left(pBH_1^* + pBK\frac{\partial H_1^*}{\partial K}\right)\Big/U & \left(pB^2H_2^* + pB^2K\frac{\partial H_2^*}{\partial K}\right)\Big/U & 0 & 0 \\ 0 & \left(pB^2A_5^* + pB^2K\frac{\partial A_5^*}{\partial K}\right)\Big/U & \left(pB^2A_1^* + pB^2K\frac{\partial A_1^*}{\partial K}\right)\Big/U & \left(pB^3A_2^* + pB^3K\frac{\partial A_2^*}{\partial K}\right)\Big/U & 0 & 0 \\ 0 & 0 & 0 & 0 & 0 & 0 \\ 0 & 0 & 0 & 0 & 0 & 0 \end{pmatrix} \tag{8.3.35}$$

As previously mentioned in Chapter 3, the flutter derivatives are experimentally obtained by running tests on sections of the deck. Consequently, mathematical expressions are unknown for the functions relating each derivative with the reduced frequency. This means that the only way to calculate the flutter derivatives in terms of the reduced frequency ($\partial H_i^*/\partial K$, $\partial A_i^*/\partial K$, $\partial P_i^*/\partial K$, i = 1,...,6) is by using the finite differential method.

After constructing the matrices expressed in (8.3.34) and (8.3.35), they are assembled for all the deck nodes, obtaining $\partial \mathbf{K}_a/\partial K_f$ and $\partial \mathbf{C}_a/\partial K_f$. Substituting these two matrices in (8.3.32), it is possible to find the derivative that is sought $\partial \mathbf{A}/\partial K_f$.

Once the derivatives $\partial \mathbf{A}/\partial x$, $\partial \mathbf{A}/\partial U_f$ and $\partial \mathbf{A}/\partial K_f$ are calculated, the indicated steps, simple operations with complex numbers, are carried out from (8.3.10) to (8.3.16) until one obtains the derivatives for the parameters U_f and K_f with respect to the chosen design variables x (I_y, I_z, J)

$$\frac{\partial U_f}{\partial x} \text{ and } \frac{\partial K_f}{\partial x} \tag{8.3.36}$$

8.4 Solving the eigenvalue problem

This section describes the numerical method used to solve the eigenvalue problem that appears when obtaining the flutter speed for the bridge. In sensitivity analysis, it is necessary to get the eigenvalues on both the right and the left for the matrix **A**. Therefore, the numerical method should provide both the eigenvalues and eigenvectors.

The eigenvalue and vector problem that must be solved in a flutter study is not straightforward because matrix **A** from the system is not symmetrical

$$\mathbf{A}=\begin{pmatrix} -\mathbf{C}_R & -\mathbf{K}_R \\ \mathbf{I} & \mathbf{0} \end{pmatrix} \tag{8.4.1}$$

The strategy worth following in this case is to convert it into one with a simpler structure: a Hessenberg matrix, which has the same eigenvalues as matrix **A**. This new one is then used in an iterative method so that the problem can be calculated. If the completed transformations have been stored, it is easy to calculate the corresponding eigenvectors using the eigenvalues as a basis.

A superior Hessenberg matrix has zeros in all the terms situated below the main diagonal, with the exception of the first sub-diagonal. For instance, in the case of a 6 × 6 matrix, the non-null elements are

$$\begin{pmatrix} \bullet & \bullet & \bullet & \bullet & \bullet & \bullet \\ \bullet & \bullet & \bullet & \bullet & \bullet & \bullet \\ & \bullet & \bullet & \bullet & \bullet & \bullet \\ & & \bullet & \bullet & \bullet & \bullet \\ & & & \bullet & \bullet & \bullet \\ & & & & \bullet & \bullet \end{pmatrix} \tag{8.4.2}$$

To reach this structure in the matrix, a series of Householder transformations are carried out, each one of which introduces zeros into the elements of the corresponding column. Householder transformation matrix **P** has the following form:

$$\mathbf{P} = \mathbf{I} - 2\mathbf{w}\mathbf{w}^T \tag{8.4.3}$$

where **w** is any vector real that leads to $|\mathbf{w}|^2 = 1$. Any matrix constructed in this way is orthogonal, that is, $\mathbf{P}^T = \mathbf{P}^{-1}$. When Householder transformations are carried out, one continues to get zeros in the desired locations within the shape

$$\mathbf{PA} = \left(\begin{array}{c|cccc} 1 & 0 & 0 & \cdots & 0 \\ - & - & - & - & - \\ 0 & & & & \\ 0 & & {}^{(n-1)}\mathbf{P}_1 & & \\ \vdots & & & & \\ 0 & & & & \end{array}\right) \left(\begin{array}{c|cccc} a_{11} & a_{12} & a_{13} & \cdots & a_{1n} \\ - & - & - & - & - \\ a_{21} & & & & \\ a_{31} & & & & \\ \vdots & & & & \\ a_{n1} & & & & \end{array}\right) = \left(\begin{array}{c|cccc} a_{11} & a_{12} & a_{13} & \cdots & a_{1n} \\ - & - & - & - & - \\ k & & & & \\ 0 & & & & \\ \vdots & & & & \\ 0 & & & & \end{array}\right) \tag{8.4.4}$$

A more efficient alternative is to carry out a process analogous to Gaussian elimination with partial-pivoting. The original matrix $\mathbf{A} = \mathbf{A}_1$ after the first steps $r - 1$ will be converted into $\mathbf{A}_r$, a superior Hessenberg matrix in the first $r - 1$ columns. Thus, step r will consist of the following:

- Finding the element in the column r under the diagonal. It may be supposed that this element is in row r'.
- Exchanging rows r' and $r + 1$. This is what is referred to as the pivot and should also be carried out with columns r' and $r + 1$.
- For $i = r + 2, r + 3, + \ldots + N$, calculating factor $n_{i,r+1}$ $a_{i,r}/a_{r+1,r}$, subtracting row $r + 1$ and row $i n_{i,r+1}$ times, and adding column i to column $r + 1$ $n_{i,r+1}$ times.
- In all, $N - 2$ steps are needed, N being the dimension of **A**.

A QR algorithm makes it possible to obtain through iteration the eigenvalues for Hessenberg matrix $\mathbf{A} = \mathbf{A}_1$, with the number of operations in the order of $O(n^2)$ per iteration. Therefore, when $s \to \infty$, $\mathbf{A}_s$ converges to a shape in which the eigenvalues are either isolated along the diagonal or are the eigenvalues of a 2 × 2 sub-matrix over the diagonal. In this method, the following relations are achieved:

$$\begin{aligned}\mathbf{Q}_s(\mathbf{A}_s - k_s\mathbf{I}) &= \mathbf{R}_s \\ \mathbf{A}_{s+1} &= \mathbf{R}_s\mathbf{Q}_s^T + k_s\mathbf{I} = \mathbf{Q}_s\mathbf{A}_s\mathbf{Q}_s^T\end{aligned} \quad (8.4.5)$$

where **Q** is orthogonal and **R** is an upper triangular. The algorithm will be done in two-by-two steps and it is particularly advisable to take what are two complex conjugate numbers for what are known as cut values, or k_s and k_{s+1}. It must be remembered that eigenvalues for an asymmetrical matrix may be complex. The following matrix is obtained:

$$\mathbf{A}_{s+2} = \mathbf{Q}_{s+1}\mathbf{Q}_s\mathbf{A}_s\mathbf{Q}_s^T\mathbf{Q}_{s+1}^T \quad (8.4.6)$$

Moreover, these relations should be achieved:

$$\mathbf{A}_s - k_s\mathbf{I} = \mathbf{Q}_s^T\mathbf{R}_s \quad (8.4.7)$$

$$\mathbf{A}_{s+1} = \mathbf{Q}_s\mathbf{A}_s\mathbf{Q}_s^T \quad (8.4.8)$$

$$\mathbf{A}_{s+1} - k_{s+1}\mathbf{I} = \mathbf{Q}_{s+1}^T\mathbf{R}_{s+1} \quad (8.4.9)$$

Using equation (8.4.8), it is possible to write equation (8.4.9) in this way

$$\mathbf{A}_s - k_{s+1}\mathbf{I} = \mathbf{Q}_s^T\mathbf{Q}_{s+1}^T\mathbf{R}_{s+1}\mathbf{Q}_s \quad (8.4.10)$$

If one defines

$$\mathbf{M} = (\mathbf{A}_s - k_{s+1}\mathbf{I})(\mathbf{A}_s - k_s\mathbf{I}) \quad (8.4.11)$$

equations (8.4.7) and (8.4.10) give

$$\mathbf{R} = \mathbf{QM} \quad (8.4.12)$$

where

$$\begin{aligned}\mathbf{Q} &= \mathbf{Q}_{s+1}\mathbf{Q}_s \\ \mathbf{R} &= \mathbf{R}_{s+1}\mathbf{R}_s\end{aligned} \quad (8.4.13)$$

Equation (8.4.6) can be written (8.4.13) as

$$\mathbf{A}_s\mathbf{Q}^T = \mathbf{Q}^T\mathbf{A}_{s+2} \quad (8.4.14)$$

To calculate the orthogonal matrix **Q** that makes the real matrix **M** triangular, a sequence of transformations are carried out through Householder $\mathbf{P}_1$ matrices (acting on the first column), $\mathbf{P}_2$ (acting on the second column), ... , $\mathbf{P}_{n-1}$. In this way,

$\mathbf{Q} = \mathbf{P}_{n-1} \ldots \mathbf{P}_2\mathbf{P}_1$, and the first row of **Q** is also the first one of $\mathbf{P}_1$, given that $\mathbf{P}_i$ is an identity matrix of $(i-1) \times (i-1)$ in its upper, left-hand corner.

The Householder $\mathbf{P}_1$ matrix is determined by the first column of $\mathbf{M}$. As $\mathbf{A}_s$ is a superior Hessenberg matrix, the first column of $\mathbf{M}$ has to have the shape $(p_1, q_1, r_1, 0, \ldots 0)^T$, in equation (8.4.11), where

$$\begin{aligned} p_1 &= a_{11}^2 - a_{11}\left(k_s + k_{s+1}\right) + k_s k_{s+1} + a_{12}a_{21} \\ q_1 &= a_{21}\left(a_{11} + a_{22} - k_s - k_{s+1}\right) \\ r_1 &= a_{21}a_{32} \end{aligned} \tag{8.4.15}$$

As a result, $\mathbf{P}_1 = \mathbf{I} - 2\mathbf{w}_1\mathbf{w}_1^T$. $\mathbf{w}_1$ is a vector in which only the first three components are distinct from zero. The matrix $\mathbf{P}_1\mathbf{A}_s\mathbf{P}_1^T$ is a superior Hessenberg one with three additional elements distinct from zero

$$\mathbf{P}_1\mathbf{A}_1\mathbf{P}_1^T = \begin{pmatrix} \bullet & \bullet & \bullet & \bullet & \bullet & \bullet & \bullet \\ \bullet & \bullet & \bullet & \bullet & \bullet & \bullet & \bullet \\ \circ & \bullet & \bullet & \bullet & \bullet & \bullet & \bullet \\ \circ & \circ & \bullet & \bullet & \bullet & \bullet & \bullet \\ & & & \bullet & \bullet & \bullet & \bullet \\ & & & & \bullet & \bullet & \bullet \\ & & & & & \bullet & \bullet \end{pmatrix} \tag{8.4.16}$$

This matrix can be reconverted into a superior Hessenberg one without altering the first row through a sequence of Householder transformations. $\mathbf{P}_2$ acts on elements 2, 3 and 4 of the first column, placing zeros in elements 3 and 4. This gives way to a matrix with the same shape as (8.4.16), with three additional non-nodal elements, which appear in the column below:

$$\mathbf{P}_2\mathbf{P}_1\mathbf{A}_1\mathbf{P}_1^T\mathbf{P}_2^T = \begin{pmatrix} \bullet & \bullet & \bullet & \bullet & \bullet & \bullet & \bullet \\ \bullet & \bullet & \bullet & \bullet & \bullet & \bullet & \bullet \\ & \bullet & \bullet & \bullet & \bullet & \bullet & \bullet \\ & \circ & \bullet & \bullet & \bullet & \bullet & \bullet \\ & \circ & \circ & \bullet & \bullet & \bullet & \bullet \\ & & & & \bullet & \bullet & \bullet \\ & & & & & \bullet & \bullet \end{pmatrix} \tag{8.4.17}$$

The operation is continued in this fashion until $\mathbf{P}_{n-1}$. In each $\mathbf{P}_r$ Householder transformation, there is a vector $\mathbf{w}_r$ with three components distinct from zero in positions r, $r+1$ and $r+2$ in column $(r-1)$, determined by the terms r, $r+1$ and $r+2$ of the matrix, which has been transformed until this point. The result is a superior Hessenberg matrix

$$\mathbf{A}_{s+2} = \mathbf{P}_{n-1}\ldots\mathbf{P}_2\mathbf{P}_1\mathbf{A}_s\mathbf{P}_1^T\mathbf{P}_2^T\ldots\mathbf{P}_{n-1}^T \tag{8.4.18}$$

and

$$\mathbf{Q} = \mathbf{P}_{n-1}\ldots\mathbf{P}_2\mathbf{P}_1 \tag{8.4.19}$$

Serving as cut values in each step are the eigenvalues of the lower, right-hand 2×2 sub-matrix within matrix $\mathbf{A}_s$

$$
\begin{aligned}
k_s + k_{s+2} &= a_{n-1,n-1} + a_{n,n} \\
k_s k_{s+1} &= a_{n-1,n-1} a_{n,n} - a_{n-1,n} a_{n,n-1}
\end{aligned}
\tag{8.4.20}
$$

The procedure converges fairly quickly and offers two ways to finish the iterations in search of an eigenvalue. The first is to look at the value of $a_{n,\,n-1}$; if it is null, a_{nn} is an eigenvalue. Row and column n of the matrix is eliminated and the following eigenvalue is searched for. The second possibility is that the other term, $a_{n-1,\,n-2}$, is made null. In this case, the eigenvalues of the lower, right-hand 2×2 sub-matrix are taken as eigenvalues, eliminating columns and rows n and $(n-1)$.

The number of iterative operations required for algorithm *QR* is in the order of $5k^2$, where k is the size of the matrix in which the following eigenvalue is searched for. The average number of iterations for each eigenvalue is around 1.8 so that the total number of operations to obtain all the eigenvalues is in the order of $3n^3$. Wilkinson[6], Smith[7] and Press[8] provide programming routines for this numerical method.

8.5 *FLAS* Code

Figure 8.5.1 gives an overview of the code which has been developed for the sensitivity analysis of flutter problems in terms of the mechanical parameters (I_y, I_z and J) for a

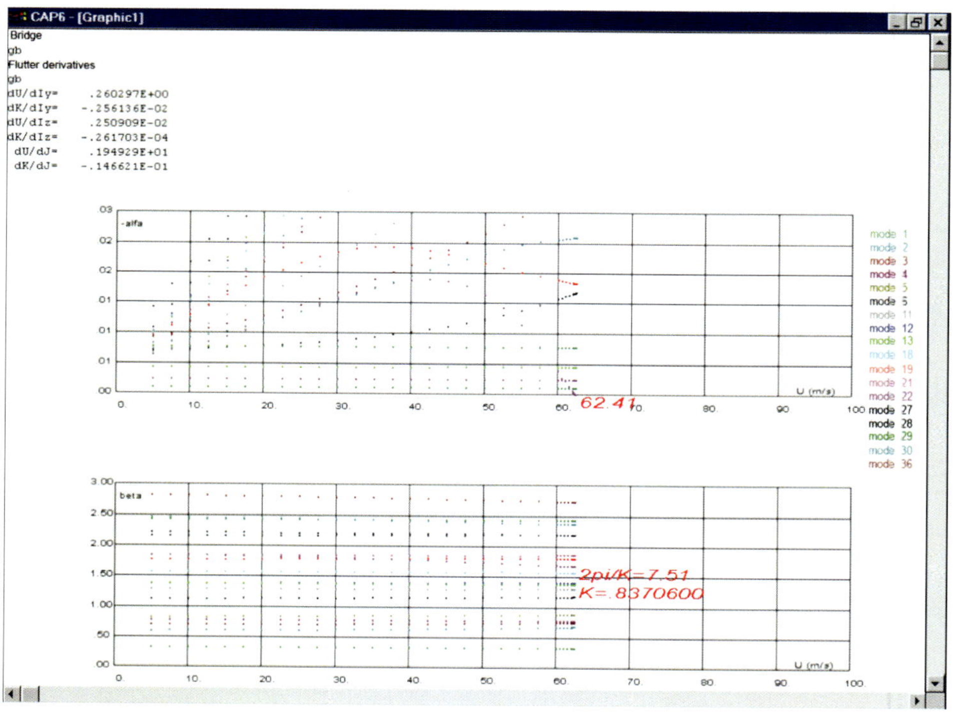

Figure 8.5.1: Overview of *FLAS* code.

bridge deck. In the first stage of this system, a finite-element model of the bridge is made. With this model, it is possible to calculate the frequencies and vibration modes.

The dynamic calculation should be carried out not only with the values of I_y, I_z and J which the bridge has, but also with the modified values of these inertias. This is so that it is later possible to use finite differentials to evaluate the sensitivity of the frequencies and modes in terms of these parameters.

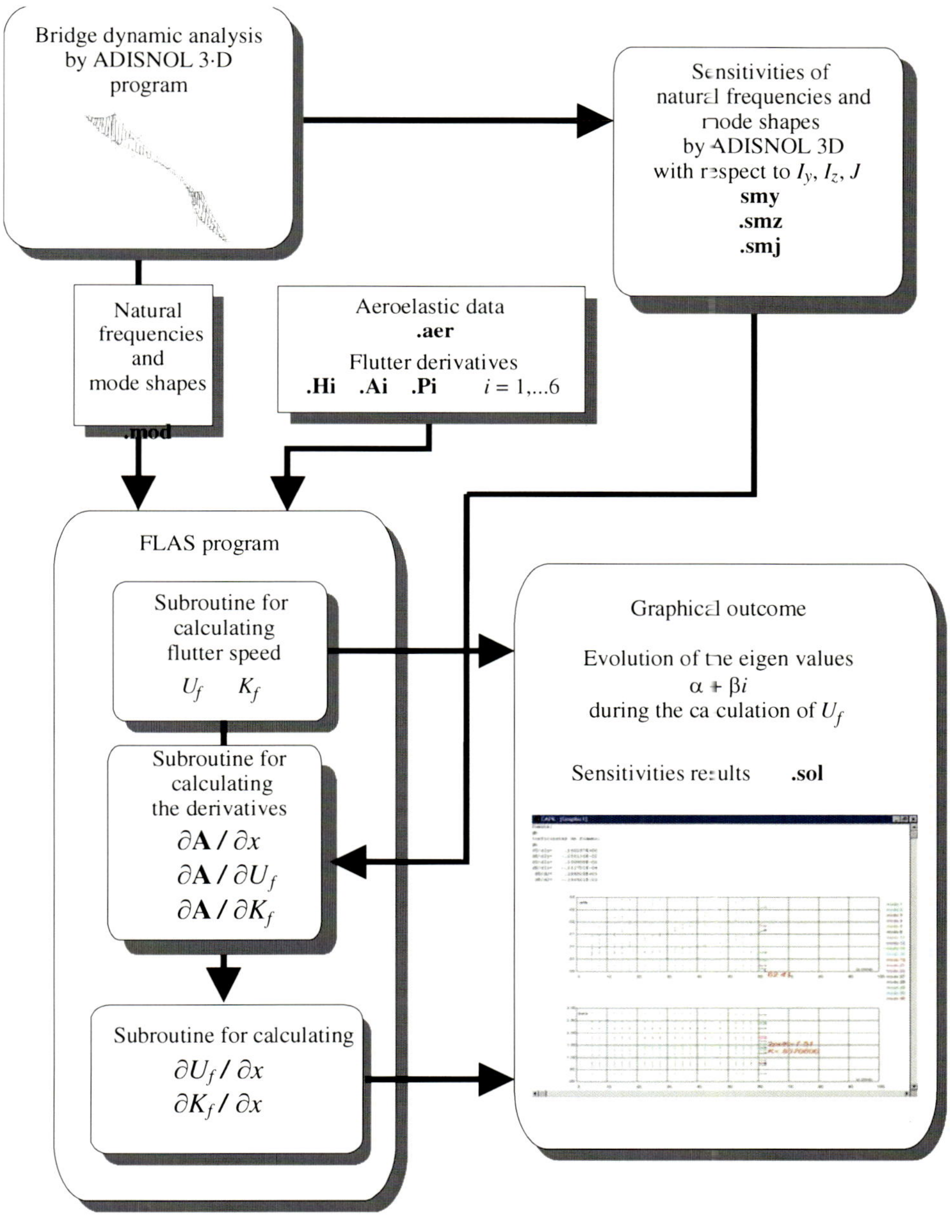

Figure 8.5.2: On-screen graphics of the *FLAS* code.

The data files of the code carry the name of the bridge and have different extensions, according to the data they hold. The **.mod** file provides the frequencies and vibration modes; the **.smy**, **.smz**, **.smj** files contain the sensibilities of the frequencies and modes; the **.aer** file defines parameters within aeroelastic calculation, such as damping, deck width, the modes used in the analysis, and so on. And lastly, it is necessary to write 18 files with the experimental flutter derivatives: **.Hi**, **.Ai**, **.Pi** with $i = 1,...,6$.

The results are stored in a file with extension **.sol** and are also presented on the computer screen. Another feature worth mentioning is how the code graphically portrays the calculation process related to flutter speed. Figure 8.5.2 shows the screen in which, as the wind speed U increases, and the search for the critical wind speed is in progress, points are drawn to represent the way in which the real and imaginary parts of the eigenvalues, α_j and β_j, respectively, evolve.

8.6 References

[1] Jurado J. A., Hernández S. [1999] *Design of ultra long span bridges with aeroelastic Constraints.* Optimization in Industry II. Anff, Alberta, Canada, June.

[2] Jurado J. A., Hernández S. [2000] *Sensitivity Analysis of Flutter Speed of Bridges.* Structures Congress ASCE, Philadelphia, USA.

[3] Nelson R. B. [1976] *Simplified calculation of eigenvector derivatives.* AIAA Journal, Vol .14, pp. 1201–1206.

[4] GENESIS [1997] *Manuals of GENESIS Version 6.0*, VMA Engineering, Colorado Spring.

[5] Haug E. J., Choi K. K., Komkov V. [1986] *Design Sensitivity Analysis of Structural System*, Academic Press, New York.

[6] Wilkinson J. H., Reinsch C. [1971] *Linear Algebra, Vol. II of Handbook for Automatic Computation*, Springer, New York.

[7] Smith B. T., Boyle J. M., Dongarra J. J., Garbow B. S., Ikebe Y., Klema V. C., Moler C. B. [1976] *Matrix eigen-system routines.* EISPACK Guide, Springer, New York.

[8] Press W., Teukolsky S., Vetterling W., Flannery B. [1992] *Numerical Recipes in C.* Cambridge University Press, Cambridge.

Chapter 9

Sensitivity of flutter response for suspension bridges under construction

9.1 Introduction

This chapter will provide another opportunity to apply the multimodal flutter methodology described in Chapter 3 as well as the analytical sensitivity analysis explained in Chapter 8. Serving as case studies are the same two suspension bridges whose construction phase was looked at in Chapter 4. These are the Great Belt and the Höga Kusten Bridges. The *ADISNOL3D* and *FLAS* codes were used to facilitate this task at each construction phase.

The sensitivity analysis of the aeroelastic response for each construction phase of these bridges entails the following stages:

1) The dynamic calculation of the structural response through free vibration without damping. Using the *ADISNOL3D* code, the natural frequencies and second-order vibration eigenmodes are obtained.
2) With the same code, a sensitivity analysis is carried out on these natural frequencies and vibration eigenmodes.
3) Then there is an aeroelastic calculation of the flutter speed for the two vibration modes – the first vertical and torsional ones. A multimodal analysis is also done, taking into account the modes that play the greatest role in aeroelastic instability. The *FLAS* code helps here, using the data provided by the *ADISNOLD3D* code.
4) The flutter parameters – that is, the critical speed and reduced frequency – are also subject to an analytical sensitivity study. Again, the FLAS code is relied on making the most of the data that came out of the dynamic sensitivity analysis achieved with *ADISNOLD3D*.
5) A comparison is made between the previously mentioned sensitivity analyses and the calculations based on centred finite differences.
6) The last step consists of a study that covers both of the bridges chosen. It includes the flutter parameters and sensitivities throughout the entire construction process: from deck placement to project closure.

Sections 4.1 and 4.3 offer a detailed description of each case study from Chapter 4. For this reason, only the construction phases considered earlier are dealt with and the chapter goes on to develop the remaining sections that have to do with sensitivity calculation. Outstanding studies on the stability of suspension bridges withstanding wind action during construction were carried out by Tanaka[1], Ge and Tanaka[2], Cobo[3], Brancaleoni[4,5] and Larsen[6].

9.2 Example 1: Höga Kusten Bridge at the construction phase

The Swedish Höga Gusten Bridge (Figure 9.2.1) has already been described in detail in Section 4.2. For this reason, Table 9.2.1 provides some of its characteristics to refresh the reader's memory, while Table 9.2.2 is an overview of its construction phases.

Figure 9.2.1: Höga Kusten Bridge.

Table 9.2.1: Geometric and mechanical properties of the Höga Kusten Bridge.

Main span length (m)	1210
Side span lengths (m)	310–280
Distance between main cables (m)	20.8
Total deck width (m)	22
Deck depth (m)	4.0
Section of each main cable (m^2)	0.234
Inertia moment under lateral bending for the deck I_z (m^4)	31.57
Torsional inertia moment for the deck J (m^4)	1.1
Deck mass (segments) (t/m)	
25 (L = 25 m)	11.0
24, 26 (L = 16 m)	8.94
23, 27 (L = 40 m)	8.033
12–22, 28–38 (L = 40 m)	7.615
11, 39 (L = 40 m)	7.788
Polar moment for deck inertia (segments) (Tm^2/m)	
25 (L = 25 m)	518.3
24, 26 (L = 16 m)	383.6
23, 27 (L = 40 m)	332.1
12–22, 28–38 (L = 40 m)	306.9
11, 39 (L = 40 m)	310.1

Table 9.2.2: Defining the construction phases studied for the Höga Kusten Bridge.

Phase	Deck segments in place	Length of completed deck (m)	Percentage of main span completed (%)	Configuration
Phase 1	22–28	217	18	Symmetrical
Phase 2	19–35	617	51	Asymmetrical
Phase 3	14–35	817	68	Symmetrical
Phase 4	10–40	1177	97	Symmetrical

Table 9.2.3: Summary of flutter speeds (m/s) obtained by various researchers studying the Höga Kusten Bridge in the construction phase.

	FLAS 2 modes	FLAS multimodal
Phase 1	32.17	32.23
Phase 2	67.45	62.63
Phase 3	72.94	61.30
Phase 4	80.40	60.45

It is also worth referring back to the results for the aeroelastic analysis of the construction phases that had been taken into consideration. Table 9.2.3 summarizes the results.

Phase 1: 18% of the main span

The aeroelastic sensitivity analysis was done through *FLAS*, which codifies the formulation outlined in Chapter 8. This process made use of the analytical sensitivity analysis for the frequencies and vibration eigenmodes implemented in the *ADISNOL3D* code. Tables 9.2.4 and 9.2.5 show the results for both aeroelastic analyses carried out earlier: using two vibration modes and a multimodal approach.

The only way to test the sensitivities is through the finite differential method. For whatever function u, its derivative in terms of a variable x can be approximated by

$$\frac{\partial u}{\partial x} \approx \frac{u(x+\Delta x)-u(x-\Delta x)}{2\Delta x} \tag{9.2.1}$$

Therefore, two steps must be done each time the sensitivity is tested through finite differentials. It is necessary to evaluate *u(x)* as well as carry out the two additional calculations for the response with the modified variable, adding or subtracting a Δx. One must take into account that the responses under evaluation are the flutter coefficients U_f and K_f, which entail lengthy calculation. This is because testing sensitivity through finite differentials means four additional calculations: two dynamic ones with *ADISNOL3D* and two aeroelastic ones with *FLAS*.

Due to the lack of accuracy inherent in finite differential calculations, two step sizes were attributed to them: $\Delta x = 0.01x$, $\Delta x = 0.025x$, that is, differentials centred between 2% and 5%, respectively.

Table 9.2.4: Comparing sensitivity analyses for the flutter coefficients with respect to I_z and J of the deck using two vibration modes for Phase 1 of the Höga Kusten Bridge.

	$\frac{\partial U_f}{\partial I_z}$	$\frac{\partial K_f}{\partial I_z}$	$\frac{\partial U_f}{\partial J}$	$\frac{\partial K_f}{\partial J}$
Analytical	0.186e–4	–0.251e–6	0.414	–0.0055
Finite differentials based on 2%	0*	1.6e–7	0.454	–0.0065
Finite differentials based on 5%	0*	–0.223e–6	0.364	–0.0052

Table 9.2.5: Comparing sensitivity analyses of the flutter coefficients with respect to I_z and J of the deck using multiple vibration modes for Phase 1 of the Höga Kusten Bridge.

	$\frac{\partial U_f}{\partial I_z}$	$\frac{\partial K_f}{\partial I_z}$	$\frac{\partial U_f}{\partial J}$	$\frac{\partial K_f}{\partial J}$
Analytical	0.187e–4	–0.252e–6	0.414	–0.00552
Finite differentials based on 2%	0*	–0.158e–6	0.454	–0.00434
Finite differentials based on 5%	0*	–0.221e–6	0.364	–0.006

Values of 0* in Tables 9.2.4 and 9.2.5 indicate that the variation in the design variable is inferior to 0.01 m/s, demonstrating the precision with which *FLAS* calculates the critical wind speed.

From Tables 9.2.4 and 9.2.5, the following conclusions may be drawn:

- There is generally a close agreement between the sensitivities calculated through finite differentials and those obtained through the analytical methodology implemented in *ADISNOL3D* and *FLAS*.
- As was the case with the flutter coefficients, both sensitivity analyses (with two or 15 modes) had almost identical results.
- The sensitivity for U_f with respect to the inertia under torsion of the deck is positive and various orders of magnitude higher than the sensitivity for the inertia under lateral bending of the deck. This suggests that for the bridge's initial design and at this stage of its construction, an increase in this variable leads to a higher flutter speed.
- The derivatives for the inertia under lateral bending of the deck are very low. Once again, there are implications for the bridge's initial design and at this stage of its construction. Any change in that variable will lead to a very slight modification in the aeroelastic behaviour of the bridge.

Phase 2: 51% of the main span

The same procedure from the last stage was followed. Tables 9.2.6 and 9.2.7 show the results for both the bimodal aeroelastic analysis and the one using 17 modes. As was the case before, to prove the analytical derivatives it is necessary to use finite differentials centred on two step sizes: $\Delta x = 0.01x$, $\Delta x = 0.025x$, or differentials of 2% and 5%, respectively.

Table 9.2.6: Comparing sensitivity analyses of the flutter coefficients with respect to I_z and J of the deck using two vibration modes for Phase 2 of the Höga Kusten Bridge.

	$\frac{\partial U_f}{\partial I_z}$	$\frac{\partial K_f}{\partial I_z}$	$\frac{\partial U_f}{\partial J}$	$\frac{\partial K_f}{\partial J}$
Analytical	0.0331	–0.769e–4	21.17	–0.0802
Finite differentials based on 2%	0.0317	–0.144e–3	21.82	–0.0766
Finite differentials based on 5%	0.0328	–0.784e–4	21.10	–0.0806

Table 9.2.7: Comparing sensitivity analyses of the flutter coefficients with respect to I_z and J of the deck using 17 vibration modes for Phase 2 of the Höga Kusten Bridge.

	$\frac{\partial U_f}{\partial I_z}$	$\frac{\partial K_f}{\partial I_z}$	$\frac{\partial U_f}{\partial J}$	$\frac{\partial K_f}{\partial J}$
Analytical	–0.021	0.190e–3	6.323	–0.0163
Finite differentials based on 2%	–0.019	0.180e–3	6.360	–0.0169
Finite differentials based on 5%	–0.021	0.187e–3	6.360	–0.0163

By studying Tables 9.2.6 and 9.2.7, it is possible to reach the following conclusions:

- On the whole, the sensitivities calculated through finite differentials correspond well with those calculated with the analytical methodology used in *ADISNOL3D* and *FLAS*.
- The results from the sensitivity analyses using 2 and 17 modes are generally different from those obtained through aeroelastic analyses.
- The derivative U_f related to the torsional inertia of the deck is positive and several orders of magnitude above the derivative for the inertia under lateral bending of the deck. However, the value obtained in the bimodal analysis is excessive and this indicates that, for the bridge's initial design, an increase in this variable will result in a higher flutter speed.
- The derivatives for the inertia under lateral bending of the deck are extremely low. This implies that for the bridge's initial design and at this construction phase any modification in that variable will lead to only a slight change in the construction's aeroelastic behaviour. It is interesting to note that even the signs of both sensitivity analyses change, although the modular values are very small.

Phase 3: 68% of the main span

Tables 9.2.8 and 9.2.9 provide the results for this third phase, which has a symmetrical configuration. These were obtained through aeroelastic analyses with 2 and 17 vibration modes, respectively. The same method described above was used to test the analytical derivatives.

Table 9.2.8: Comparing sensitivity analyses of the flutter coefficients with respect to I_z and J of the deck using two vibration modes for Phase 3 of the Höga Kusten Bridge.

	$\frac{\partial U_f}{\partial I_z}$	$\frac{\partial K_f}{\partial I_z}$	$\frac{\partial U_f}{\partial J}$	$\frac{\partial K_f}{\partial J}$
Analytical	–0.0084	0.166e–4	18.26	–0.123
Finite differentials based on 2%	0*	0.216e–4	18.18	–0.122
Finite differentials based on 5%	–0.0079	0.172e–4	18.36	–0.124

Table 9.2.9: Comparing sensitivity analyses of the flutter coefficients with respect to I_z and J of the deck using 17 vibration modes for Phase 3 of the Höga Kusten Bridge.

	$\frac{\partial U_f}{\partial I_z}$	$\frac{\partial K_f}{\partial I_z}$	$\frac{\partial U_f}{\partial J}$	$\frac{\partial K_f}{\partial J}$
Analytical	0.0155	–0.856e–4	9.933	–0.0184
Finite differentials based on 2%	0.0158	–0.817e–4	9.540	–0.0189
Finite differentials based on 5%	0.0156	–0.837e–4	9.820	–0.0175

The 0* value in tables means that the variation of U_f is lower than 0.01 m/s which is the precision of FLAS program.

The data in Tables 9.2.8 and 9.2.9 lead to the following conclusions:

- As was the case in the previous phases, the agreement between the sensitivities calculated through finite differentials and those obtained from the analytical methodology applied in *ADISNOL3D* and *FLAS* is fairly good.
- There is, however, a noticeable difference between the results from the sensitivity analyses using 2 and 17 modes and those got through aeroelastic analyses.
- Yet again, the derivative for U_f with respect to the torsional inertia of the deck is positive and various orders of magnitude above the derivative for its inertia under lateral bending. Nevertheless, like before, the value obtained through the bimodal analysis is excessive.
- The derivatives for the deck's inertia under lateral bending are very low. It is again the case that for the bridge's initial design and at this stage in its construction, changing this variable in any way will hardly alter the structure's aeroelastic behaviour. It has to be added that, once again, the signs in both sensitivity analyses change.

Phase 4: 97% of the main span

In this fourth phase, the cables have been hung along almost the entire length of the deck. Tables 9.2.10 and 9.2.11 have the results for the aeroelastic analyses using 2 and 17 vibration modes, once again comparing them with the finite differentials.

Table 9.2.10: Comparing sensitivity analyses of the flutter coefficients with respect to I_z and J of the deck using two vibration modes for Phase 4 of the Höga Kusten Bridge.

	$\frac{\partial U_f}{\partial I_z}$	$\frac{\partial K_f}{\partial I_z}$	$\frac{\partial U_f}{\partial J}$	$\frac{\partial K_f}{\partial J}$
Analytical	–0.621	0.00124	68.57	–0.2106
Finite differentials based on 2%	–0.630	0.00135	68.63	–0.2110
Finite differentials based on 5%	–0.622	0.00127	68.72	–0.2107

Table 9.2.11: Comparing sensitivity analyses of the flutter coefficients with respect to I_z and J of the deck using 17 vibration modes for Phase 4 of the Höga Kusten Bridge.

	$\frac{\partial U_f}{\partial I_z}$	$\frac{\partial K_f}{\partial I_z}$	$\frac{\partial U_f}{\partial J}$	$\frac{\partial K_f}{\partial J}$
Analytical	0.159	–0.625e–4	6.80	–0.0048
Finite differentials based on 2%	0.174	–0.739e–4	6.81	–0.0051
Finite differentials based on 5%	0.161	–0.638e–4	6.73	–0.0049

Tables 9.2.10 and 9.2.11 lead to conclusions similar to the ones reached in earlier sections. The only point worth making is that the sensitivities for the inertia under lateral bending of the deck are lower than the ones for the other variable considered, by about two orders of magnitude. It is possible to observe how the signs are inverted according to the type of analysis considered, with bi- or multimodal (17-mode) analysis.

Next, by way of a summary, Figures 9.2.2–9.2.5 chart the evolution of the sensitivity analyses carried out as the deck steadily was being laid, following the construction phases considered earlier.

By studying the sensitivity analyses for the flutter coefficients, it is easy to see how essential it is to consider a sufficient number of modes within them. Generally, the differences between the bi- and mulitmodal sensitivity analyses are more marked than they are within the aeroelastic analyses. One can observe that the torsional inertia of the deck is the variable most likely to affect the flutter coefficient across the construction phases considered. With the multimodal analysis, the values for the sensitivities of flutter speed U_f with respect to this variable grow until 50% of the deck has been completed, and are later maintained with few variations. In terms of the inertia under lateral bending of the deck, it is possible to detect that its influence on the flutter coefficients is practically null until 68% of the main span has been completed. This is in agreement with the results provided by the multimodal analysis. That influence becomes somewhat noticeable during the last construction phases. During these very stages, nevertheless, what does stand out is the disparity between those sensitivities according to the number of modes considered in the analysis.

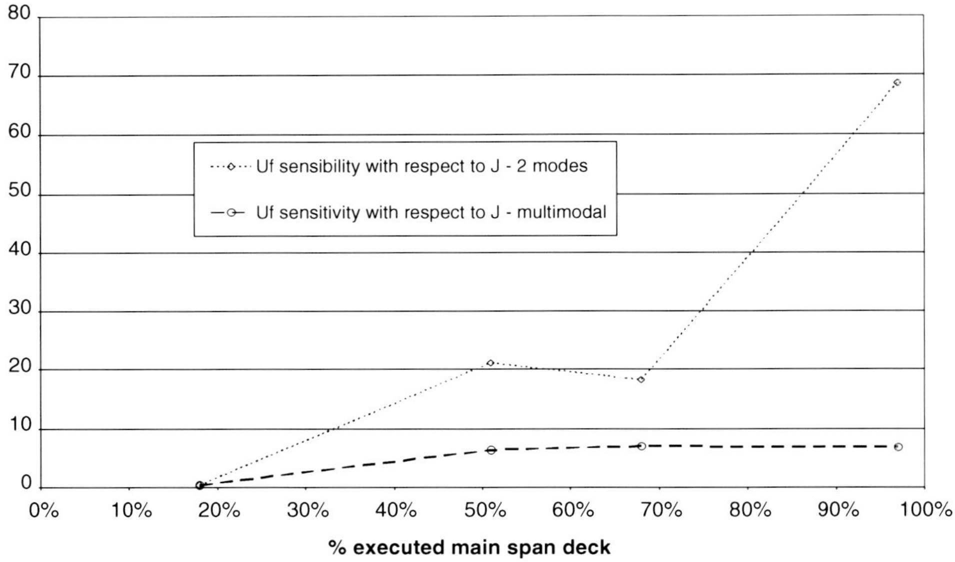

Figure 9.2.2: Evolution of the sensitivities for flutter speed U_f with respect to the torsional inertia J of the deck, as it was being laid for the Höga Kusten Bridge.

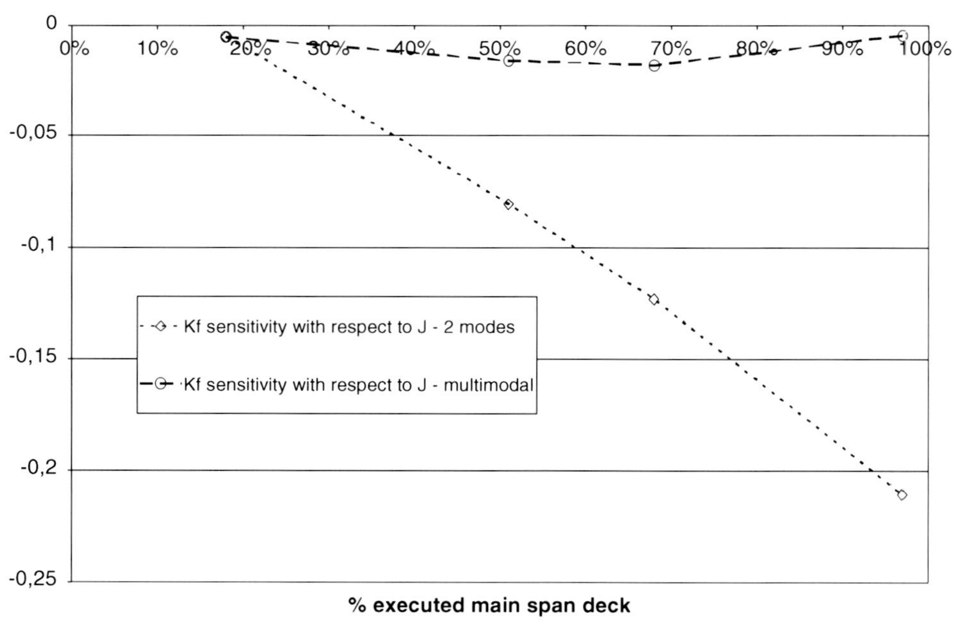

Figure 9.2.3: Evolution of the sensitivities for the reduced frequency K_f with respect to the inertia under torsion J of the deck, as it was being laid for the Höga Kusten Bridge.

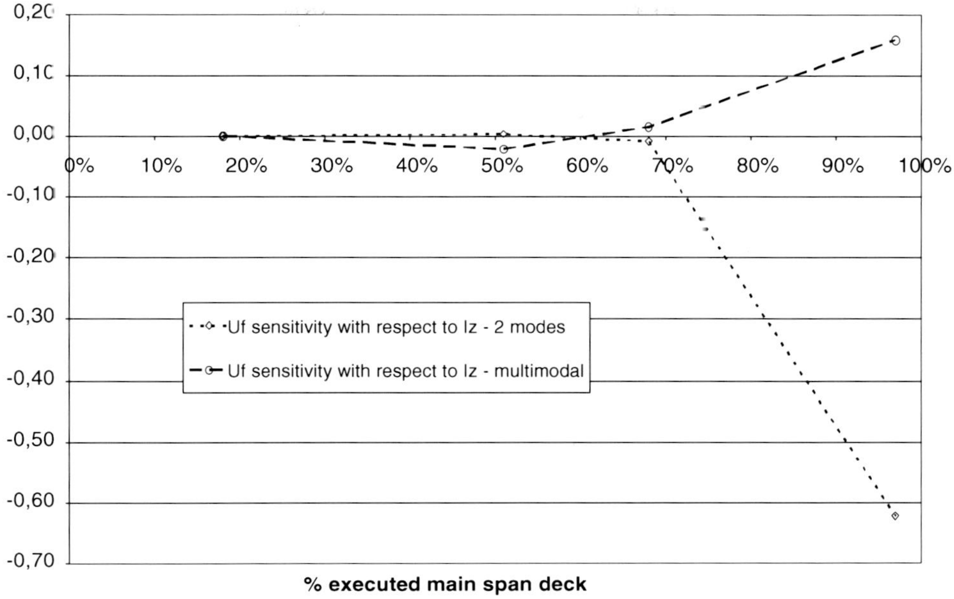

Figure 9.2.4: Evolution of the sensitivities for the flutter speed U_f with respect to the inertia under lateral bending I_z of the deck, as it was being laid for the Höga Kusten Bridge.

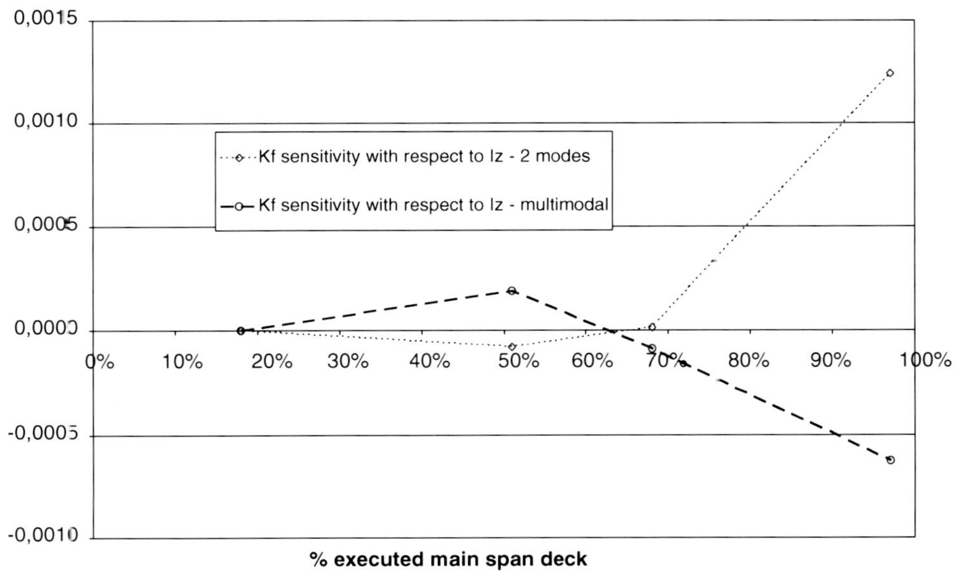

Figure 9.2.5: Evolution of the sensitivities for the reduced frequency K_f with respect to the inertia under lateral bending I_z of the deck, as it was being laid for the Höga Kusten Bridge.

9.3 Example 2. Great Belt suspension bridge under construction

As Chapters 4 and 5 already provide a full description of the Great Belt Bridge, only a general overview of its main geometric and mechanical features can be found in Table 9.3.1.

As can be seen in Figure 9.3.1, the construction process for the bridge deck followed a sequence of phases, starting from the centre of the span and working out towards the towers.

Figure 9.3.1: Laying the deck for the Great Belt Bridge.

Table 9.3.1: Geometrical and mechanical properties of the Great Belt Bridge.

Main span length (m)	1624
Side span length (m)	535
Main cable sag (m)	180
Distance between main cables (m)	27
Total width of deck (m)	31
Deck depth (m)	4.4
Cross section of each main cable (m^2)	0.44
Mass for each main cable (t/m)	3.74
Inertia moment under lateral bending for deck I_z (m^4)	100.0
Inertia moment under torsonal bending for deck J (m^4)	7.6
Deck mass (t/m)	15.26
Polar moment of inertia for deck (tm^2/m)	1106.8

In this example, the phases described in Table 9.3.2 were studied. All of these had a symmetrical configuration. The results from the aeroelastic analysis for each phase are explained in Chapter 4. However, Table 9.3.3 provides a reminder of the flutter speed values for each phase through bi- and multimodal analyses.

Below, one finds an overview of the analytical sensitivities for the flutter coefficients with respect to the deck's mechanical parameters I_z and J (inertia under lateral bending and inertia under torsion). There is also a comparison between these results and those obtained through finite differentials. The sensitivity analysis was done using the same methodology previously applied with the *FLAS* code, implementing the formulation described in Chapter 8. Use was made of the sensitivity analysis based on frequencies and vibration eigenmodes and carried out through the *ADISNOL3D* code.

Table 9.3.2: Construction phases studied for the Great Belt Bridge.

Phase	Central span length (m)	Central span percentage (%)
Phase 1	150	9.25
Phase 2	250	15.4
Phase 3	350	21.6
Phase 4	450	27.8
Phase 5	649.6	40
Phase 6	974.4	60
Phase 7	1299.2	80
Phase 8	1461.6	90

Table 9.3.3: Flutter analysis of the construction phases of the Great Belt Bridge.

		FLAS 2 modes	FLAS multimodal
Phase 1	U_f (m/s)	33.83	35.00
9.25 %	K_f	0.64312	0.61620
Phase 2	U_f (m/s)	37.25	38.12
15.4%	K_f	0.52244	0.50664
Phase 3	U_f (m/s)	42.14	42.04
21.6 %	K_f	0.49182	0.48633
Phase 4	U_f (m/s)	46.33	46.28
27.8%	K_f	0.48903	0.48393
Phase 5	U_f (m/s)	52.95	56.15
40%	K_f	0.53887	0.47888
Phase 6	U_f (m/s)	75.62	58.88
60%	K_f	0.44825	0.63306
Phase 7	U_f (m/s)	87.26	54.54
80%	K_f	0.42573	0.92167
Phase 8	U_f (m/s)	90.26	58.79
90%	K_f	0.426013	0.868412

It is yet again the case that in order to test these sensitivities, the finite deferential method must be used. Table 9.3.4 lists the results obtained through aeroelastic analyses using two and multiple vibration modes; both of these were explained in Chapter 4.

Table 9.3.4: Comparing the sensitivities of the flutter coefficients with respect to I_z and J of the deck through analyses using two vibration modes in Phase 1 of the Great Belt Bridge.

		$\frac{\partial U_f}{\partial I_z}$	$\frac{\partial K_f}{\partial I_z}$	$\frac{\partial U_f}{\partial J}$	$\frac{\partial K_f}{\partial J}$
Phase 1	Analytical	0.272e−4	−0.386e−6	−0.009	0.143e−3
2 modes	Finite differentials 2%	0*	−0.240e−6	−0.0657	0.690e−3
9.25%	Finite differentials 5%	0*	−0.350e−6	−0.0105	0.238e−3
Phase 1	Analytical	0.31e−4	−0.422e−6	−0.011	0.161e−3
Multimodal	Finite differentials 2%	0*	−0.285e−6	0*	−0.740e−4
9.25%	Finite differentials 5%	0*	−0.342e−6	−0.0105	0.208e−3
Phase 2	Analytical	0.644e−4	−0.62e−6	−0.541e−3	0.297e−4
2 modes	Finite differentials 2%	0*	−0.50e−6	0*	0.204e−4
14.4%	Finite differentials 5%	0*	−0.52e−6	0*	0.203e−4
Phase 2	Analytical	0.717e−4	−0.68e−6	−0.16e−2	0.37e−4
Multimodal	Finite differentials 2%	0*	−0.50e−6	0*	0.105e−4
14.4%	Finite differentials 5%	0*	−0.52e−6	0*	0.28e−4
Phase 3	Analytical	0.129e−3	−0.898e−6	0.0438	−0.288e−3
2 modes	Finite differentials 2%	0*	−0.105e−5	0.0657	−0.217e−3
21.6%	Finite differentials 5%	0*	−0.112e−5	0.0526	−0.343e−3
Phase 3	Analytical	0.743e−5	−0.575e−7	0.0651	−0.46e−3
Multimodal	Finite differentials 2%	0*	−0.430e−7	0.0657	−0.37e−3
21.6	Finite differentials 5%	0*	−0.450e−7	0.0780	−0.57e−3
Phase 4	Analytical	0.178e−3	−0.928e−6	0.107	−0.623e−3
2 modes	Finite differentials 2%	0*	−1.01e−6	0.131	−0.793e−3
27.8%	Finite differentials 5%	0*	−1.16e−6	0.105	−0.560e−3
Phase 4	Analytical	0.272e−4	−0.193e−6	0.117	−0.698e−3
Multimodal	Finite differentials 2%	0*	−0.200e−6	0.131	−0.720e−3
27.8%	Finite differentials 5%	0*	−0.200e−6	0.105	−0.719e−3
Phase 5	Analytical	0.107e−3	−0.31e−6	0.201	−0.00116
2 modes	Finite differentials 2%	0*	−0.28e−6	0.216	−0.00124
40%	Finite differentials 5%	0*	−0.27e−6	0.205	−0.00126
Phase 5	Analytical	0.177e−4	−0.124e−6	0.225	−0.00133
Multimodal	Finite differentials 2%	0*	−0.137e−6	0.241	−0.00141
40%	Finite differentials 5%	0*	−0.134e−6	0.231	−0.00138
Phase 6	Analytical	0.482e−3	−0.777e−6	0.565	−0.243e−2

continued

Table 9.3.4: *Continued.*

2 modes	Finite differentials 2%	0*	−0.792e−6	0.571	−0.251e−2
60%	Finite differentials 5%	0*	−0.801e−6	0.568	−0.253e−2
Phase 6	Analytical	0.727e−3	−0.609e−5	0.484	−0.291e−2
Multimodal	Finite differentials 2%	0*	−0.728e−5	0.493	−0.302e−2
60%	Finite differentials 5%	0*	−0.715e−5	0.482	−0.315e−2
Phase 7	Analytical	−0.774e−3	0.791e−6	1.128	−0.00604
2 modes	Finite differentials 2%	0*	0.884e−6	1.215	−0.00671
80%	Finite differentials 5%	0*	0.812e−6	1.143	−0.00682
Phase 7	Analytical	0.362e−3	−0.484e−5	0.574	−0.00502
Multimodal	Finite differentials 2%	0*	−0.502e−5	0.571	−0.00513
80%	Finite differentials 5%	0*	−0.526e−5	0.568	−0.00526
Phase 8	Analytical	−0.059	0.756e−4	1.793	−0.0083
2 modes	Finite differentials 2%	−0.061	0.735e−4	1.842	−0.0088
90%	Finite differentials 5%	−0.060	0.722e−4	1.815	−0.0089
Phase 8	Analytical	0.194e−2	−0.199e−4	1.188	−0.0113
Multimodal	Finite differentials 2%	0*	−0.203e−4	1.176	−0.0115
90%	Finite differentials 5%	0.2e−2	−0.201e−4	1.182	−0.0115

The 0* value in tables means that the variation of U_f is lower than 0.01 m/s which is the precision of FLAS program.

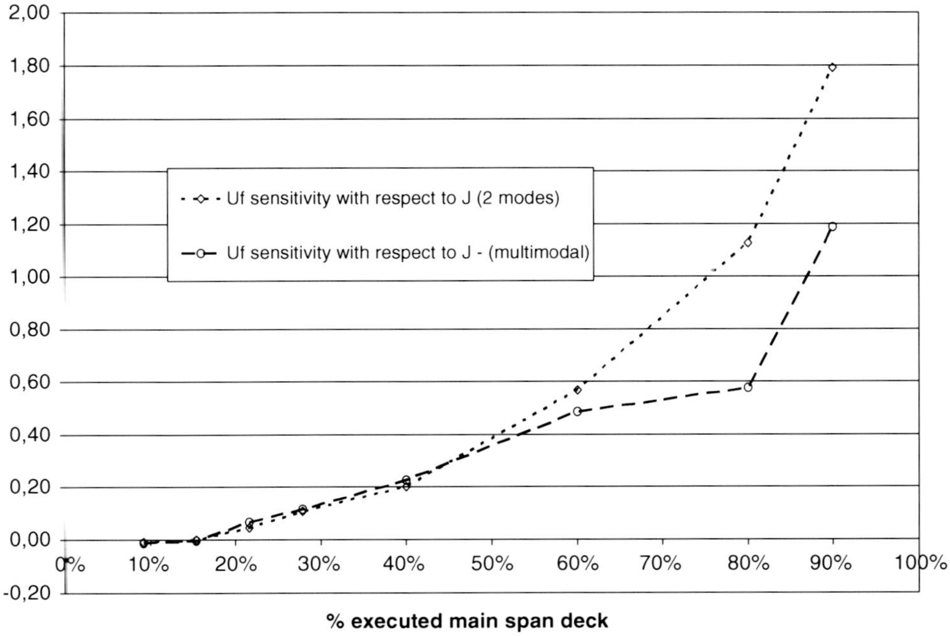

Figure 9.3.2: Evolution of the sensitivity of U_f with respect to inertia under torsion J for the deck, as it was being constructed for the Great Belt Bridge.

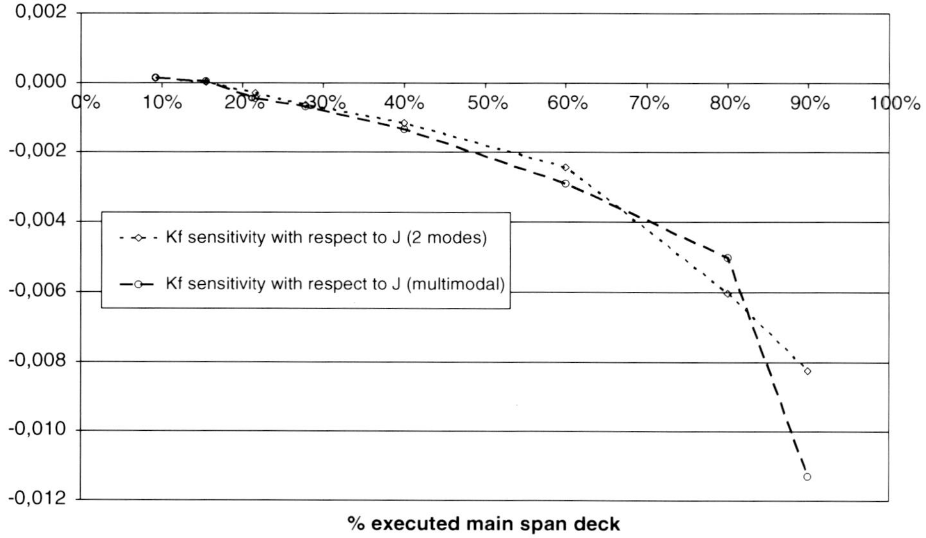

Figure 9.3.3: Evolution of the sensitivity of K_f with respect to inertia under torsion J for the deck, as it was being constructed for the Great Belt Bridge.

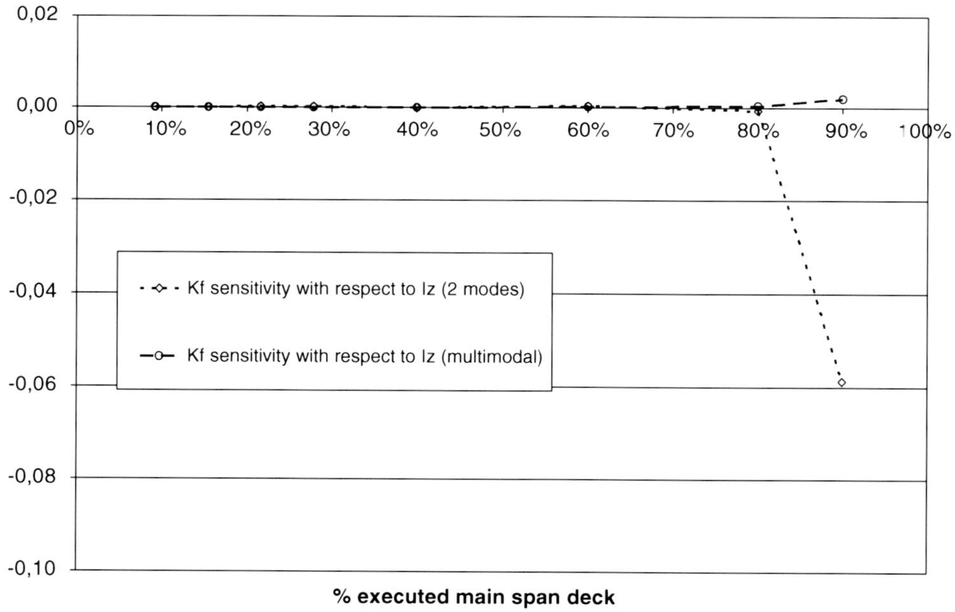

Figure 9.3.4: Evolution of the sensitivity of U_f with respect to the inertia under lateral bending I_z of the deck, as it was being constructed for the Great Belt Bridge.

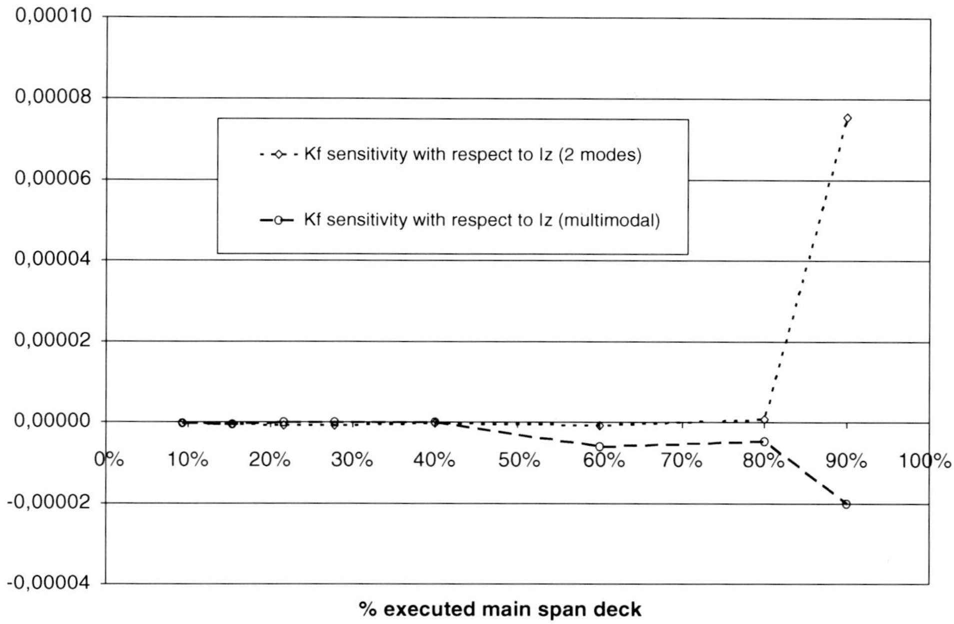

Figure 9.3.5: Evolution of the sensitivity of K_f with respect to the inertia under lateral bending I_z of the deck, as it was being constructed for the Great Belt Bridge.

Figures 9.3.2–9.3.5 chart the evolution of the sensitivity analysis carried out as the deck was being laid during the construction phases described earlier. Results are shown for both types of analyses: using two and multiple vibration modes.

What stands out from these results is the lack of precision underlying those obtained through the analysis using two vibration modes. This imprecision becomes more noticeable in the later construction phases. It is again the case that the inertia under torsion for the deck is the variable that has greatest impact on the flutter coefficient across all the construction phases. With this bridge, the sensitivity of the flutter speed U_f with respect to this variable is positive and rises from the beginning to the end of the deck's construction process. The sensitivity of U_f in terms of the inertia under lateral bending of the deck is virtually null during the whole construction process. The only exception is the final phases (when 90% of the central span had been completed); a slight influence was felt. It is possible to conclude that for this bridge's design, the inertia under lateral bending of the deck had practically no influence on its flutter coefficients while it was being constructed.

9.4 References

[1] Tanaka H. [1998] *Aeroelastic stability of suspension bridges during erection*. Structural Engineering International. Journal of the Intenational Association for Bridge and Structural Engineering (IABSE). Vol. 8, No. 2, pp. 118–12[illegible].

[2] Ge Y. J., Tanaka H. [2000] *Aerodynamic stability of long-span suspension bridges under erection*. Journal of Structural Engineering. Vol. 126, No 12, pp.1404–1412.

[3] Cobo del Arco D. [1998] *An analysis of wind stability. Improvements to the response of suspension bridges*. Doctoral Thesis from the Universitat Politécnica de Catalunya. Escola Técnica Superior D'Enginyers de Camins, Canals i Ports. (Director: Prof. Ángel Aparicio)

[4] Brancaleoni F. [1992] *The construction phase and its aerodynamic issues*. Aerodynamics of Large Bridges, A. Larsen (ed.) Balkema, Rotterdam.

[5] Brancaleoni F. [1988] *Behaviour of long span suspension bridges under construction*. 13th IABSE Congress Report. Helsinki, pp. 905–910.

[6] Larsen A. [1993] *Aerodynamic aspects of the final design of the 1624 m. suspension bridge across the Great Belt*. Journal of Wind Engineering and Industrial Aerodynamics, Vol. 48, pp. 261–285.

Chapter 10

Flutter response sensitivity of completed cable-supported bridges

The objective here is to demonstrate this methodology for sensitivity analysis of flutter in cable-supported bridges that have already been completed. Serving as case studies are the same four bridges examined in Chapter 5: the Great Belt, Akashi Strait, original Tacoma and Vasco de Gama Bridges. Because their characteristics have been detailed in Chapter 5, they are only summarized here alongside the results for flutter speed and reduced frequency.

10.1 Example 1. Great Belt Bridge

Table 10.1.1 outlines the most important properties of the suspension bridge over the Great Belt (Figure 10.1.1), whose main span is 1624 m. These properties were obtained from Larsen[1] and Cobo[2].

The first analysis made on the Great Belt considers the coupling of the two classic modes in the phenomenon of flutter, both of which are symmetrical: the vertical bending and torsion modes. The latter is responsible for flutter instability, which is reached with $U_f = 89.91$ and a reduced frequency of $K_f = 0.4494$, corresponding to $w = 1.303$ rad/s.

In multimodal analysis, this calculation was made with 18 vibration modes, including all the significant ones that appear; even the third mode of pure tension was considered (mode 36). Consequently, vibration modes of every kind came into play for this calculation and it seems that the result of the aeroelastic analysis provides the most accurate representation of the bridge's real behaviour under flutter. The critical flutter speed obtained is $U_f = 62.41$ m/s, 30% lower than the one calculated through the bimodal analysis, which produced unreliable results. The reduced response frequency from this analysis is $K_f = 0.8371$, corresponding with1.68 rad/s.

Chapter 5 has already provided an in-depth commentary on these results and here these are only touched upon before the sensitivity analysis is carried out.

These case studies will be used for the sensitivity analysis of the critical flutter speed U_f with respect to the deck's mechanical parameters, like inertia under vertical and lateral bending – I_y, and I_z respectively – or inertia under torsion J. To this end, they will be subjected to the analytical method developed in Chapter 7, which is implemented in the *FLAS* code. Different vibration modes can be included in the aeroelastic analysis and it is possible to obtain the sensitivities immediately. In contrast, the differential method as a way of testing was only done for calculations using two and 18 modes, due to the time-intensive nature of this method.

Table 10.1.1: Geometrical and mechanical properties of Great Belt Bridge.

Main span length (m)	1624
Side span length (m)	535
Hanger sag (m)	180
Width between cables (m)	27
Total deck width (m)	31
Deck depth (m)	4.4
Section of each main cable (m^2)	0.44
Mass of each main cable (t/m)	3.74
Moment of inertia under vertical bending I_y (m^4)	4.0
Moment of inertia under lateral bending I_z (m^4)	100.0
Moment of inertia under torsion J (m^4)	7.6
Deck mass (t/m)	15.26
Polar moment of inertia for deck (tm^2/m)	1106.8

Figure 10.1.1: Great Belt Bridge.

The analytical method of sensitivity analysis entails knowing the modes and natural frequencies, their sensibilities vis à vis the mechanical parameters (I_y, I_z or J) as well as the values for the flutter speed U_f and reduced frequency K_f obtained in the aeroelastic analysis. Directly from these data it is possible to get the following sensitivities:

$$\frac{\partial U_f}{\partial I_y}, \frac{\partial K_f}{\partial I_y}, \frac{\partial U_f}{\partial I_z}, \frac{\partial K_f}{\partial I_z}, \frac{\partial U_f}{\partial J}, \frac{\partial K_f}{\partial J} \tag{10.1.1}$$

One can prove these derivatives through the centred finite differential method, the same one used in the case studies related to bridges under construction. This is done for two step sizes: $\Delta x = 0.01x$, $\Delta x = 0.05x$, that is, by calculating the centred finite differential of 1% and 5% for each design variable x.

10.1.1 Sensitivity of the aeroelastic analysis with 2 modes for the Great Belt

This analysis takes into account the two vibration modes described in Table 10.1.2

The results of these aeroelastic analyses, carried out using these two modes for design variable $x = I_y$, are shown in Table 10.1.3. Table 10.1.4 compares sensitivities in relation to I_y obtained by using the *FLAS* code and finite differential calculations.

Relying on the same methodology, one can test the sensitivity analysis for the other design variables: inertia under lateral bending I_z and under torsion J. The results are analysed in Tables 10.1.5–10.1.8.

Table 10.1.2: Frequencies and modes for bimodal analysis of Great Belt

Mode no.	ω (rad/s)	Type
2	0.613	VS
22	1.859	TS

Table 10.1.3: Aeroelastic analysis of the Great Belt using two modes and variation of I_y.

% I_y	I_y (m^4)	U_f (m/s)	K_f
95	3.80	89.96	0.4488854
99	3.96	89.92	0.4492886
100	4.00	89.91	0.4494074
101	4.04	89.91	0.4494954
105	4.20	89.87	0.4498980

Table 10.1.4: Comparing sensitivities in terms of I_y from bimodal study of Great Belt.

Sensitivity	$\partial U_f/\partial I_y$	$\partial K_f/\partial I_y$
FLAS	−0.208340	0.0025370
Differential del 1%	−0.125	0.0025850
Differential del 5%	−0.2105	0.0025315

Table 10.1.5: Aeroelastic analysis of Great Belt using two 2 modes and variation of I_z.

%I_z	I_z(m^4)	U_f(m/s)	K_f
95	95	89.91	0.4491259
99	99	89.91	0.4493918
100	100	89.91	0.4494074
101	101	89.91	0.4494245
105	105	89.92	0.4493862

Table 10.1.6: Compaing sensitivities with respect to I_z in Great Belt using two modes.

Sensitivity	$\partial U_f/\partial I_z$	$\partial K_f/\partial I_z$
FLAS code	0.0050738	−0.00000186
Deference of 1%	0.0(*)	0.00001635
Differential of 5%	0.001	−0.00000253

Table 10.1.7: Aeroelastic analyses of Great Belt using two modes and variation of *J*.

% *J*	*J* (m^4)	U_f (m/s)	K_f
95	7.220	89.09	0.4509694
99	7.524	89.75	0.4497437
100	7.600	89.91	0.4494074
101	7.676	89.07	0.4491285
105	7.980	89.71	0.4479059

Table 10.1.8: Comparing sensitivities in terms of *J* using two modes for Great Belt.

Sensitivity	$\partial U_f/\partial J$	$\partial K_f/\partial J$
FLAS code	2.13544	−0.0040078
Differential of 1%	2.10526	−0.0040467
Differential of 5%	2.13157	−0.0040309

(*) The null value from the 1% differential occurs because the variation of I_z causes a variation below 0.01 m/s, the accuracy with which the flutter speed is calculated.

In the previous tables, it is possible to notice how the derivative values obtained through finite differentials bear a fairly close resemblance to the ones ahieved with the method developed by the *FLAS* code. It is worth noting the derivatives calculated through finite differentials with increments of 5% are the closest to the ones produced by *FLAS*, in which 1% increments were used.

The derivatives of U_f related to inertia under torsion J are considerably higher than those obtained from inertia under bending I_y and I_z. This suggests that it is the design variable with greatest influence on flutter speed. From what can be observed, the derivative related to I_y is one order of magnitude less than J, while the values that are almost null for derivatives in terms of I_z indicate, that, when this parameter is varied, the bridge response is hardly altered. Looking at the derivatives' signs, it is possible to see how the increments of J favourably raise the flutter speed. However, a surprising outcome is that the derivatives of U_f with respect to I_y are negative. This result indicates that, for the design in which the sensitivity is being calculated, an increase in the inertia under vertical bending leads to a decrease in the flutter speed.

10.1.2 Sensitivity of the aeroelastic analysis of the Great Belt using 18 modes

This section will follow the same methodology used earlier, although this time18 modes will be employed in the aeroelastic analysis, as displayed in Table 10.1.9. Table 10.1.10–10.1.15 show how the flutter speed U_f and reduced frequency K_f change once the deck properties I_y, I_z or J have been modified, and the corresponding sensitivity been done.

Table 10.1.9: Frequencies and modes in the analysis of the Great Belt using 18 modes

Mode no.	ω (rad/s)	Type
1	0.331	LS
2	0.613	VS
3	0.704	VA
4	0.789	LA
5	0.835	VS
6	1.134	VA
11	1.293	VA
12	1.388	VS
13	1.391	LS
18	1.575	VS
19	1.777	VA
21	1.850	VA
22	1.859	TS
27	2.164	VS
28	2.223	LTA
29	2.438	LS
30	2.470	TA
36	2.830	TS

Table 10.1.10: Aeroelastic analysis of Great Belt using 18 modes and variation of I_y.

% I_y	I_y (m^4)	U_f (m/s)	K_f
95	3.80	62.37	0.8374646
99	3.96	62.40	0.8371599
100	4.00	62.41	0.8370600
101	4.04	62.42	0.8369642
105	4.20	62.47	0.8364283

Table 10.1.11: Comparing sensitivities related to I_y in the Great Belt, using 18 modes.

Sensitivity	$\partial U_f/\partial I_y$	$\partial K_f/\partial I_y$
FLAS code	−0.260296	0.00256136
Differential of 1%	−0.25	0.00244625
Differential of 5%	−0.25	0.00259075

Table 10.1.12: Aeroelastic analyses of Great Belt using 18 modes anda variation of I_z.

% I_z	I_z (m^4)	U_f (m/s)	K_f
95	95	62.40	0.8372286
99	99	62.41	0.8371251
100	100	62.41	0.8370600
101	101	62.41	0.8370727
105	105	67.00	0.8369633

Table 10.1.13: Comparing sensitivities with respect to I_z in the Great Belt using 18 modes.

Sensitivity	$\partial U_f/\partial I_z$	$\partial K_f/\partial I_z$
FLAS code	0.00250909	−0.00002617
Differential of 1%	0.0(*)	−0.00002620
Differential of 5%	0.002	−0.00002653

Table 10.1.14: Aeroelastic analysis of Great Belt using 18 modes and a variation of J.

% J	J (m^4)	U_f (m/s)	K_f
95	7.220	61.66	0.8428039
99	7.524	62.25	0.8383710
100	7.600	62.41	0.8370600
101	7.676	62.57	0.8358302
105	7.980	63.22	0.8304440

Table 10.1.15: Comparing sensitivities with respect to J in the Great Belt using18 modes.

Sensitivity	$\partial U_f/\partial J$	$\partial K_f/\partial J$
FLAS code	1.94929	−0.0146621
Differential of 1%	2.10526	−0.0167157
Differential of 5%	2.05263	−0.0162630

The tables above reveal how the sensitivities calculated through finite differentials are fairly close to the ones produced with the *FLAS* code. In terms of the magnitude for the derivatives, values in the wind speed derivatives U_f, exceeding the reduced frequency K_f value can be seen. As was the case with the bimodal analysis, it is also possible to detect that the derivatives related to J are one magnitude higher than ones related to I_y, while the derivatives referring to I_z are almost null.

As for the sign of the derivative, it can once again be confirmed that an increase in the inertia under vertical bending has a negative influence on flutter speed: that is, as I_y goes up, U_f goes down.

10.1.3 Comparison of the sensitivity analyses for the Great Belt

It has already been stated that testing through finite differentials entails a great deal of dynamic and aeroelastic calculations for the bridge. On the other hand, the method used by *FLAS* to obtain the sensitivities is straightforward: it provides, on par with each aeroelastic analysis, the derivatives values corresponding with the deck's mechanical parameters. Its rapid results allow one to compare the aeroelastic and sensitivity analyses, considering all the possible vibration modes one would like to include.

Therefore, 17 sensitivity analyses could be carried out; these are compared in Tables 10.1.16–10.1.18. In each of these, a new vibration mode is included, specifically the one corresponding to the lowest natural frequency among the modes left out of the previous calculation.

A perusal of the results in the tables above reveals how the flutter changes when a symmetrical mode is introduced into the calculation. As an example, when mode 18 *VS* is presented, 80.48 increases to 62.18 m/s. In contrast, when asymmetrical modes are incorporated into the calculation, the results for speed or frequency do not change. As for sensitivity values, these can also be seen to alter when symmetrical modes are introduced. A case in point is, when mode 12 *VS* is presented, all the sensitivity values undergo a noticeable change. In the Great Belt Bridge, flutter is produced with the coupling of the symmetrical modes, and it is therefore essential to include as many symmetrical modes as possible in these aeroelastic and sensitivity analyses.

10.1.4 Flutter speed in modified designs for the Great Belt Bridge

Interesting data produced by the sensitivity analysis that would be of great utility to design engineers can be taken from the bank of hypothetical values reached by the flutter speed once the design variables have been changed.

Table 10.1.16: Sensitivities from the aeroelastic analyses for the Great Belt related to I_y.

Analysis, modes included	U_f	K_f	$\partial U_f/\partial I_y$	$\partial K_f/\partial I_y$
(1°) modes 2VS y 22TS	89.91	0.4494074	−0.208340	0.002537
(2°)=(1°) + mode 1LS	89.91	0.4494087	−0.208345	0.002537
(3°)=(2°)+mode 3VA	89.91	0.4494087	−0.208345	0.002537
(4°)=(3°)+mode 4LS	89.91	0.4494087	−0.208345	0.002537
(5°)=(4°)+mode 5VS	89.43	0.4568964	−0.229649	0.002790
(6°)=(5°)+mode 6VA	89.43	0.4568964	−0.229649	0.002790
(7°)=(5°)+mode 11VA	89.43	0.4568964	−0.229649	0.002790
(8°)=(7°)+mode 12VS	80.44	0.5859156	−0.270669	0.031215
(9°)=(8°)+mode 13LS	80.48	0.5857527	−0.270323	0.031138
(10°)=(9°)+mode 18VS	62.18	0.8407247	−0.195918	0.001567
(11°)=(10°)+mode 19VA	62.18	0.8407247	−0.195918	0.001567
(12°)=(11°)+mode 21VA	62.18	0.8407247	−0.195918	0.001567
(13°)=(12°)+mode 27VS	62.21	0.8401298	−0.190123	0.001469
(14°)=(13°)+mode 28LTA	62.21	0.8401298	−0.190123	0.001469
(15°)=(14°)+mode 29LS	62.21	0.8401237	−0.190438	0.001484
(16°)=(15°)+mode 30TA	62.21	0.8401237	−0.190438	0.001484
(17°)=(16°)+mode 36TS	62.41	0.8370600	−0.260296	0.002561

Table 10.1.17: Sensitivities from the aeroelastic analyses of Great Belt with respect to I_z.

Analysis, modes included	U_f	K_f	$\partial U_f/\partial I_z$	$\partial K_f/\partial I_z$
(1°) modes 2VS y 22TS	89.91	0.4494074	0.001507	−0.0000018
(2°)=(1°) + mode 1LS	89.91	0.4494087	0.001506	−0.0000018
(3°)=(2°)+mode 3VA	89.91	0.4494087	0.001506	−0.0000018
(4°)=(3°)+mode 4LS	89.91	0.4494087	0.001495	−0.0000018
(5°)=(4°)+mode 5VS	89.43	0.4568964	0.001495	−0.0000017
(6°)=(5°)+mode 6VA	89.43	0.4568964	0.001495	−0.0000017
(7°)=(6°)+mode 11VA	89.43	0.4568964	0.001495	−0.0000017
(8°)=(7°)+mode 12VS	80.44	0.5859156	0.000924	0.0000008
(9°)=(8°)+mode 13LS	80.48	0.5857527	0.002275	−0.0000037
(10°)=(9°)+mode 18VS	62.18	0.8407247	0.002499	−0.0000262
(11°)=(10°)+mode 19VA	62.18	0.8407247	0.002499	−0.0000262
(12°)=(11°)+mode 21VA	62.18	0.8407247	0.002499	−0.0000262
(13°)=(12°)+mode 27VS	62.21	0.8401298	0.002497	−0.0000261
(14°)=(13°)+mode 28LTA	62.21	0.8401298	0.002497	−0.0000261
(15°)=(14°)+mode 29LS	62.21	0.8401237	0.002492	−0.0000261
(16°)=(15°)+mode 30TA	62.21	0.8401237	0.002492	−0.0000261
(17°)=(16°)+mode 36TS	62.41	0.8370600	0.002509	−0.0000261

Table 10.1.18: Sensitivities from the aeroelastic analyses for Great Belt with respect to J.

Analysis, included modes	U_f	K_f	$\partial U_f/\partial J$	$\partial U_f/\partial J$
(1°) modes 2VS y 22TS	89.91	0.4494074	2.13544	−0.0040078
(2°)=(1°) + mode 1LS	89.91	0.4494087	2.13543	−0.0040078
(3°)=(2°)+mode 3VA	89.91	0.4494087	2.13543	−0.0040078
(4°)=(3°)+mode 4LS	89.91	0.4494087	2.13543	−0.0040078
(5°)=(4°)+mode 5VS	89.43	0.4568964	2.35261	−0.0070026
(6°)=(5°)+mode 6VA	89.43	0.4568964	2.35261	−0.0070026
(7°)=(6°)+mode 11VA	89.43	0.4568964	2.35261	−0.0070026
(8°)=(7°)+mode 12VS	80.44	0.5859156	1.52570	−0.0024791
(9°)=(8°)+mode 13LS	80.48	0.5857527	1.52669	−0.0025252
(10°)=(9°)+mode 18VS	62.18	0.8407247	2.98192	−0.0185495
(11°)=(10°)+mode 19VA	62.18	0.8407247	2.18192	−0.0185495
(12°)=(11°)+mode 21VA	62.18	0.8407247	2.18192	−0.0185495
(13°)=(12°)+mode 27VS	62.21	0.8401298	2.18733	−0.0186054
(14°)=(13°)+mode 28LTA	62.21	0.8401298	2.18733	−0.0186055
(15°)=(14°)+mode 29LS	62.21	0.8401237	2.18640	−0.0185933
(16°)=(15°)+mode 30TA	62.21	0.8401237	2.18640	−0.0185933
(17°)=(16°)+mode 36TS	62.41	0.8370600	1.94929	−0.0146621

Given an initial stage defined by (I_y, I_z, J), to which the critical speed U_f, corresponds, the flutter speed related to a modified design defined by $(I_y + DI_y, I_z + DI_z, J + DJ)$ is expressed as U_{fm} and, if a linear approximation is acceptable, can be written as

$$U_{fm} \approx U_f + \sum_{i=1}^{3} \frac{\partial U_f}{\partial x_i} \Delta x_i \qquad (10.1.2)$$

Table 10.1.19: Modified flutter speed in m/s changing a design varible.

U_{fm} (m/s)	$\partial U_f/\partial I_y = -0.260296$	$\partial U_f/\partial I_z = 0.002509$	$\partial U_f/\partial J = 1.94929$
– 20%	62.62	62.36	59.44
– 10%	62.51	62.38	60.93
– 5%	62.46	62.39	61.67
0%	62.41	62.41	62.41
+ 5%	62.35	62.42	63.15
+ 10%	62.30	62.43	63.89
+ 20%	62.20	62.46	65.37

Table 10.1.20: Modified flutter speed in m/s changing the three design variables.

DI_y	DI_z	DJ	U_{fm}
20%	20%	20%	65.21
–20%	20%	20%	**65.63**
20%	–20%	20%	65.11
20%	20%	–20%	59.29
–20%	–20%	20%	65.53
–20%	20%	–20%	59.71
20%	–20%	–20%	59.19
–20%	–20%	–20%	59.61

where

$$\Delta x_1 = \Delta I_y, \Delta x_2 = \Delta I_z, \Delta x_3 = \Delta J \tag{10.1.3}$$

In Table 10.1.19 new flutter speeds were calculated for various increments in each of the design variables until values of ±20% were reached. This was done while working with the aeroelastic analysis using 18 vibration modes for the Great Belt Bridge, which is the most complete study among those presented here.

The results on the table show how the variations in the inertias under bending, I_y or I_z, fail to produce perceptible changes in the flutter speed, and for this reason the design is not very sensitive to those parameters. As for the inertia moment under bending J, the sensitivities are more significant, to the order of ±5%.

Combining the largest increments for each one of the design parameters, one can create a new table containing the most extreme values speeds that have been modified from U_{fm} for the intervals of variation given earlier. These values appear in Table 10.1.20.

The highest speed in the table takes a value of 65.63 m/s, a 5% difference from the one obtained in the real design. By way of comment, it can be said that, overall, the Great Belt Bridge only ends up being sensitive to major changes in its inertia under torsion and the numerical values for modified flutter speed are reduced.

10.2 Example 2. Akashi Strait Bridge

The properties for the Akashi Strait Bridge, pictured in Figure 10.2.1, are presented again in Table 10.2.1. Among the most noteworthy of aeroelastic studies is one by Katsuchi[3]

Figure 10.2.1: Akashi Strait Bridge.

Table 10.2.1: Geometrical and mechanical properties of Akashi Strait Bridge.

Main span length (m)	1990
Side span length (m)	960
Tower height (m)	297
Width between cables (m)	35.5
Total deck width (m)	35.5
Deck depth (m)	14
Section of each main cable (m^2)	0.79
Moment of inertia under vertical bending Iy (m^4)	24
Moment of inertia under lateral bending Iz (m^4)	130
Moment of inertia under torsion J (m^4)	17.8
Deck mass (t/m)	28.7
Polar moment of inertia for deck (tm^2/m)	5800

at John Hopkins University. Others include the experimental testing on a full model of the bridge on a scale of 1/100, done by Miyata *et al.*[4]. The deck's mechanical properties have been taken from Cobo[2].

For this example the sensitivity analyisis is also obtained for the flutter speed U_f and the reduced frequency K_f related to the mechanical parameters of the deck (I_y, I_z o J).

$$\frac{\partial U_f}{\partial I_y}, \frac{\partial K_f}{\partial I_y}, \frac{\partial U_f}{\partial I_z}, \frac{\partial K_f}{\partial I_z}, \frac{\partial U_f}{\partial J}, \frac{\partial K_f}{\partial J} \tag{10.2.1}$$

Testing these derivatives will be done through finite differentials, with step sizes of $\Delta x = 0.01x$ e $\Delta x = 0.05x$: in other words, the differentials will be calculated by varying between 1% and 5% any design variable x.

10.2.1 Sensitivity of aeroelastic analysis using two modes for the Akashi Strait Bridge

One must recall that this analysis was done by taking into account two vibration modes described in Table 10.2.2.

The results from the aeroelastic analyses performed with these two modes for different values for design variable $x = I_y$, can be found in Table 7.2.3. Table 10.2.4 compares the sensitivities related to I_y, obtained with finite differential calculations by applying the *FLAS* code.

The null value (*) for the finite differential of 1% is due to the fact that the flutter speed is obtained through an incremental process while searching for a critical value. In this process 0.005 m/s of speed is being added in the last stages of the search. Consequently, the final values finales calculated for the critical speed have only significant two decimals. When the speed variations with respect to the mechanica parameters are below 0.01 m/s, the finite differential offers no data and becomes null.

By using the same methodology it is possible to test the sensitivity analysis for the other design variables: inertia under lateral bending $x = I_z$ and inertia under torsion $x = J$. The results are analysed in Tables 10.2.5–10.2.8.

Table 10.2.2: Frequencies and modes for the bimoda analysis of the Akashi Strait Bridge.

Mode no.	ω (rad/s)	Type
2	0.416	VS
12	1.027	TLS

Table 10.2.3: Aeroelastic analyses for Akashi Bridge using two modes and a variation of I_y.

% I_y	I_y (m^4)	U_f (m/s)	K_f
95	22.80	77.68	0 4555339
99	23.76	77.69	0 4554671
100	24.00	76.69	0 4554658
101	24.24	77.69	0 4554339
105	25.20	77.70	0 4553983

Table 10.2.4: Comparing sensitivities with respect to I_y in the Akashi Strait Bridge by means of a bimodal analysis.

Sensitivity	$\partial U_f / \partial I_y$	$\partial K_f / \partial I_y$
FLAS	−0.0093861	−0.00006265
Differential based on 1%	0.0 (*)	−0.00006916
Differential based on 5%	−0.0083333	−0.00005650

Table 10.2.5: Aeroelastic analyses for the Akashi Strait Bridge using two modes and a variation of I_z

% I_z	I_z(m^4)	U_f (m/s)	K_f
95	123.5	73.63	0.4772100
99	128.7	76.44	0.4620242
100	130.0	77.69	0.4554658
101	131.3	79.30	0.4471693
105	136.5	92.55	0.3871338

Table 10.2.6: Comparing sensitivities related to I_z on the Akashi Bridge with bimodal study.

Sensitivity	$\partial U_f / \partial I_z$	$\partial K_f / \partial I_z$
FLAS	1.08682	−0.00567140
Differential based on 1%	1.1	−0.00571342
Differential based on 5%	1.45538	−0.00692893

Table 10.2.7: Aeroelastic analyses of the Akashi Strait Bridge using two modes and a variation of *J*.

%*J*	*J*(m^4)	U_f (m/s)	K_f
95	16.910	82.47	0.4274294
99	17.622	78.34	0.4512766
100	17.800	77.69	0.4554658
101	17.978	77.15	0.4590769
105	18.690	75.84	0.4692287

Table 10.2.8: Comparing sensitivities related to *J* in the Akashi Bridge in a bimodal study.

Sensitivity	$\partial U_f / \partial I_z$	$\partial K_f / \partial I_z$
FLAS	−3.32425	0.0219147
Differential based on 1%	−3.34269	0.0219109
Differential based on 5%	−3.72472	0.0234827

An analysis of the tables above brings to light how the derivatives from the *FLAS* code are similar to those calculated through finite differentials. With this example, it seems that the method developed here offers values closer to those provided by the differentials of 1%.

Glancing at the magnitudes of the derivatives, one can see once again that those related to *J* are higher, although its negative sign does not appear to be very logical. In any case, it is important to realize that the aeroelastic analysis of a structure like the Akashi Strait

Bridge, whose main span is 2000 m, cannot be considered correctly executed if only two vibration modes have been taken into account.

10.2.2 Sensitivities from the 17-mode aeroelastic analysis of the Akashi Strait Bridge

Here, the aim is to work with the 17 modes described in the section on the bridge's dynamic analysis. The natural frequencies are given in Table 10.2.9.

In the following tables it is possible to discern how the speed U_f and reduced frequency K_f change once the deck properties I_y, I_z or J are modified and the related sensitivity analyses are carried out.

In Table 10.2.10, the reduced frequency value takes a considereable leap when it reaches $I_y = 24.24$ m^4 and after $I_y = 24.48$ m^4 is used. This occurs because the flutter phenomenon changes and other modes become responsible for the instability. Therefore it makes no sense to work the derivative through a finite centred differential, with a 2% or higher step, because such a derivative has yet to exist. It is, however, possible to obtain directional derivatives or, in other words, use the finite differential from further back or further forward to calculate the derivative to the right or to the left of those values considered for the Akashi Strait Bridge. This lack of continuity is represented in Figure 10.2.2.

Table 10.2.9: Frequencies and modes for the 17-mode analysis of the Akashi Strait Bridge.

Mode no.	ω (rad/s)	Type	Mode no.	ω (rad/s)	Type
1	0.2836	LS	14	1.0939	VA
2	0.4160	VS	24	1.3797	TA
3	0.4623	VA	25	1.4263	VS
4	0.5339	VA	26	1.5019	TA
5	0.5943	LA	27	1.573	LA
6	0.7708	VS	28	1.640	TS
10	0.9934	LTS	29	1.771	VA
11	1.0196	VS	38	2.0172	TS
12	1.0275	TLS			

Table 10.2.10: Aeroelastic analysis of the Akashi Strait Bridge using17 modes and a variation of I_y

% I_y	I_y (m^4)	U_f (m/s)	K_f
97	23.28	93.33	0.5064281
98	23.52	93.33	0.5064268
99	23.76	93.33	0.5064256
100	24.00	93.33	0.5064244
101	24.24	93.33	0.5064232
102	24.48	93.33	0.3607419
103	24.72	93.23	0.3611787

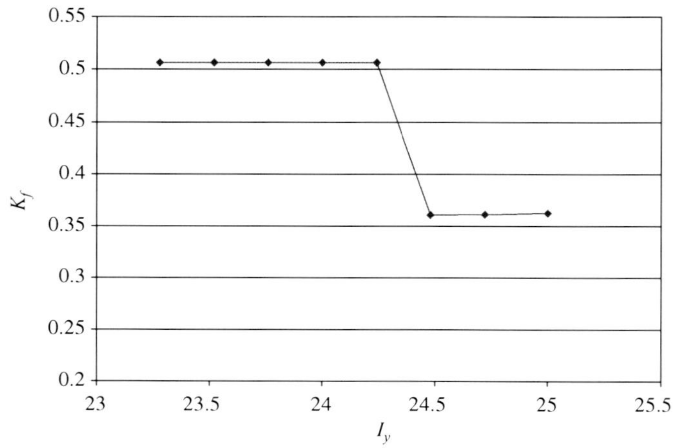

Figure 10.2.2 Values for reduced frequency according to I_y.

Table 10.2.11: Comparing sensitivities related to I_y in the Akashi Bridge using 17 modes.

Sensitivity	$\partial U_f / \partial I_y$	$\partial U_f / \partial I_y$
FLAS code $I_y = 24$ m^4	−0.000787	−0.0000049
FLAS code $I_y = 24.48$ m^4	−0.422404	0.00184
Differential further back from 1%	0.0 (*)	−0.0000050
Differential further back from 2%	0.0 (*)	−0.0000050
Differential forward from 1%	−0.416666	0.00182

Table 10.2.12: Aeroelastic analysis of the Akashi Strait Bridge using 17 modes and a variation of I_z.

% I_z	I_z (m^4)	U_f (m/s)	K_f
97	126.1	93.33	0.5064029
98	127.4	93.33	0.5064101
99	128.7	93.33	0.5064173
100	130.0	93.33	0.5064244
101	131.3	93.33	0.5064244
102	132.6	93.33	0.3609317
103	133.9	93.26	0.3612977

Table 10.2.11 demonstrates how the finite differentials from further back produce values close to the semianalytical derivative in I_y, = 24 m^4. The finite differential from further forward gives a value that approaches the one obtained in *FLAS* by calculating the derivative in 1.02. $I_y = 24.48$ m^4. The values for the derivative of U_f marked by the asterisk (*) cannot be obtained through differentials because the variation falls below 0.01 m/s.

Similarly to what occurred with I_y, Table 10.2.12 shows how, when I_z is increased by 2%, there is a sharp rise in the reduced frequency value, indicating a change in the form taken by flutter.

In Table 10.2.13 the finite differentials from further back give values that are similar to the derivative from the *FLAS* code in I_z = 130 m^4, while the finite differential from further forward produces a value that closely resembles the one obtained by calculating the derivative in 1.02. I_z = 132.6 m^4. (*) As is the case with previous derivatives, the variations of U_f below 0.01 m/s cannot be tested through finite differentials.

In Table 10.2.14, it is possible to see how, when J is increased, the reduced frequency value is around 0.50; when it is decreased, that value is closer 0.37. Consequently, the change in flutter response produced is very close to the value chosen for the bridge's inertia under torsion J = 17.8 m^4.

In Table 10.2.15, it can be observed the derivatives from finite differentials that have been moved forward coincide with the semi-analytical derivative from J considered for

Table 10.2.13: Comparing sensitivities related to I_z In the Akashi Bridge using 17 modes.

Sensitivity	$\partial U_f / \partial I_z$	$\partial K_f / \partial I_z$
FLAS code I_z = 130 m^4	0.000573	0.0000019
FLAS code I_z = 132.6 m^4	−0.045866	0.0002714
Differential further back from 1%	0.0 (*)	0.0000054
Differential further back from 2%	0.0 (*)	0.0000055
Differential forward from 1%	−0.053846	0.0002815

Table 10.2.14: Aeroelastic analyses of the Akashi Strait Bridge using 17 modes and a variation of J.

% J	J(m^4)	U_f (m/s)	K_f
97	17.266	89.01	0.3777937
98	17.444	90.45	0.3719550
99	17.622	91.82	0.3665753
100	17.800	93.33	0.5064244
101	17.978	93.64	0.5063586
102	18.156	93.95	0.5062911
103	18.334	94.26	0.5062216

Table 10.2.15: Comparing sensitivities related to J in the Akashi Bridge using 17 modes.

Sensitivity	$\partial U_f / \partial J$	$\partial K_f / \partial J$
FLAS code J = 17.8 m^4	1.73575	−0.0003315
FLAS code J = 17.444 m^4	7.87026	−0.0314479
FLAS code J = 18.156 m^4	1.73594	−0.0003525
Differential further back from 1%	7.69662	−0.0302230
Differential further back from 2%	7.89325	−0.0315123
Differential forward from 1%	1.74157	−0.0003696
Differential forward from 2%	1.74157	−0.0003744
Differential forward from 3%	1.74157	−0.0003851

this bridge and in values that are higher than this one ($J + \Delta J$) However, the derivatives that have been moved back correspond to those found in lower inertias ($J - \Delta J$). This table also reveals how the derivatives from finite differentials just about coincide with the values got through the *FLAS* code. This excellent result is due to the fact that the derivatives' magnitude in terms of the moment of inertia under torsion far exceeds those under bending I_y or I_z.

It is only logical that the sign for the derivative related to J is positive in this 17-mode analysis. This result brings to light the deceptive nature of the derivative sign for this design variable obtained from the analysis with two vibration modes. It has already been mentioned that, thanks to the bridge's exceptional length, the deformation cannot be adequately represented with these two modes alone. For this reason, the very derivative is affected.

The sensitivities from this 17-mode aeroelastic analysis have been fairly similar in magnitude and sign to the ones obtained for the Great Belt Bridge. It is again the case that the derivatives for J are positive, suggesting that an increase in this parameter will in turn raise the flutter speed. The derivatives for I_y are negative and have lower orders of magnitude. Thus, calculating backward or forward influences their value. Finally, this example resembles the last one in that the derivatives for I_z are mostly from a lesser modulus.

10.2.3 Comparing the sensitivity analyses for the Akashi Strait Bridge

As was done with the Great Belt example, diverse sensitivity analyses are presented so that it is possible to consider other combinations of vibration modes for the bridge.

By analyzing Tables 10.2.16–10.2.18, it is easy to observe the vast differences between the values resulting from the bimodal analyses and those coming out of the analyses using multiple vibration modes. Comparing the three mulitmodal aeroelastic analyses, one can

Table 10.2.16: Sensitivities for I_y from the aeroelastic analyses of the Akashi Strait Bridge.

Modes	U_f	K_f	$\partial U_f / \partial I_y$	$\partial K_f / \partial I_y$
2, 12	77.69	0.4554658	0.009386	−0.000062
3, 24	96.06	0.4906476	0.005696	−0.000037
Symmetrical	93.55	0.3597848	−0.435841	0.001892
Asymmetrical	93.34	0.5063988	−0.000785	−0.000005
17 modes	93.33	0.5064244	−0.000787	−0.000005

Table 10.2.17: Sensitivities for I_y from the aeroelastic analyses of the Akashi Strait Bridge.

Modes	U_f	K_f	$\partial U_f / \partial I_z$	$\partial K_f / \partial I_z$
2, 12	77.69	0.4554658	1.086820	−0.005671
3, 24	96.06	0.4906476	0.000805	0.000001
Symmetrical	93.55	0.3597848	−0.089170	0.000472
Asymmetrical	93.34	0.5063988	−0.000611	0.000005
17 modes	93.33	0.5064244	−0.000573	0.000002

Table 10.2.18: Sensitivities for J from the aeroelastic analyses of the Akashi Strait Bridge.

Modes	U_f	K_f	$\partial U_f / \partial J$	$\partial K_f / \partial J$
2, 12	77.69	0.4554658	−3.324250	0.021914
3, 24	96.06	0.4906476	1.773175	−0.000247
Symmetrical	93.55	0.3597848	10.56924	−0.041167
Asymmetrical	93.34	0.5063988	1.738016	−0.000342
17 modes	93.33	0.5064244	1.735750	−0.000331

trace a similar flutter speed, 93 m/s. In the calculation considering symmetrical modes, the reduced frequency is 0.35. However, the value is 0.50 in the other two analyses. Both of these are multimodal, with one using asymmetrical modes and the other, 17 modes. This points to the existence of two forms of flutter in the Akashi Strait Bridge. In both, the flutter speed is similar, but, in one, the symmetrical modes are coupled and the same is true of the asymmetrical modes in the other. Between them, the response frequency changes.

The sensitivity analyses done through the *FLAS* code, either considering all the vibration modes or only working with the asymmetrical ones, are almost identical. This seems to confirm that the forms of symmetrical vibration have hardly any influence on the calculation using all the modes.

10.2.4 Flutter speed in modified designs of the Akashi Strait Bridge

The next procedure mirrors what was done in the previous example: sensitivity analyses to calculate the approximate values which the flutter speed would have when the design variables are modified. Once again this expression is employed:

$$U_{fm} \approx U_f + \sum_{i=1}^{3} \frac{\partial U_f}{\partial x_i} \Delta x_i \tag{10.2.2}$$

where

$$\Delta x_1 = \Delta I_y , \Delta x_2 = \Delta I_z , \Delta x_3 = \Delta J \tag{10.2.3}$$

Working with the 17-vibration mode aeroelastic analysis of the Akashi Strait Bridge, the new flutter speeds have been calculated in Table 10.2.19 for various increases in the each of the design variables.

The table demonstrates how increasing or decreasing the design variables produces different variations in the flutter speed because there are various derivatives according to the direction in which those variables are altered.

The Akashi Strait Bridge is like its counterpart in the Great Belt Changes produced in the flutter speed by I_y and I_z are much lower than those caused by variations in J, whose impact is notable. Thus, decreasing inertia under torsion by 20% reduces flutter speed by 30%, while an increase of the same value manages to raise that speed, but on a lesser scale by only 7%.

Modifying at the same time the three design variables through increments of ±20%, one gets the values for U_{fm} indicated in Table 10.2.20.

There is a similarity between the values in this table and those in Table 7.2.21. It can be said that the Akashi Strait Bridge is fairly insensitive to small variations in I_y and I_z and just the opposite with variations in J. By way of contrast with the results from the Great Belt Bridge, it seems that the truss cross section is more sensitive to J, for span lengths close to m, than is the case with the aerodynamic box girder section.

10.3 Example 3. Original Tacoma Bridge

Because of its historical importance and its presence in Chapter 5, the original Tacoma Bridge, Figure 10.3.1, has been included among the case studies for sensitivity analysis. It main features are provided in Table 10.3.1. The geometric dimensions were taken from the plans made by the Federal Work Agency[5] and an article by Herzog[6].

Just as before, the *FLAS* code is used to see how the flutter speed U_f and reduced frequency K_f vary in terms of the deck's mechanical parameters (I_y, I_z or J), that is

$$\frac{\partial U_f}{\partial I_y}, \frac{\partial K_f}{\partial I_y}, \frac{\partial U_f}{\partial I_z}, \frac{\partial K_f}{\partial I_z}, \frac{\partial U_f}{\partial J}, \frac{\partial K_f}{\partial J} \tag{10.3.2}$$

Following the same procedure used with the other case studies here, these derivatives will be tested through finite differentials with varying step sizes, $\Delta x = 0.01x$, $\Delta x = 0.05x$.

Table 10.2.19: Modified flutter speed in m/s changing a design variable.

U_{fm} (m/s)	$\partial U_f/\partial I_y = -0.000787$	$\partial U_f/\partial I_z = 0.000573$	$\partial U_f/\partial J = 7.6966$
– 20%	93.33	93.32	65.93
– 10%	93.33	93.33	79.64
– 5%	93.33	93.33	86.48
0%	93.33	93.33	93.33
U_{fm} (m/s)	$\partial U_f/\partial I_y = -0.41666$	$\partial U_f/\partial I_z = -0.053846$	$\partial U_f/\partial J = 1.74157$
+ 5%	92.83	92.98	94.89
+ 10%	92.33	92.63	96.43
+ 20%	91.33	91.93	99.53

Table 10.2.20: Modified flutter speed in m/s changing the three design variables.

ΔI_y	ΔI_z	ΔJ	U_{fm}
20%	20%	20%	96.13
–20%	20%	20%	98.13
20%	–20%	20%	97.52
20%	20%	–20%	62.53
–20%	–20%	20%	**99.52**
–20%	20%	–20%	64.53
20%	–20%	–20%	63.92
–20%	–20%	–20%	65.92

Figure 10.3.1 Original Tacoma Bridge.

Table 10.3.1: Geometrical and mechanical properties for old Tacoma Bridge.

Main span length (m)	853.4
Side span length (m)	335.3
Tower height (m)	71
Width between cables (m)	11.9
Total deck width (m)	11.9
Deck depth (m)	2.3
Cross section of each main cable (m^2)	0.124
Mass of each main cable (t/m)	1.05
Moment of inertia under vertical bending I_y (m^4)	0.154
Moment of inertia under lateral bending I_z (m^4)	5.69
Moment of inertia under torsion J (m^4)	6.07×10^{-6}
Deck mass (t/m)	6.22
Polar moment polar for deck (tm^2/m)	106.5

10.3.1 Sensitivity from bimodal aeroelastic analysis of the Tacoma Bridge

One may recall how this analysis is done: by taking into account the two vibration modes described in Table 10.3.2.

Table 10.3.3 gives the results of the aeroelastic analyses that use these two modes for different values of the design variable $x = I_y$. Chapter 5 explains these calculations in detail. Table 10.3.4 compares the sensitivities related to I_y that have been obtained by applying the *FLAS* code with those calculated through centred finite differentials.

Table 10.3.2: Frequencies and modes in bimodal analysis of Tacoma Bridge.

Mode no.	ω (rad/s)	Type
2	0.817	VA
5	1.505	TA

Table 10.3.3: Aeroelastic analyses of the Tacoma Bridge using two modes and variable I_y.

% I_y	I_y (m^4)	U_f (m/s)	K_f
95	0.1463	11.49	1.5592532
99	0.15246	11.49	1.5592449
100	0.154	11.49	1.5592428
101	0.15554	11.49	1.5592407
105	0.1617	11.49	1.5592335

Table 10.3.4: Comparing sensitivities related to I_y in the Tacoma Bridge using two modes.

Sensitivity	$\partial U_f / \partial I_y$	$\partial K_f / \partial I_y$
FLAS code	−0.0088573	−0.0014306
Differential based on 1%	0.0(*)	−0.0013636
Differential based on 5%	0.0(*)	−0.0012792

Table 10.3.5: Aeroelastic analyses of the Tacoma Bridge using two modes and variable.

% I_z	I_z(m^4)	U_f (m/s)	K_f
95	5.4055	11.47	1.5621270
99	5.6331	11.49	1.5598118
100	5.69	11.49	1.5592428
101	5.7469	11.50	1.5580018
105	5.9745	11.53	1.5544476

(*) It has already been observed in previous case studies that, when the increase in speed ΔU_f produced by ΔI_y is less than 0.01 m/s, it is impossible to calculate the derivative $\partial U_f/\partial I_y$ through finite differentials.

By applying the same methodology here, one can test the sensitivity analysis for the other design variables: inertia under lateral bending $x = I_z$ and inertia under torsion $x = J$. The results are analysed in Tables 10.3.5–10.3.8.

In the Tacoma Bridge, the value for the inertia under torsion J is so small that the variations of 1% or 5% in its value lead to insignificant results. For this reason, it will be necessary to also evaluate the finite differential with $J = 10^{-3}$, or alter its value by various orders of magnitude.

Table 10.3.6: Comparing sensitivities related to I_z in the Tacoma Bridge using two modes.

Sensitivity	$\partial U_f / \partial I_z$	$\partial K_f / \partial I_z$
FLAS code	0.112831	−0.0133368
Differential of 1%	0.087873	−0.0159050
Differential of 5%	0.105448	−0.0134950

Table 10.3.7: Aeroelastic analyses of the Tacoma Bridge using two modes and variable J.

%J	J (m^4)	U_f (m/s)	K_f
95	5.7665×10^{-6}	11.49	1.5592407
99	6.0093×10^{-6}	11.49	1.5592428
100	6.07×10^{-6}	11.49	1.5592428
101	6.1307×10^{-6}	11.49	1.5592438
105	6.3735×10^{-6}	11.49	1.5592459
	10^{-3}	11.54	1.5605943

Table 10.3.8: Comparing sensitivities in terms of J for the Tacoma Bridge using two modes.

Sensitivity	$\partial U_f / \partial J$	$\partial K_f / \partial J$
FLAS code	49.8084	1.77195
Differential based on 1%	0.0(*)	8.23723
Differential based on 5%	0.0(*)	8.56672
Differential forwarded with $J_f = 10^{-3}$	50.3053	1.35975

Due to the almost null value of the inertia under torsion of the Tacoma Bridge, the finite differentials based on 1% and 5% provide poor results from derivatives. Nevertheless, when the finite differential is brought forward, increasing by a few orders of magnitude the inertia under torsion $J + DJ = 10^{-3}$, the values for the derivatives obtained through differentials come sufficiently close to the ones got through the *FLAS* code.

Here, the magnitude of the derivative related to J is quite considerable, undoubtedly because raising the inertia under torsion of the deck would guarantee an improvement in the bridge's behaviour. The other derivatives are very small; this suggests that altering their values make little difference. Once again it is important to look at the negative sign of derivative related to I_y, which indicates that a decreased inertia under bending would raise the critical speed, although the low value for the derivative might mean that it would be necessary to modify a great deal.

10.3.2 Sensitivity from the aeroelastic analysis using 10 modes for the Tacoma Bridge

As in the previous section, the same methodology will be followed here. This time, however, the first 10 modes of vibration will be used. Their natural frequencies are listed in Table 10.3.9.

Tables 10.3.10–10.3.15 include the flutter speed U and reduced frequency K after the deck properties I_y, I_z or J have been modified, as well as those from the corresponding sensitivity analyses. (*) In a situation similar to the one for the bimodal analysis, the increase in speed ΔU produced by the variations of I_y is below 0.01 m/s, and the derivative $\partial U/\partial Iy$ cannot be calculated through finite differentials.

Table 10.3.9: Frequencies and modes used in 10-mode analysis of the Tacoma Bridge.

Mode no.	ω (rad/s)	Type
1	0.568	LS
2	0.817	VA
3	1.189	VS
4	1.296	LA
5	1.505	TA
6	1.608	TS
7	1.705	VA
8	1.792	VS
12	2.179	VS
16	2.321	TA

Table 10.3.10: Aeroelastic analyses for the Tacoma Bridge using 10 modes and variable I_y.

% I_z	I_y(m^4)	U_f (m/s)	K_f
95	0.1463	11.49	1.5592473
99	0.15246	11.49	1.5592390
100	0.154	11.49	1.5592369
101	0.15554	11.49	1.5592348
105	0.1617	11.49	1.5592276

Table 10.3.11: Comparing sensitivities related to I_y in the Tacoma Bridge using 10 modes.

Sensitivity	$\partial U_f / \partial I_y$	$\partial K_f / \partial I_y$
FLAS code	−0.008858	−0.001429
Differential of 1%	0.0(*)	−0.001363
Differential of 5%	0.0(*)	−0.001279

Table 10.3.12: Aeroelastic analyses of the Tacoma Bridge using 10 modes and variable I_y.

% I_z	I_z(m^4)	U_f (m/s)	K_f
95	5.4055	11.46	1.5628039
99	5.6331	11.49	1.5598062
100	5.69	11.49	1.5592369
101	5.7469	11.50	1.5579956
105	5.9745	11.53	1.5551143

Table 10.3.13: Comparing sensitivities related to I_z in the Tacoma Bridge using 10 modes.

Sensitivity	dU_f / dI_z	dK_f / dI_z
FLAS code	0.112692	−0.0133232
Differential del 1%	0.087873	−0.0159103
Differential del 5%	0.123023	−0.0135112

Table 10.3.14: Aeroelastic analyses of the Tacoma Bridge using 10 modes and variable J.

%J	J (m^4)	U_f (m/s)	K_f
95	5.7665×10^{-6}	11.49	1.5592348
99	6.0093×10^{-6}	11.49	1.5592369
100	6.07×10^{-6}	11.49	1.5592369
101	6.1307×10^{-6}	11.49	1.5592380
105	6.3735×10^{-6}	11.49	1.5592400
	10^{-3}	11.54	1.5605890

Table 10.3.15: Comparing sensitivities related to J for the Tacoma Bridge using 10 modes.

Sensitivity	$\partial U_f / \partial J$	$\partial K_f / \partial J$
FLAS code	49.8246	1.77047
Differential based on 1%	0.0(*)	9.06095
Differential based on 5%	0.0(*)	8.56672
Forwarded differential with $J_f = 10^{-3}$	50.3053	1.36035

(*) Yet again the derivatives obtained through differentials with an almost null value from J are inexact. Nevertheless, a satisfactory result is achieved with the finite differential brought forward, increasing the inertia under torsion to $J + \Delta J = 10^{-3}$.

The sensitivity derivatives in the 10-mode analysis closely resemble in sign and magnitude the ones obtained in the bimodal analysis. This similarity may partly be explained by the fact that this bridge has a shorter central span (853.4 m) when compared with the bridges featured in the other case studies. On the other hand, another reason is that mode 5 under torsion is chiefly responsible for the flutter, which already appears in the bimodal study, while the remaining modes hardly play a role.

10.3.3 Comparing sensitivity analyses for the Tacoma Bridge

Equally true with the other case studies, it is easy to compare the various sensitivity analyses by considering other combinations of vibration modes for the bridge, always relying on the *FLAS* code.

By studying Tables 10.3.16–10.3.18, one can discern how the flutter speed and reduced frequency almost do not change in terms of number of modes considered.

Table 10.3.16: Sensitivities with respect to I_y from the aeroelastic analyses of the Tacoma Bridge.

Modes	U_f	K_f	$\partial U_f / \partial I_y$	$\partial K_f / \partial I_y$
2, 4	11.49	1.5592428	−0.00885737	0.00143060
1,2,3,4,5	11.49	1.5592368	−0.00885881	0.00142923
1,2,3,4,5,6	11.49	1.5592368	−0.00885880	0.00142925
1,2,3,4,5,6,7	11.49	1.5592368	−0.00885879	0.00142922
1,2,3,4,5,6,7,8	11.49	1.5592368	−0.00885881	0.00142923
1,2,3,4,5,6,7,8,12	11.49	1.5592368	−0.00885880	0.00142923
1,2,3,4,5,6,7,8,12,16	11.49	1.5592369	−0.00885878	0.00142927

Table 10.3.17: Sensitivities with respect to I_z from the aeroelastic analyses of the Tacoma Bridge.

Modes	U_f	K_f	$\partial U_f / \partial I_z$	$\partial K_f / \partial I_z$
2, 4	11.49	1.5592428	0.112831	−0.0133368
1,2,3,4,5	11.49	1.5592368	0.112693	−0.0133232
1,2,3,4,5,6	11.49	1.5592368	0.112693	−0.0133232
1,2,3,4,5,6,7	11.49	1.5592368	0.112693	−0.0133232
1,2,3,4,5,6,7,8	11.49	1.5592368	0.112693	−0.0133232
1,2,3,4,5,6,7,8,12	11.49	1.5592368	0.112693	−0.0133232
1,2,3,4,5,6,7,8,12,16	11.49	1.5592369	0.112692	−0.0133232

Table 10.3.18: Sensitivities with respect to J from the aeroelastic analyses of the Tacoma Bridge.

Modes	U_f	K_f	$\partial U_f / \partial J$	$\partial K_f / \partial J$
2, 4	11.49	1.5592428	49.8084	0.00177195
1,2,3,4,5	11.49	1.5592368	49.8247	0.00177043
1,2,3,4,5,6	11.49	1.5592368	49.8247	0.00177043
1,2,3,4,5,6,7	11.49	1.5592368	49.8246	0.00177044
1,2,3,4,5,6,7,8	11.49	1.5592368	49.8248	0.00177042
1,2,3,4,5,6,7,8,12	11.49	1.5592368	49.8247	0.00177043
1,2,3,4,5,6,7,8,12,16	11.49	1.5592369	49.8246	0.00177047

The sensitivities also show little variation across the different analyses. This behaviour is caused by the way in which the asymmetrical mode of torsion, present in all the analyses, leads to flutter in the Tacoma Bridge. The other modes have little to do with this phenomenon.

10.3.4 Flutter speed within modified designs of the Tacoma Bridge

Thanks to the sensitivity analysis and following the same pattern occurring in the previous case studies, it is possible to know the values adopted by the flutter speed when any of the design variables are changed. To this end, the expression below is applied

Table 10.3.19: Modified flutter speed in m/s changing a design variable.

U_{fm} (m/s)	$\partial U_f / \partial I_y = -0.008857$	$\partial U_f / \partial I_z = 0.112831$	$\partial U_f / \partial J = 49.8246$
– 20%	11.49	11.36	11.49
– 10%	11.49	11.42	11.49
– 5%	11.49	11.49	11.49
+ 5%	11.49	11.49	11.49
+ 10%	11.49	11.55	11.49
+ 20%	11.49	11.62	11.49

Figure 10.4.1: Vasco da Gama Bridge.

$$U_{fm} \approx U_f + \sum_{i=1}^{3} \frac{\partial U_f}{\partial x_i} \Delta x_i \tag{10.3.2}$$

where

$$\Delta x_1 = \Delta I_y, \Delta x_2 = \Delta I_z, \Delta x_3 = \Delta J \tag{10.3.3}$$

Table 10.3.19 has the results of working with the aeroelastic analysis that uses 10 vibration modes for the Tacoma Bridge. The new flutter speeds are calculated with several increases for each one of the design variables.

All the flutter speeds are insignificant when the three design variables I_y, I_z and J are modified. It seems surprising that, even though the derivative $\partial U_f / \partial J$ is higher than it is in previous examples, U_f remains unchanged when J is altered. This is because the value for the moment of inertia under torsion in this bridge is so insignificant ($J = 6.7 \times 10^{-6}$ m^4), that 20% variations lead to almost null values insufficient for raising U_f.

10.4 Example 4. Vasco Da Gama Bridge

Chapter 5 has already provided an aeroelastic study of the Vasco da Gama Bridge in Figure 10.4.1, which is a cable-stayed structure. In this section, the *FLAS* code will be used to carry out a sensitivity analysis and draw out any differences it may have from the other case studies, which deal with suspension bridges. Table 10.4.1 has the main geometrical properties and mechanical parameters for this bridge. Obtaining and testing the sensitivities

Table 10.4.1: Geometrical and mechanical properties for the Vasco da Gama.

Main span length (m)	420
Side span length (m)	62+70.7+70
Tower height (m)	150
Width between cables (m)	29.2
Total deck width (m)	30.9
Deck depth (m)	2.65
Cross section of each cable (m^2)	0.019
Moment of inertia under vertical bending I_y (m^4)	10.5
Moment of inertia under lateral bending I_z (m^4)	2300
Moment of inertia under torsion J (m^4)	6.2
Deck mass (t/m)	53.9
Polar moment of inertia for deck (tm^2/m)	6800

Table 10.4.2: Frequencies and modes for the bimodal analysis of the Vasco da Gama.

Mode no.	ω (rad/s)	tipo
2	1.704	VS
4	2.977	TS

Table 10.4.3: Aeroelastic analyses of the Vasco de Gama using two modes and a variation of I_y.

% I_y	I_z (m^4)	U_f (m/s)	K_f
95	9.975	80.21	1.1141672
99	10.395	80.17	1.1146778
100	10.500	80.16	1.1148242
101	10.605	80.15	1.1149702
105	11.025	80.11	1.1155560

Table 10.4.4: Comparing sensitivities in terms of I_y for the Vasco da Gama using two modes.

Sensitivity	$\partial U_f / \partial I_y$	$\partial K_f / \partial I_y$
FLAS code	−0.0894887	0.00130807
Differential of 1%	−0.0952381	0.00139238
Differential of 5%	−0.0952381	0.00132256

for this bridge are done in the same way as before. The geometric dimensions were taken from Capra[7], as well as the description of the bridge made by Machado and Branco[8].

10.4.1 Sensitivity from the bimodal aeroelastic analysis of the Vasco da Gama Bridge

Table 10.4.2 shows how this analysis was done using just two vibration modes. Table 10.4.2 summarises the results for the aeroelastic analyses, which employed these two modes with different values of the design variable $x = I_y$. Table 10.4.4 compares the sensitivities in terms of I_y obtained by applying the method developed in *FLAS* with those calculated through finite differentials.

Turning to the same methodology, one can test the sensitivity analysis for the other design variables: inertia under lateral bending $x = I_z$ and inertia under torsion $x = J$. An analysis of the results is shown in Tables 10.4.5–10.4.8.

From the tables above, it is possible to trace a close correlation between the differences calculated with the *FLAS* code and those obtained through finite differentials. Those of the greatest magnitude are yet again the derivatives from J. Those from I_y are negative and with a lower order of magnitude, a trend seen in the other case studies. Those from I_z are almost null.

Table 10.4.5: Aeroelastic analyses of the Vasco da Gama using two modes and a variation of I_z.

$\% I_z$	I_z (m^4)	U_f (m/s)	K_f
95	2185	80.18	1.1146778
99	2277	80.17	1.1148020
100	2300	80.16	1.1148242
101	2323	80.16	1.1148438
105	2415	80.14	1.1148991

Table 10.4.6: Comparing sensitivities in terms of I_z for the Vasco da Gama using two modes.

Sensitivity	$\partial U_f / \partial I_z$	$\partial K_f / \partial I_z$
FLAS code	−0.00017212	0.00000024
Differential of 1%	−0.00021739	0.00000090
Differential of 5%	−0.00017391	0.00000028

Table 10.4.7: Aeroelastic analyses of the Vasco da Gama using two modes and a variation of J.

$\%J$	J (m^4)	U_f (m/s)	K_f
95	5.890	79.61	1.1144070
99	6.138	80.05	1.1147625
100	6.200	80.16	1.1148242
101	6.262	80.27	1.1149416
105	6.510	80.69	1.1152433

Table 10.4.8: Comparing sensitivities in terms of J for the Vasco da Gama using two modes.

Sensitivity	$\partial U_f / \partial J$	$\partial K_f / \partial J$
FLAS code	1.74109	0.0013381
Differential of 11%	1.77419	0.0014443
Differential of 5%	1.74193	0.0013488

10.4.2 11-mode sensitivity aeroelastic analysis for the Vasco da Gama Bridge

The same methodology takes the spotlight in this section. Now, however, the focus is on the first 11 vibration modes, whose natural frequencies are detailed in Table 10.4.9. Tables 10.4.10–10.4.15 chart the changes in flutter speed U_f and reduced frequency K_f that take place when the deck properties I_y, I_z or J, are modified and also include the corresponding sensitivity analyses.

The results for the sensitivities in the 11-mode aeroelastic analysis are very similar, whether they are from calculations done with the *FLAS* code or an evaluation made through finite differentials. On comparing the derivative values obtained in this analysis with the results of the bimodal study, one does not come across changes in signs.

Table 10.4.9: Frequencies and modes in the 11-mode analysis of the Vasco da Gama.

Mode no.	ω (rad/s)	Type
1	1.490	LS
2	1.704	VS
3	1.945	VA
4	2.977	TS
5	3.513	TA
6	3.523	VS
7	4.179	TA
8	4.294	VA
9	4.877	VS
10	5.387	VA
11	5.434	TS

Table 10.4.10: Aeroelastic analyses of the Vasco da Gama using 11 modes and a variation of I_y.

% I_y	I_y(m^4)	U_f(m/s)	K_f
95	9.975	83.06	1.0736813
99	10.395	82.95	1.0750889
100	10.500	82.93	1.0754324
101	10.605	82.90	1.0758445
105	11.025	82.80	1.0772207

Table 10.4.11: Comparing sensitivities in terms of I_y for the Vasco da Gama using 11 modes.

Sensitivity	$\partial U_f / \partial I_y$	$\partial K_f / \partial I_y$
FLAS code	−0.230455	0.003163
Differential of 1%	−0.238095	0.003598
Differential of 5%	−0.247619	0.003370

Table 10.4.12: Aeroelastic analyses of the Vasco da Gama using 11 modes and a variation of I_z.

% I_z	$I_z(m^4)$	U_f (m/s)	K_f
95	2185	82.94	1.0754491
99	2277	82.93	1.0754819
100	2300	82.93	1.0754324
101	2323	82.92	1.0754491
105	2415	82.90	1.0754937

Table 10.4.13: Comparing sensitivities in terms of I_z for the Vasco da Gama using 11 modes.

Sensitivity	$\partial U_f / \partial I_z$	$\partial K_f / \partial I_z$
FLAS code	−0.000190	0.00000041
Differential of 1%	−0.000217	0.00000071
Differential of 5%	−0.000173	0.00000019

Table 10.4.14: Aeroelastic analyses of the Vasco da Gama using11 modes and a variation of *J*.

%*J*	$J(\mathrm{m}^4)$	$U_f(\mathrm{m/s})$	K_f
95	5.890	82.24	1.0766437
99	6.138	82.79	1.0757355
100	6.200	82.93	1.0754324
101	6.262	82.06	1.0752507
105	6.510	83.59	1.0742337

Table 10.4.15: Comparng sensitivities in terms of *J* for the Vasco da Gama using 11 modes.

Sensitivity	$\partial U_f / \partial J$	$\partial K_f / \partial J$
FLAS code	2.18247	−0.003977
Differential of 1%	2.17741	−0.003909
Differential of 5%	2.17741	−0.003887

However, there seems to be a slight change in the derivative value. Thus, the derivatives related to I_y more than double, and those from *J* are also somewhat bigger.

10.4.3 Comparing the sensitivity analyses for the Vasco da Gama Bridge

One can trace the same trends that were obvious in the other case studies and run tests by laying the data alongside those obtained through finite differentials from the *FLAS* code. After, it is easy to compare the results from the diverse sensitivity analyses considering other combinations of vibration modes for the bridge.

A study of Tables 10.4.16–10.4.18 reveals that the flutter speed hardly changes in terms of the number of modes considered. U_f lies between 80.16 and 82.93 m/s. The same occurs with the reduced frequency, whose values are of the order of $K_f \approx 1.1$. In contrast, the sensitivities do have a fairly strong influence on the number of modes that come into play.

Table 10.4.16: Sensitivities for I_y from the aeroelastic analyses of the Vasco da Gama.

Modes	U_f	K_f	$\partial U_f / \partial I_y$	$\partial K_f / \partial I_y$
2, 4	80.16	1.1148242	−0.089488	0.001308
4 first ones	80.16	1.1148997	−0.089592	0.001309
5 first ones	80.16	1.1148996	−0.089593	0.001309
6 first ones	82.78	1.0774689	−0.232292	0.003198
7 first ones	82.78	1.0774688	−0.232292	0.003198
8 first ones	82.78	1.0774688	−0.232293	0.003198
9 first ones	82.92	1.0755922	−0.228903	0.003142
10 first ones	82.92	1.0755922	−0.228897	0.003142
11 first ones	82.93	1.0754324	−0.230455	0.003163

Table 10.4.17: Sensitivities for I_z from the aeroelastic analyses of the Vasco da Gama.

Modes	U_f	K_f	$\partial U_f / \partial I_z$	$\partial K_f / \partial I_z$
2, 4	80.16	1.1148242	−0.0001721	0.00000024
4 first ones	80.16	1.1148997	−0.0001715	0.00000023
5 first ones	80.16	1.1148996	−0.0001713	0.00000023
6 first ones	82.78	1.0774689	−0.0001875	0.00000037
7 first ones	82.78	1.0774688	−0.0001875	0.00000037
8 first ones	82.78	1.0774688	−0.0001875	0.00000037
9 first ones	82.92	1.0755922	−0.0001912	0.00000042
10 first ones	82.92	1.0755922	−0.0001912	0.00000042
11 first ones	82.93	1.0754324	−0.0001904	0.00000041

Table 10.4.18: Sensitivities for J from the aeroelastic analyses of the Vasco da Gama.

Modes	U_f	K_f	$\partial U_f / \partial J$	$\partial K_f / \partial J$
2, 4	80.16	1.1148242	1.74109	0.0013381
4 first ones	80.16	1.1148997	1.74099	0.0013382
5 first ones	80.16	1.1148996	1.74093	0.0013391
6 first ones	82.78	1.0774689	2.15683	−0.0036821
7 first ones	82.78	1.0774688	2.15683	−0.0036822
8 first ones	82.78	1.0774688	2.15683	−0.0036822
9 first ones	82.92	1.0755922	2.17813	−0.0039256
10 first ones	82.92	1.0755922	2.17813	−0.0039257
11 first ones	82.93	1.0754324	2.18247	−0.0039779

On revising the tables, one can see that, when the symmetrical modes (modes: 6, 9 y 11) are included, there are marked variations in the sensitivities. For example, the sixth *VS* mode makes the derivative of U_f in terms of I_y go from $\partial U_f / \partial I_y = -0.089$ to $\partial U_f / \partial I_y = -0.23$. In other words, it is has grown nearly three times in size.

Moreover, the derivative of K_f related to J changes in sign, going from $\partial K_f / \partial J = 0.0013$ to $\partial K_f / \partial J = -0.0036$. Mode 9 *VS* produces a slighter change in the sensitivity values. Although mode 11 *TS* also changes them, it does so in a much more subtle manner.

What is happening is that the symmetrical modes play the most dynamic role in flutter for this bridge. Therefore, these modes have to be taken into account in any sensitivity analysis. One might conclude by saying an effective sensitivity analysis entails working with many more modes than are needed just to obtain flutter speed. It is always essential to include those that participate most in instability. In the Vasco da Gama Bridge example, these are the symmetrical modes.

10.4.4 Flutter speed in the modified design of the Vasco da Gama Bridge

For this case study, a sensitivity analysis may also be used to calculate flutter speeds when any percentages within the design variables are altered. Once again this expression makes an appearance

$$U_{fm} \approx U_f + \sum_{i=1}^{3} \frac{\partial U_f}{\partial x_i} \Delta x_i \tag{10.4.2}$$

where

$$(\Delta x_1 = \Delta I_y, \Delta x_2 = \Delta I_z, \Delta x_3 = \Delta J) \tag{10.4.3}$$

Employed here is the same aeroelastic analysis using 11 vibration modes that was applied to the Tacoma Bridge. Table 10.4.19 has the calculations for the new flutter speeds with each increment of the design variables.

Again it is the case that the variable with the greatest impact on flutter speed is inertia under torsion J, with a 4% difference. The changes produced by modifying I_y and I_z are modest. By altering all three design variables and applying 20% increments, one can get the values for U_{fm} that are given in Table 10.4.20.

Table 10.4.19: Modified flutter speed in m/s changing a design variable.

U_{fm} (m/s)	$\partial U_f / \partial I_y = -0.230455$	$\partial U_f / \partial I_z = -0.00019$	$\partial U_f / \partial J = 2.18247$
– 20%	83.41	83.02	80.22
– 10%	83.17	82.97	81.57
– 5%	83.05	82.95	82.25
0%	82.93	82.93	82.93
+ 5%	82.81	82.91	83.61
+ 10%	82.69	82.88	84.28
+ 20%	82.45	82.84	85.63

Table 10.4.20: Modified flutter speed in m/s changing three design variables.

ΔI_y	ΔI_z	ΔJ	U_{fm}
20%	20%	20%	85.06
−20%	20%	20%	86.03
20%	−20%	20%	85.24
20%	20%	−20%	79.65
−20%	−20%	20%	*86.21*
−20%	20%	−20%	80.62
20%	−20%	−20%	79.83
−20%	−20%	−20%	80.80

The results obtained show percentage variations similar to the ones achieved for the Great Belt Bridge, despite the fact that the Vasco da Gama is a cable-stayed structure with a moderate span length.

10.5 References

[1] Larsen A. [1993] *Aerodynamic aspects of the final design of the 1624* m *suspension bridge across the Great Belt.* Journal of Wind Engineering and Industrial Aerodynamics, Vol. 48, pp. 261–285.

[2] Cobo D. [1998] *An analysis of wind stability. Improvements to the response of suspension bridges*. Tesis doctoral de La Universitat Politécnica de Catalunya. Escola Técnica Superior D'Enginyers de Camins, Canals i Ports.

[3] Katsuchi T. [1997] *An Analytical Study on Flutter and Buffeting of the Akashy-Kaikyo Bridge.* Eassy submitted in conformity with the requirements for Master of Science in Engineering.

[4] Miyata T,. Yokoyama K., Yasuda M., Hikami Y. [1992] *Wind effects and full model wind tunnel tests.* Proceedings of the Eirst International Symposium on *Aerodynamics of Large Bridges.* A.A.Balkema, Copenhagen, Denmark, pp. 217–236.

[5] Ammann O. [1941] The Failure of the Tacoma Narrows Bridge. Report to the Federal Work Agency, March 1940, pp. 37–126.

[6] Herzog M. [1990] *Die aerostatische Stabilität der Hängebrücken.* Schweizer Ingenieur und Architekt Nr.9, 1 März, pp. 223–227.

[7] Capra A., Leveille A. [1997] *Vasco da Gama main bridge seismic design.* International Conference New Technologies in Structural Engineering. Vol. 1. Lisbon, Portugal.

[8] Machado L., Branco F. A. [1997] *The Vasco da Gama bridge proyect.* International Conference New Technologies in Structural Engineering. Vol. 2. Lisbon, Portugal.

Chapter 11

A formulation of optimization in bridge aeroelasticity

11.1 Introduction

Structural design, especially in bridges, must complete different phases. Along the design process, the structure is being constantly improved until it becomes a material entity fully defined. At each design phase, decisions must be taken by the designer in order to choose from the available alternatives aspects such as the adopted material, the structural elements joints, or the location and number of foundations. The design process concludes when an efficient design has been worked out from the designer's point of view. Generally speaking, the design process can be divided into three main stages

- *Settlement of the requirements plan.* It establishes the targets to be satisfied along the structure service life. Some of those targets are fixed characteristics such as deck high compatible with the aim of allowing ship traffic, number of vehicles to be absorbed or the minimum wind speed to avoid aeroelastic instabilities. Along this stage, other structure properties as the kind of material to be used can be also defined.
- *Election and analysis of the structure.* The requirements plan can usually be satisfied by means of different designs. Therefore, along this stage, several questions such as the structural typology or the foundation method among others must be answered. This entails the study and analysis of a number of solutions based upon their adequate performance in similar design problems. Thus, modifications are made in the considered alternatives and some of them are discarded until an only feasible solution is obtained. But ahead, this final solution can be improved with the aim of reaching a more efficient design.
- *Design of details.* Some aspects such as joints visual appearance which have a limited interest and do not affect the general structural design are defined.

Considering the aforementioned three design stages, the second one requires the bigger number of designer's decisions. The design process is considered finished when the chosen solution shows a "good enough" performance.

11.2 Conventional design method

The structural conventional design, whose conceptual flow chart is presented in Figure 11.2.1, relies on heuristic rules based upon the experience and expertise of the

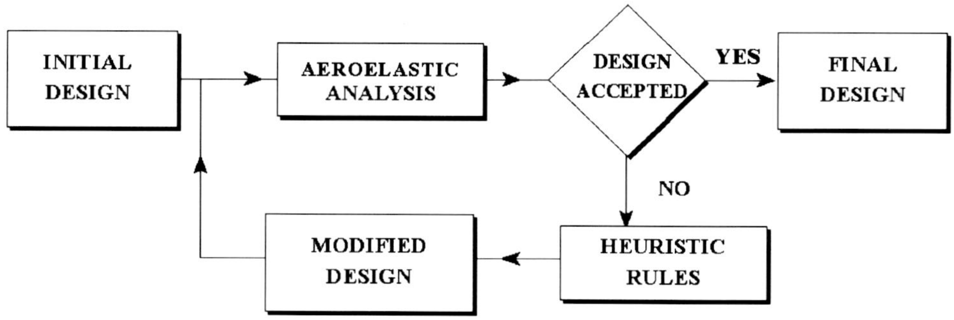

Figure 11.2.1: Conventional design flow chart.

designer. These heuristic rules are used with the aim of modifying an initial design until a series of constraints are satisfied and, therefore, that modified design can be considered acceptable.

This method presents important drawbacks such as the possibility of the introduced modifications to produce worse designs than the former ones as there is not guaranty about a modified design performing better than the previous one. Another key issue is the election of the design variables (structural parameters suitable to change) to be modified in each iteration. Moreover, once reached the final design, concerns can exist regarding unattended design variables which could have lead to a better design.

It can be said that the conventional design is a subjective process and that its final result relies on the efficiency and experience of the designer team. The design target can be defined as a good enough solution which satisfies the requirements plan. As a matter of fact, reaching a valid solution using the conventional design method can be almost impossible when dealing with a very complex problem or an inexperienced design team is responsible for the design.

11.3 Sensitivity analysis

According to Choi and Kim[1], design sensitivity analysis is used to compute the rate of performance measure change with respect to design variable changes. Generally speaking, a structural performance measure depends on the design; for example, a change in the cross-sectional area of a beam would affect the structural weight. Sensitivities play a key role in gradient-based optimization methods however; they offer very valuable information by themselves as they show the design variables which are worthy to change as well as the sign of the modifications in order to improve the studied structural response. Mathematically, sensitivities are obtained by means of the evaluation of the derivative of the structural response which is being analyzed with regard to the considered design variable.

In Chapters 9 and 10 of this book, several examples about the analytical evaluation of the sensitivity of the flutter speed with regard to geometrical properties of the deck have been presented. Those results are very interesting as they improve the designer's knowledge of the structural behaviour, avoiding unfruitful design modifications. In Figure 11.3.1, the flow chart of the design method based upon sensitivity analysis is shown. It can be seen that the heuristic rules of the conventional design method have been substituted by the sensitivity analyses, which offer objective information and, therefore, they are a guide for the designer in the way towards a feasible design.

11.4 Optimum design

Optimum design differs from the conventional design process at the design modification stage. Instead of heuristic rules, mathematical algorithms are used to produce a modified design which is guaranteed that is going to perform better than the previous one. In general, the most efficient optimization methods are those which use the information offered by sensitivity analysis. In Figure 11.4.1, the conceptual flow chart of the gradient-based optimization method is presented.

Optimum design, also known as automated design, has been for decades successfully and frequently employed in engineering realms such as automobile and aircraft industries. In civil engineering, optimum design techniques have been used, although in a lesser extent. Some interesting works are the following: Hernández[2], Hernández and

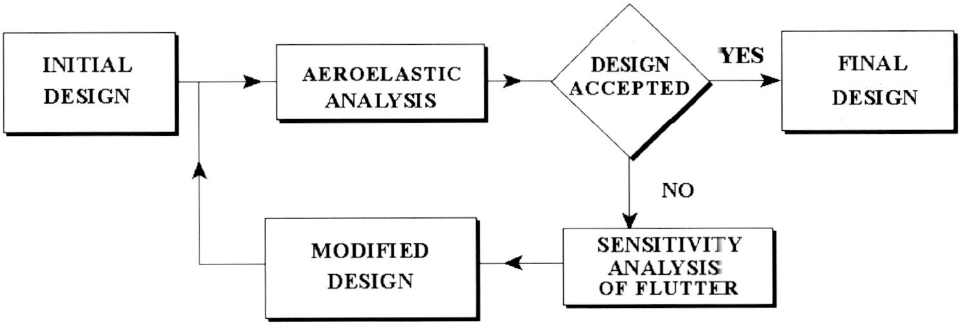

Figure 11.3.1: Design based upon sensitivity analysis flow chart.

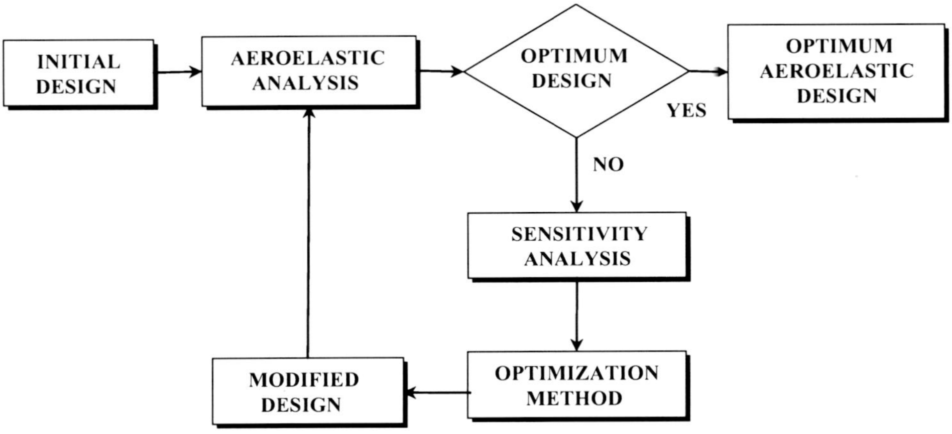

Figure 11.4.1: Gradient-based optimum design flow chart.

Fontán[3], Ohkubo and Taniwaki[4], Navarrina[5], Navarrina *et al.*[6] or Perezzán[7]. The main advantages of optimum design with regard to conventional design are as follows:

- The modified design obtained after each iteration is always better than the previous one.
- When the final design is reached, it can be guaranteed that it is the best available solution if the Kuhn–Tucker[8] criterion is satisfied.

Mathematically, the optimum design problem can be posed as follows[9]: an objective function F, for instance the weight of a structure, which is a function of a set of design variables $\mathbf{x} = (x_1, \ldots, x_n)$ being n the number of selected design variables, must be minimized, or maximized if that is the case:

$$\min F(\mathbf{x}) \tag{11.4.1}$$

subject to

$$g_j(\mathbf{x}) \leq 0 \quad j = 1, \ldots, l \tag{11.4.2}$$

where $g_j(\mathbf{x})$ is the set of chosen constraints and l is the total number of constraints.

In the former formulas, it has been expressed mathematically what the designer does somehow along the conventional design process. Thus, the structural parameters that can suffer changes are the design variables contained in vector $\mathbf{x}$. Moreover, the requirements plan includes limitations which cannot be avoided, for instance stresses limitations, maximum allowable movements under certain load conditions or the minimum flutter speed. Those constraints, many times stipulated by the construction codes, are expressed mathematically by the set of constraints $g_j(\mathbf{x})$. Finally, the design characteristic to be improved along the design process can be expressed as the objective function $F(\mathbf{x})$.

A graphical representation of a two design variables (x_1, x_2) minimization problem is presented in Figure 11.4.2, where the feasible region is bounded in grey and the objective function is represented by straight lines each corresponding to designs with equal objective function values.

The design problem begins with an initial design $\mathbf{x}_1$; due to heuristic rules a modified design $\mathbf{x}_2$, better than the former, is reached. However, after another iteration a worse design $\mathbf{x}_3$ is obtained. It is possible also that once a modified design is obtained $\mathbf{x}_4$, a new modified design $\mathbf{x}_5$ does not improve the design at all, in terms of the objective function. Therefore, the uncertainty associated with the conventional design process is clear.

In Figure 11.4.3, a graphical example of the performance of the optimum design process is shown. As in the former case, the initial design is $\mathbf{x}_1$; however each obtained modified design improves its previous one until an optimum design is reached.

11.5 Suspension bridges optimum design

To date, optimum design techniques, also called automated design techniques, have not been extensively used in bridge engineering, especially in long-span bridges. Therefore, although cable-supported bridges are projected using wind-tunnel testing and complex computer analysis, the decision regarding the structural elements to be modified relies

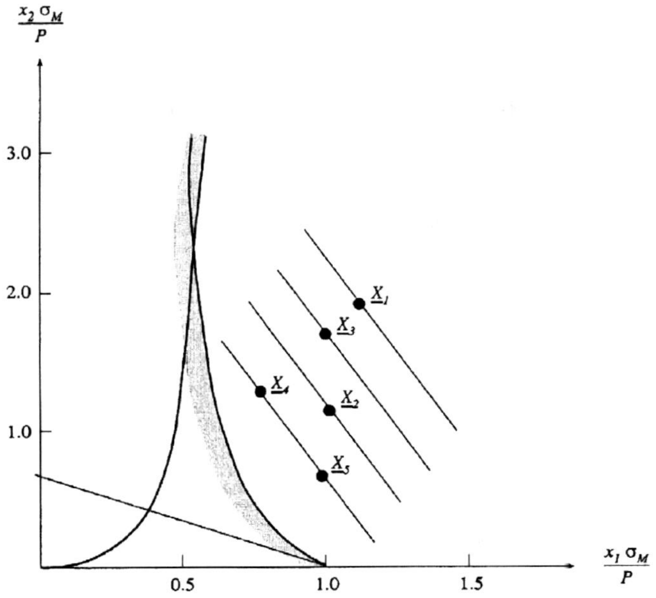

Figure 11.4.2: Graphic interpretation of the conventional design process.

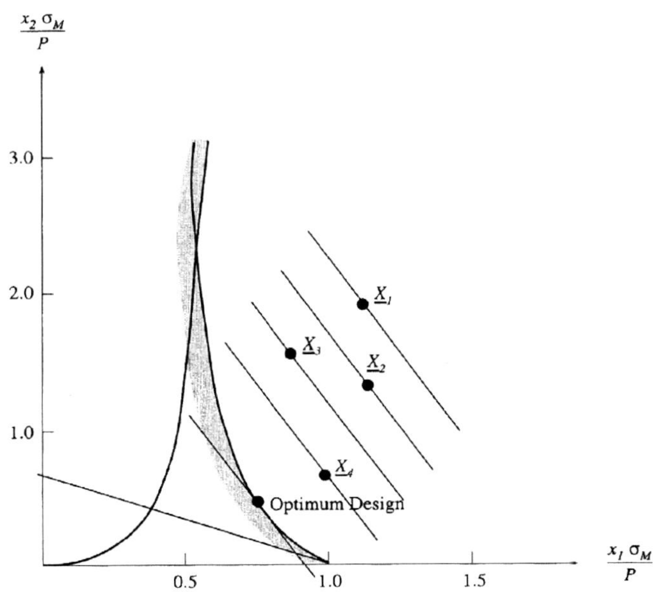

Figure 11.4.3: Graphic interpretation of the optimum design process.

exclusively in the experience and expertise of the engineering team. In other words, state-of-the-art projects such as the Great Belt, Akashi Strait or Messina Bridges have not been designed using at any stage automated design techniques. It is easy to guess that optimum design methods could improve the final design.

One of the main concerns when dealing with long-span bridges is their aeroelastic performance. Thus, long-span bridges optimum design formulation must include among the set of considered constraints at least one regarding the bridge flutter wind speed. Moreover, service limit state, in general defined by construction codes, must also be included as long-span bridges decks are usually made of steel and therefore movements under static loads bound the final design more than the material stress level. Another key issue is to identify the characteristic or the set of characteristics in multiobjective optimization that the designer wishes to improve. A typical property whose improvement is pursued is the structure weight as decreasing weight means to decrease material costs as well.

11.5.1 Formulation of the optimum design problem

The optimization problem to be solved is to obtain the minimum weight bridge deck compatible with a stipulated minimum flutter critical wind speed and a maximum allowed deck vertical displacement. It is clear that as the deck weight is going to be minimized, along the design process the deck mass is going to be considered as non-constant. The former circumstance implies certain extensions in the analytical sensitivity analysis formulation presented in Chapter 7 (see Section 11.5.2). Additionally, the bridge deck weight is going to depend on the cross-section area as it is assumed that the deck length does not change, neither the deck material, which is steel.

Additionally, the formulation and examples that are going to be presented are referred to symmetric aerodynamic box section decks. The deck design protocol begins with detailed studies of the deck cross-section perimeter shape with the aim of obtaining an efficient performance under wind-induced effects. Once the box cross-section behaviour is considered adequate, new studies, as the ones that are going to be presented, can be carried out in order to improve the design. However, it must be bore in mind that at this stage, the external cross-section shape is fixed and so do the flutter derivatives required to evaluate the flutter response. Taking into account the former consideration, the consequence is that the deck weight, that is, the box cross-section area, is a function of the steel thicknesses which are the chosen design variables. Design variables may change continuously inside a range between a lower limit and an unbounded upper limit. Therefore the optimum design problem to be solved is

$$\min F = A(e_1,\ldots,e_n) \tag{11.5.1}$$

subject to

$$e_{\min} \leq e_i \leq e_{\max} \qquad i = 1,\ldots,n \tag{11.5.2}$$

$$g_1 : \quad \frac{U_{f,cr}}{U_f} - 1 \leq 0 \tag{11.5.3}$$

$$g_2 : \quad \frac{w_c}{w_{c,\max}} - 1 \leq 0 \tag{11.5.4}$$

In the former expressions, A is the area of the deck cross section. The deck weight is directly related with the cross-section area; therefore, the minimum area deck corresponds with the minimum weight deck. When the suspension bridge deck is a box girder, then it is clear that the cross-section area is dependent upon the material thicknesses ($e_1, \ldots, e_n$). Material thicknesses must be included inside a certain range whose lower bound $e_{\min}$ is the minimum thickness due to manufacturing or constructive considerations. In practical applications, the upper bound $e_{\max}$ can be neglected. In Section 11.5.4, the relationship between a box cross-section area and material thicknesses will be surveyed in depth. Furthermore, U_f is the wind flutter speed and $U_{f,cr}$ is the minimum wind speed that the bridge must support before flutter occurs, w_c the vertical displacement of the centre of the main span, while $w_{c,\max}$ is the maximum allowable vertical displacement in the centre of the main span which is usually fixed by structural codes.

In the optimization problem presented in equations (11.5.1)–(11.5.4), the design variables are the thicknesses of the symmetric box cross section. The automated design problem can be efficiently solved by means of a gradient-based method; thus the derivatives of the objective function as well as the behaviour constraints with regard to the aforementioned thicknesses must be evaluated. Careful reflection is worthy to demonstrate that those design variables are the core ones governing the bridge deck aeroelastic behaviour.

The aeroelastic response of a long-span bridge depends upon the geometric properties of the deck cross section I_y, I_z, J, A as these parameters are included in the deck mass and stiffness matrixes as much in box as in truss girders. Therefore, the aforementioned geometric properties are going to play a key role in the aeroelastic performance of the bridge as can be inferred from the dynamic equilibrium equation for the bridge deck (3.4.10). Mathematically that dependency can be expressed as

$$U_f = U_f(I_y, I_z, J, A) \tag{11.5.5}$$

Evaluation of the derivatives of the flutter speed U_f with respect to any of the cross-section geometric properties can be done by means of the formulas presented in Chapter 8.

For box girders, the geometric properties of the deck cross section depend upon the lengths and thicknesses of the cross-section composite parts (see Section 11.5.4). This can be seen in Figure 11.5.1, where a simplified symmetric box cross section is shown.

It is assumed that the box cross-section external shape does not change along the optimum design process. That is, the lengths of the cross-section composite parts are fixed, although thicknesses are freed to change. This approach is consistent with the practice of first developing an extensive wind-tunnel testing campaign in order to fulfil an efficient bridge deck aerodynamic behaviour, followed by a comprehensive design phase in which the external deck geometry is considered unchangeable. Mathematically, the dependencies of the cross-section geometric properties with regard to the thicknesses can be written as

$$\begin{aligned} I_y &= I_y(e_1, e_2, \ldots, e_n) \\ I_z &= I_z(e_1, e_2, \ldots, e_n) \\ J &= J(e_1, e_2, \ldots, e_n) \\ A &= A(e_1, e_2, \ldots, e_n) \end{aligned} \tag{11.5.6}$$

Thus, the flutter speed is a function of the deck cross-section thicknesses

$$U_f = U_f(I_y(e_1,e_2,...,e_n), I_z(e_1,e_2,...,e_n), J(e_1,e_2,...,e_n), A(e_1,e_2,...,e_n)) \quad (11.5.7)$$

The sensitivity of the flutter speed with regard to any of the thicknesses of the cross-section composite parts is

$$\frac{\partial U_f}{\partial e_i} = \frac{\partial U_f}{\partial I_y}\frac{\partial I_y}{\partial e_i} + \frac{\partial U_f}{\partial I_z}\frac{\partial I_z}{\partial e_i} + \frac{\partial U_f}{\partial J}\frac{\partial J}{\partial e_i} + \frac{\partial U_f}{\partial A}\frac{\partial A}{\partial e_i} \quad (11.5.8)$$

A matrix expression may be used to obtain the flutter speed derivatives:

$$\begin{Bmatrix} \frac{\partial U_f}{\partial e_1} \\ \vdots \\ \frac{\partial U_f}{\partial e_n} \end{Bmatrix} = \begin{pmatrix} \frac{\partial I_y}{\partial e_1} & \frac{\partial I_z}{\partial e_1} & \frac{\partial J}{\partial e_1} & \frac{\partial A}{\partial e_1} \\ \vdots & \vdots & \vdots & \vdots \\ \frac{\partial I_y}{\partial e_n} & \frac{\partial I_z}{\partial e_n} & \frac{\partial J}{\partial e_n} & \frac{\partial A}{\partial e_n} \end{pmatrix} \cdot \begin{Bmatrix} \frac{\partial U_f}{\partial I_y} \\ \frac{\partial U_f}{\partial I_z} \\ \frac{\partial U_f}{\partial J} \\ \frac{\partial U_f}{\partial A} \end{Bmatrix} \quad (11.5.9)$$

where $\frac{\partial U_f}{\partial I_y}, \frac{\partial U_f}{\partial I_z}, \frac{\partial U_f}{\partial J}, \frac{\partial U_f}{\partial A}$ can be evaluated using (8.3.16). On the other hand, $\frac{\partial I_y}{\partial e_i}, \frac{\partial I_z}{\partial e_i}, \frac{\partial J}{\partial e_i}$ y $\frac{\partial A}{\partial e_i}$ with $i=1,...,n$ are the elements of the Jacobi matrix and must be calculated for each specific deck cross section. For a generic simplified symmetric box cross section like the one of Figure 11.5.1, the analytical derivatives of the geometric properties I_y, I_z, J, A with regard to the three thicknesses e_1, e_2 and e_3 can be found in Section 11.5.4.

A similar expression can be obtained for the derivatives of the reduced frequency K with regard to the box cross-section thicknesses:

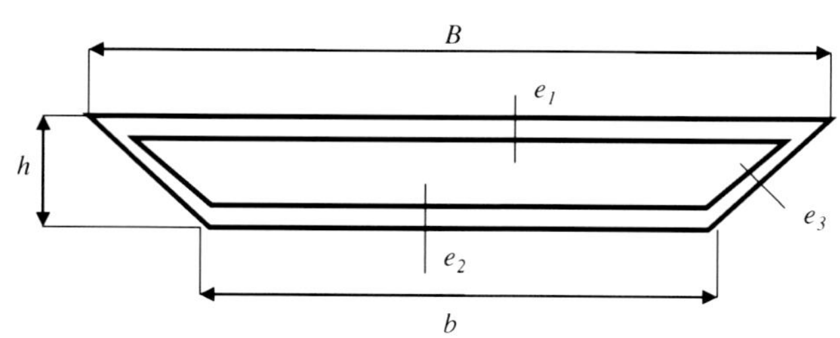

Figure 11.5.1: Simplified symmetric box cross section.

$$\begin{Bmatrix} \dfrac{\partial K_f}{\partial e_1} \\ \vdots \\ \dfrac{\partial K_f}{\partial e_n} \end{Bmatrix} = \begin{pmatrix} \dfrac{\partial I_y}{\partial e_1} & \dfrac{\partial I_z}{\partial e_1} & \dfrac{\partial J}{\partial e_1} & \dfrac{\partial A}{\partial e_1} \\ \vdots & \vdots & \vdots & \vdots \\ \dfrac{\partial I_y}{\partial e_n} & \dfrac{\partial I_z}{\partial e_n} & \dfrac{\partial J}{\partial e_n} & \dfrac{\partial A}{\partial e_n} \end{pmatrix} \cdot \begin{Bmatrix} \dfrac{\partial K_f}{\partial I_y} \\ \dfrac{\partial K_f}{\partial I_z} \\ \dfrac{\partial K_f}{\partial J} \\ \dfrac{\partial K_f}{\partial A} \end{Bmatrix} \tag{11.5.10}$$

11.5.2 Extensions of the sensitivity analysis formulation due to the assumption of variable mass

From equations (11.5.9) and (11.5.10), it can be inferred that sensitivity analyses with regard to the cross-section thicknesses are obtained once and that the sensitivities of the flutter speed or the reduced frequency with regard to the cross-section geometric properties (I_y, I_z, J and A) have been evaluated. Nevertheless, special care must be adopted when tackling with variable mass structural elements. For analysis purposes, the deck mass per unit of length can be expressed as

$$\underline{m}_x = \underline{m}_y = \underline{m}_z = \rho A + \underline{m}^{ns} \tag{11.5.11}$$

where the symbol _ highlights that the parameter is expressed by unit of length, ρ is the deck material density, A is the deck cross-section area and $\underline{m}^{ns}$ is the non-structural mass per unit of length, which represents the mass contribution of bulkheads, stiffening members or the bearing surface.

Similarly, the deck polar moment of inertia $\underline{m}_{\varphi x}$ can be evaluated as

$$\underline{m}_{\varphi x} = \rho(I_y + I_z) + \underline{m}_{\varphi x}^{ne} \tag{11.5.12}$$

where I_y is the deck cross-section vertical moment of inertia, I_z is the deck cross-section lateral moment of inertia and $\underline{m}_{\varphi x}^{ns}$ is the non-structural elements contribution to the polar moment of inertia per unit of length. Along the optimum design process, non-structural contributions in (11.5.11) and (11.5.12) are considered to remain fixed.

Therefore, the analytical derivative of the mass matrix required to evaluate the sensitivities of the natural frequencies (7.2.2) and vibration modes (7.2.3) with regard to a generic design variable x_i can be written as follows:

$$\frac{\partial \mathbf{M}}{\partial x_i} = \mathbf{IE}\left(\frac{\partial \mathbf{M}^b}{\partial x_i}\right) \tag{11.5.13}$$

where **IE** represents the assembly operation and $\mathbf{M}^b$ is the element mass matrix in the global coordinates system. Depending upon the considered design variable, that is, the deck cross-section geometric properties, the elements of the derivative of the mass matrix can be null or not.

Additionally, the derivatives of vectors **p**, death loads, and **w**, vector linking axial force and movements, in equation (7.2.20) with regard to the deck cross-section area A are not null. Those derivatives are mandatory for the evaluation of the derivative of the non-linear stiffness matrix and the sensitivities of the bridge displacements. Thus, derivatives of natural frequencies and vibration modes as well as sensitivities of flutter speed considering variable mass differ from those evaluated assuming constant mass.

11.5.3 Solving the optimum design problem: description of the *DIOPTICA* code

The optimization problem formulated in (11.5.1)–(11.5.4) requires evaluation of the objective function, constraints and gradients of the objective function, as well as the constraints because the modified feasible directions method, a gradient-based method, is going to be implemented. Analytical evaluation of sensitivity analyses of flutter is a very computer time demanding process. In fact, it involves evaluation of natural frequencies and mode shapes prior to the evaluation of sensitivity analysis of natural eigenresponses as well as determination of the flutter speed by solving a non-linear eigenvector problem and evaluation of flutter derivatives with regard to each considered design variable[10]. Aeroelastic analysis of long-span suspension bridges may require including a large number of vibration modes, and in addition a great number of design variables could also be adopted, multiplying the problem size. Therefore, the optimization problem to be solved can be considered as a *grand challenge problem* due to its enormous computer power requirements. Other grand challenge problems can be found in diverse disciplines as climate modelling, turbulent fluid mechanics as well as in the optimum design realm[11].

However, the demanded computer time is a factor that can be shortened using distributed computing. Distributed programming solves computational problems using simultaneously a set of computers that work together in a coordinated manner[12]–[14]. In the sensitivity analysis problem, evaluation of natural eigenresponses is a serial problem; nevertheless, calculus of the aeroelastic eigenresponses can be carried out in parallel, as well as analytic sensitivity analysis, because the derivatives with regard to each design variable can be carried out independently and simultaneously using a set of networked computers[15]. The results that are going to be presented in the following chapter have been computed using a cluster of personal computers, which offers an important computational power, keeping hardware acquisition and maintenance costs in low levels. Generally speaking, the required computer time for solving the non-linear eigenvector problem associated with the calculation of flutter speed may be divided by q, q being the number of existing cluster nodes. Moreover, if the number of cluster nodes is greater than the number of selected design variables, sensitivity analysis with regard to each design variable can be carried out simultaneously dividing the computer time required in sequential programming by n, n being the number of design variables.

The computer code developed in order to solve the optimum design of suspension bridges with box cross-section deck considering the flutter and cinematic constraints has been called DIOPTICA. This piece of software includes several special-purpose modules as well as a distributed computation version of codes FLAS and ADISNOL 3D (see Chapters 4 and 7) assuming variable mass. The optimization algorithms have not been programmed as the commercial code ADS[16] has been joined to DIOPTICA through in-home routines. In fact two programming languages have been used: the user interface, the communications between computers, the main program flows and the calls to the ADS optimization algorithms have been programmed using VISUAL BASIC, while the complex mathematical algorithms have been programmed in FORTRAN as dynamic link libraries.

In Figure 11.5.2, the DIOPTICA code flowchart is shown. It can be noted that its main structure is similar to that of Figure 11.4.1, as the aim of the piece of software is to solve an optimum design problem applying a gradient-based method. Additionally, the DIOPTICA flowchart includes detailed information regarding the main task, which are solved. Furthermore, in blue colour are shadowed the serial tasks (S in the text) while in pink colour are shadowed those tasks which have been distributed among the cluster nodes (P in the text).

- *Pre-analysis module (S)*. DIOPTICA employs the COSMOS/M pre-processor which is called GEOSTAR to generate the bridge BEAM3D finite elements model. Additionally, using GEOSTAR the bars orientation, materials and sections properties and assignments, as well as boundary conditions are defined. Furthermore the selected design variables for the optimization problem, aeroelastic data and flutter derivatives are introduced. All these data are stored in text files which will be accessed by DIOPTICA when needed. In Figure 11.5.3, an example of a bridge beam elements model generated using GEOSTAR is shown.

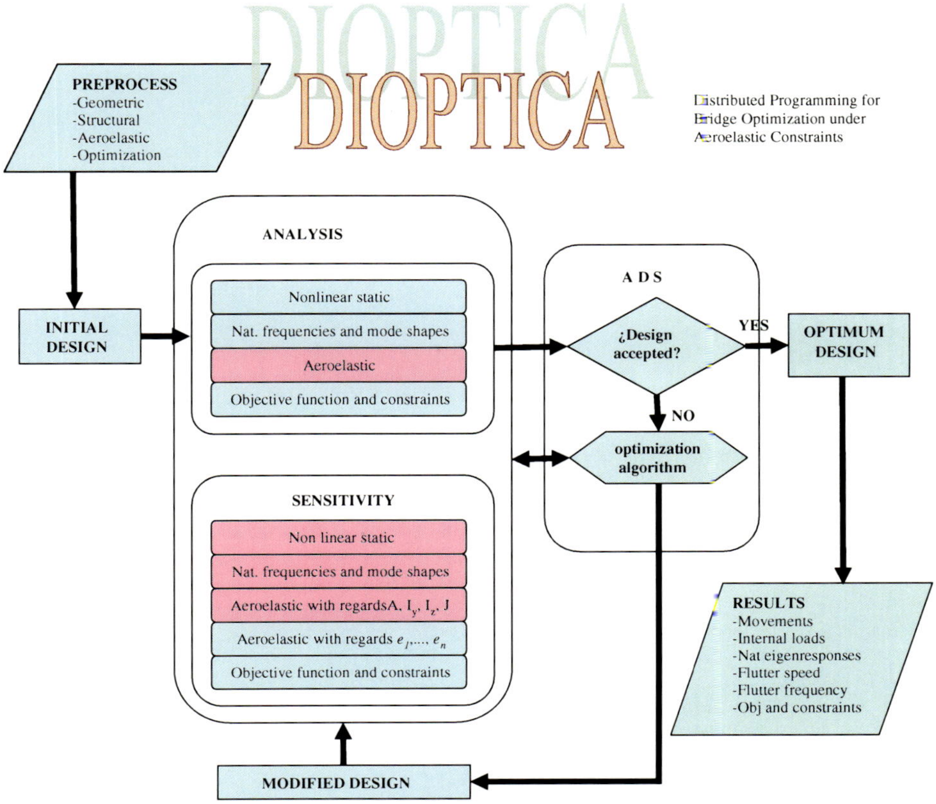

Figure 11.5.2: DIOPTICA code flowchart.

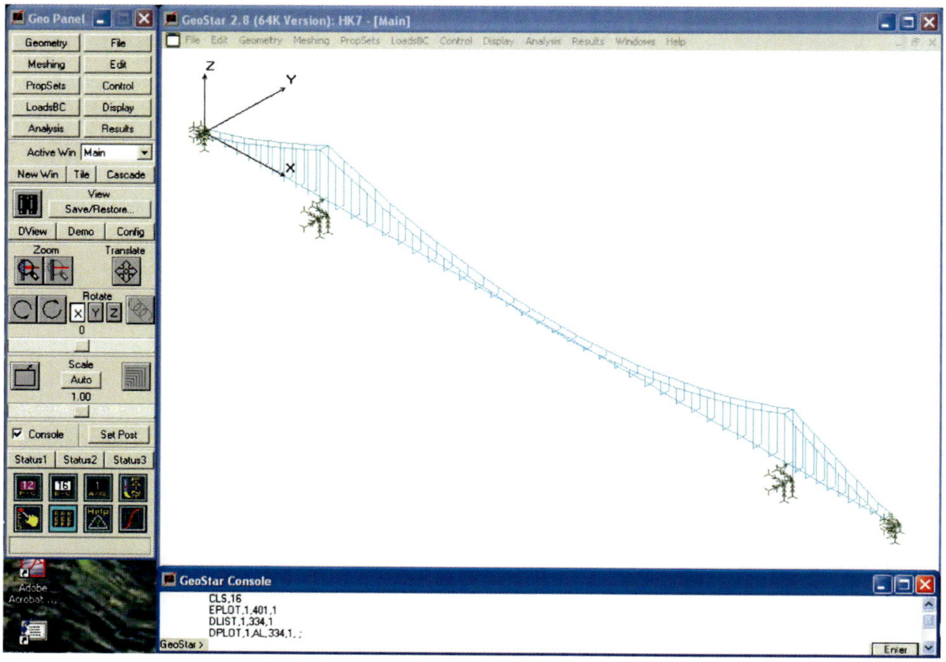

Figure 11.5.3: Bridge model generation using GEOSTAR.

- *Non-linear static analysis module (S)*. This module solves the static structural problem considering second-order theory. Mathematically, the problem to be solved is the following:

$$\mathbf{p} = \mathbf{K}_{\text{nolin}} \mathbf{u} \tag{11.5.14}$$

  where $\mathbf{p}$ is a vector containing the nodal loads, $\mathbf{u}$ is the nodal displacements vector and $\mathbf{K}_{\text{nolin}}$ is the stiffness matrix considering the geometric rigidity. The displacements obtained after solving this problem will be used to evaluate the cinematic constraint in the optimum design problem.
- *Natural eigenresponses module (S)*. In this part of the code, the bridge free vibration problem is solved considering second-order theory. Mathematically, the eigenvalue problem to be solved in order to obtain the natural frequencies and proper modes is the following:

$$\left[\mathbf{K}_{\text{nolin}} - \omega_n^2 \cdot \mathbf{M}\right] \cdot \phi_n = \mathbf{0} \tag{11.5.15}$$

  $\mathbf{M}$ being the bridge mass matrix, ω_n the n-esime natural frequency and ϕ_n the n-esime proper mode. The natural eigenresponses obtained are mass normalized.
- *Aeroelastic analysis module (P)*. The flutter problem for the studied bridge is solved applying the formulation presented in Chapter 3. The non-linear eigenvalues problem to be solved is

$$(\mathbf{A} - \mu \mathbf{I})\mathbf{w}_\mu = \mathbf{0} \tag{11.5.16}$$

The problem (11.5.16) can be solved iteratively by means of distributed computing. Matrix **A** is a function of the flutter derivatives, which depend on the reduced frequency K. Nevertheless, for a given wind speed, the reduced frequency associated to each mode is not known until problem (11.5.16) is solved for each mode. For each chosen mode, the solution process is independent for a given wind speed; thus each cluster node is able to cope simultaneously with the solution of the non-linear eigenvalues problem for a given aeroelastic mode and wind speed.
The process ends when the flutter wind speed is identified.

- *Objective function and constraints evaluation module (S)*. The aim of this code module is to evaluate the optimization problem objective function and constraints for the initial and modified designs by means of expressions (11.5.1)–(11.5.4).
- *Static sensitivity analysis module (P)*. The task which is solved is the analytical evaluation of the derivative of the vertical displacement in the centre of the main span of the bridge under a selected load combination, that is,

$$\frac{\partial \mathbf{u}}{\partial x_i} \quad x_i = I_y, I_z, J, A \tag{11.5.17}$$

The former derivatives are evaluated by means of equation (7.2.20). The sensitivity analyses with regard to each geometric property of the deck cross section can be carried out independently and simultaneously in different cluster nodes applying distributed programming.

- *Natural eigenresponses sensitivity analysis module (P)*. This computer code module allows the evaluation of the derivatives of the bridge natural frequencies and mode shapes with regard to the deck cross-section geometric properties by means of equations (7.2.2) and (7.2.3).

$$\frac{\partial\left(\omega_n^2\right)}{\partial x_i}, \frac{\partial \phi_n}{\partial x_i} \quad x_i = I_y, I_z, J, A \tag{11.5.18}$$

As in the former case, parallel programming techniques can be implemented in order to decrease computer times evaluating sensitivities with regard to each variable in different cluster nodes.

- *Flutter speed sensitivity analysis module (P)*. Analytical evaluation of the derivative of the wind flutter speed with regard to the deck geometric properties (8.3.16) is performed following formulation presented in Chapter 8.

$$\frac{\partial U_f}{\partial x_i} \quad x_i = I_y, I_z, J, A \tag{11.5.19}$$

Again, sensitivities with regard to each considered variable can be calculated in different cluster nodes.

- *Flutter speed sensitivity analysis with regard to the box deck cross-section thicknesses (S)*. The design variables adopted in the optimization problem have been the thicknesses of the deck box cross section. Thus derivatives of the flutter speed with regard to those design variables must be obtained in order to be able to evaluate the gradient of the flutter constraint

$$\frac{\partial U_f}{\partial x_i} \qquad x_i = e_1,\ldots,e_n \tag{11.5.20}$$

Once the sensitivities of the flutter wind speed with regard to the deck geometric properties have been obtained (11.5.19), the ones which are interesting at this stage can be evaluated by means of expression (11.5.9).

- *Objective function and constraints sensitivity analysis module (S)*. The optimum design problem is going to be solved by means of a gradient-based method; thus derivatives of the objective function and constraints with regard to the box deck thicknesses must be supplied, that is

$$\frac{\partial F}{\partial x_i},\ \frac{\partial g_l}{\partial x_i} \qquad x_i = e_1,\ldots,e_n \tag{11.5.21}$$

which can be obtained by means of the Jacobi matrix that appears in (11.5.9) and (11.5.10). Using the gradient operator ∇, it can be written as

$$\nabla_{\mathbf{e}} F = \mathbf{J}\cdot\nabla_{\mathbf{I}} F \tag{11.5.22}$$

$$\nabla_{\mathbf{e}} g_l = \mathbf{J}\cdot\nabla_{\mathbf{I}} g_l \qquad l = 1,2 \tag{11.5.23}$$

- *Results module (S)*. The final result of the carried out optimization problem is the values of the thicknesses of the box deck cross section that produce the minimum weight deck compatible with the selected set of constraints. Additionally, a series of results are obtained for each modified design along the optimization processes. Some of them are the following:
  - nodal movements $\mathbf{u}$;
  - loads in bars $\mathbf{p}^{b}$;
  - natural vibration frequencies ω_n;
  - proper mode shapes ϕ_n;
  - natural frequencies sensitivities with regard to the deck geometric properties $\partial(\omega_n^2)/\partial x_i \quad x_i = I_y, I_z, J, A$;
  - natural mode shapes sensitivities with regard to the deck geometric properties $\partial\phi_n/\partial x_i \quad x_i = I_y, I_z, J, A$.
  - flutter wind speed U_f and reduced frequency K_f
  - flutter wind speed and reduced frequency derivatives with regard to the deck geometric properties $\partial U_f/\partial x_i$, $\partial K_f/\partial x_i \quad x_i = I_y, I_z, J, A$;
  - objective function and constraints values for each design step min $F = A(e_1,\ldots, e_n)$, $g_1 : U_{f,cr}/U_f - 1 \le 0$ y $g_2 : w_c/w_{c,\max} - 1 \le 0$;
  - design variables values for each design step $e_1,\ldots, e_n$;
  - objective function and constraints gradients with regard to the thicknesses of the deck $\nabla_e F$ y $\nabla_e g_l$ $(l = 1, 2)$;
  - design variables, objective function and constraints values for the optimum design.

11.5.4 Symmetric box cross section: geometric properties and analytical derivatives with regard to thicknesses

For a simplified symmetric box cross section as the one showed in Figure 11.5.1, a series of auxiliary magnitudes can be defined.

$$s = \frac{B-b}{2} \tag{11.5.24}$$

$$t = h - e_1 - e_2 \tag{11.5.25}$$

$$q = \frac{ts}{h} = \frac{(h-e_1-e_2)(B-b)}{2h} \tag{11.5.26}$$

$$p = \frac{e_3\sqrt{s^2+h^2}}{h} + e_1\frac{s}{h} \tag{11.5.27}$$

$$r = B - 2(q+p) = B - \frac{1}{h}\left[(h-e_2)(b-h) + e_3\sqrt{(B-b)^2+4h^2}\right] \tag{11.5.28}$$

The formulae for calculating the location of the centroid and the geometrical properties of the cross section are the following:

$$z_g = \frac{\left[\frac{Bh^2}{2} - \frac{sh^2}{3} - qt\left(e_2 + \frac{2}{3}t\right) - rt\left(e_2 + \frac{t}{2}\right)\right]}{\left[(B-s)h - (q+r)t\right]} \tag{11.5.29}$$

$$A = Bh - sh - qt - rt \tag{11.5.30}$$

$$\begin{aligned} I_y &= \frac{1}{12}Bh^3 + Bh\left(z_g - \frac{h}{2}\right)^2 \\ &-2\left[\frac{1}{36}sh^3 + \frac{1}{2}sh\left(z_g - \frac{h}{3}\right)^2 + \frac{1}{36}qt^3 + \frac{1}{2}qt\left(e_2 + \frac{2}{3}t - z_g\right)^2\right] \\ &-\left[\frac{1}{12}rt^3 + rt\left(e_2 + \frac{1}{2}t - z_g\right)^2\right] \end{aligned} \tag{11.5.31}$$

$$I_z = \frac{1}{12}hB^3 - 2\left[\frac{1}{36}hs^3 + \frac{1}{2}hs\left(\frac{B}{2} - \frac{s}{3}\right)^2 + \frac{1}{36}tq^3 + \frac{1}{2}tq\left(\frac{r}{2} + \frac{q}{3}\right)^2\right] - \frac{1}{2}tr^3 \tag{11.5.32}$$

$$J = \frac{\left[(B-s)h - \frac{1}{2}Be_1 - \frac{1}{2}be_2 - e_3\sqrt{s^2+h^2}\right]}{\left(\frac{B}{e_1} + \frac{b}{e_2} + \frac{2\sqrt{s^2+h^2}}{e_3}\right)} \tag{11.5.33}$$

The derivatives of the cross-section geometric parameters with regard to the thicknesses can be evaluated analytically by means of the expressions ahead:

$$\frac{\partial z_g}{\partial e_1}=\frac{(2q+r)(t+e_2)A-(2q+r)\left[B\frac{h^2}{2}-s\frac{h^2}{3}-2\left(e_2+\frac{2}{3}t\right)\left(\frac{q}{2}t\right)-\left(e_2+\frac{t}{2}\right)rt\right]}{A^2} \tag{11.5.34}$$

$$\frac{\partial z_g}{\partial e_2}=\frac{re_2A-r\left[B\frac{h^2}{2}-s\frac{h^2}{3}-2\left(e_2+\frac{2}{3}t\right)\left(\frac{q}{2}t\right)-\left(e_2+\frac{t}{2}\right)rt\right]}{A^2} \tag{11.5.35}$$

$$\frac{\partial z_g}{\partial e_3}=\frac{\frac{t}{h}\left(e_2+\frac{t}{2}\right)\sqrt{4s^2+4h^2}\,A-\frac{t}{h}\sqrt{4s^2+4h^2}\left[B\frac{h^2}{2}-s\frac{h^2}{3}-2\left(e_2+\frac{2}{3}t\right)\left(\frac{q}{2}t\right)-\left(e_2+\frac{t}{2}\right)rt\right]}{A^2} \tag{11.5.36}$$

$$\begin{aligned}\frac{\partial I_y}{\partial e_1}&=2\left(z_g-\frac{h}{2}\right)Bh\frac{\partial z_g}{\partial e_1}-\\ &-2\left[sh\left(z_g-\frac{h}{3}\right)\frac{\partial z_g}{\partial e_1}-q\frac{t^2}{9}-\left(e_2+\frac{2}{3}t-z_g\right)\left(\frac{2}{3}+\frac{\partial z_g}{\partial e_1}\right)qt-q\left(e_2+\frac{2}{3}t-z_g\right)^2\right]+\\ &+r\frac{t^2}{4}+2\left(e_2+\frac{t}{2}-z_g\right)\left(\frac{1}{2}+\frac{\partial z_g}{\partial e_1}\right)rt+\left(e_2+\frac{t}{2}-z_g\right)^2 r\end{aligned} \tag{11.5.37}$$

$$\begin{aligned}\frac{\partial I_y}{\partial e_2}&=2\left(z_g-\frac{h}{2}\right)Bh\frac{\partial z_g}{\partial e_2}-\\ &-2\left[sh\left(z_g-\frac{h}{3}\right)\frac{\partial z_g}{\partial e_2}-q\frac{t^2}{9}-\left(e_2+\frac{2}{3}t-z_g\right)\left(\frac{1}{3}-\frac{\partial z_g}{\partial e_2}\right)qt-q\left(e_2+\frac{2}{3}t-z_g\right)^2\right]-\\ &-\left[q\frac{t^2}{6}-r\frac{t^2}{4}+2\left(e_2+\frac{t}{2}-z_g\right)\left(\frac{1}{2}-\frac{\partial z_g}{\partial e_2}\right)rt+(2q-r)\left(e_2+\frac{t}{2}-z_g\right)^2\right]\end{aligned} \tag{11.5.38}$$

$$\begin{aligned}\frac{\partial I_y}{\partial e_3}&=2\left(z_g-\frac{h}{2}\right)Bh\frac{\partial z_g}{\partial e_3}-\\ &-2\left[sh\left(z_g-\frac{h}{3}\right)\frac{\partial z_g}{\partial e_3}-\left(e_2+\frac{2}{3}t-z_g\right)\left(\frac{\partial z_g}{\partial e_3}\right)qt\right]+\\ &+\frac{t^3\sqrt{4s^2+4h^2}}{12h}+2\left(e_2+\frac{t}{2}-z_g\right)\left(\frac{\partial z_g}{\partial e_3}\right)rt+\frac{\left(e_2+\frac{t}{2}-z_g\right)^2 t\sqrt{4s^2+4h^2}}{h}\end{aligned} \tag{11.5.39}$$

$$\frac{\partial I_z}{\partial e_1} = -2\left[-\frac{q^3}{9} - \frac{q^2}{3}\left(\frac{r}{2}+\frac{q}{3}\right) - q\left(\frac{r}{2}+\frac{q}{3}\right)^2\right] + \frac{r^3}{12} \tag{11.5.40}$$

$$\frac{\partial I_z}{\partial e_2} = -2\left[-\frac{q^3}{9} - \frac{2}{3}q^2\left(\frac{r}{2}+\frac{q}{3}\right) - q\left(\frac{r}{2}+\frac{q}{3}\right)^2\right] - \frac{1}{12}\left(-r^3 + 6r^2 q\right) \tag{11.5.41}$$

$$\frac{\partial I_z}{\partial e_3} = -2\left[-\frac{qt\sqrt{4s^2+4h^2}\left(\frac{r}{2}+\frac{q}{3}\right)}{2h}\right] + \frac{r^2 t\sqrt{s^2+h^2}}{2h} \tag{11.5.42}$$

$$\frac{\partial J}{\partial e_1} = \frac{8\left(Bh - sh - \frac{Be_1}{2} - \frac{be_2}{2} - \sqrt{s^2+h^2}\,e_3\right)\left(\frac{-B}{2}\right)\left(\frac{B}{e_1}+\frac{b}{e_2}+2\frac{\sqrt{s^2+h^2}}{e_3}\right)}{\left(\frac{B}{e_1}+\frac{b}{e_2}+2\frac{\sqrt{s^2+h^2}}{e_3}\right)^2} + \frac{\frac{B}{e_1^2}\left[4\left(Bh - sh - \frac{Be_1}{2} - \frac{be_2}{2} - \sqrt{s^2+h^2}\,e_3\right)^2\right]}{\left(\frac{B}{e_1}+\frac{b}{e_2}+2\frac{\sqrt{s^2+h^2}}{e_3}\right)^2} \tag{11.5.43}$$

$$\frac{\partial J}{\partial e_2} = \frac{8\left(Bh - sh - \frac{Be_1}{2} - \frac{be_2}{2} - \sqrt{s^2+h^2}\,e_3\right)\left(\frac{-b}{2}\right)\left(\frac{B}{e_1}+\frac{b}{e_2}+2\frac{\sqrt{s^2+h^2}}{e_3}\right)}{\left(\frac{B}{e_1}+\frac{b}{e_2}+2\frac{\sqrt{s^2+h^2}}{e_3}\right)^2} + \frac{\frac{b}{e_2^2}\left[4\left(Bh - sh - \frac{Be_1}{2} - \frac{be_2}{2} - \sqrt{s^2+h^2}\,e_3\right)^2\right]}{\left(\frac{B}{e_1}+\frac{b}{e_2}+2\frac{\sqrt{s^2+h^2}}{e_3}\right)^2} \tag{11.5.44}$$

$$\frac{\partial J}{\partial e_3} = \frac{8\left(Bh - sh - \frac{Be_1}{2} - \frac{be_2}{2} - \sqrt{s^2+h^2}\,e_3\right)\left(-\sqrt{s^2+h^2}\right)\left(\frac{B}{e_1}+\frac{b}{e_2}+2\frac{\sqrt{s^2+h^2}}{e_3}\right)}{\left(\frac{B}{e_1}+\frac{b}{e_2}+2\frac{\sqrt{s^2+h^2}}{e_3}\right)^2} + \frac{\frac{2\sqrt{s^2+h^2}}{e_3^2}\left[4\left(Bh - sh - \frac{Be_1}{2} - \frac{be_2}{2} - \sqrt{s^2+h^2}\,e_3\right)^2\right]}{\left(\frac{B}{e_1}+\frac{b}{e_2}+2\frac{\sqrt{s^2+h^2}}{e_3}\right)^2} \tag{11.5.45}$$

$$\frac{\partial A}{\partial e_1} = 2q + r \tag{11.5.46}$$

$$\frac{\partial A}{\partial e_2} = r \tag{11.5.47}$$

$$\frac{\partial A}{\partial e_3} = \frac{t}{h}\sqrt{2s^2 + 4h^2} \tag{11.5.48}$$

11.6 References

[1] Choi K. K., Kim N. H. [2005] *Structural Sensitivity Analysis and Optimization 1.* Springer, Berlin.

[2] Hernández S. [1990] *Métodos de diseño óptimo de estructuras.* Colegio de Ingenieros de Caminos, Canales y Puertos. Madrid. In Spanish.

[3] Hernández S., Fontán A. [2002] *Practical Applications of Design Optimization.* WIT Press, Southampton.

[4] Ohkubo S., Taniwaki H. [1991] *Shape and Sizing Optimization of Steel Cable-Stayed Bridges.* Optimization of Structural Systems and Industrial Applications. Computational Mechanics Publications co-published with Elsevier Applied Science.

[5] Navarrina F. [1987] *Una Metodología General para Optimización Estructural en Diseño Asistido por Ordenador.* ETS de Ingenieros de Caminos, Canales y Puertos, Doctoral Thesis from Universidad Politécnica de Cataluña. In Spanish.

[6] Navarrina F., Valera A., París J., Colominas I., Casteleiro M. [2005] *Optimización de Torres de Alta Tensión mediante un Método de Programación Lineal Secuencial con Búsqueda Unidireccional Cuadrática.* Congreso de Métodos Numéricos en Ingeniería. 4–7 Julio, Granada, España. In Spanish.

[7] Perezzán J. C. [1999] *Análisis de Sensibilidad y Optimización de Estructuras de Barras de Nudos Rígidos en Teoría Lineal y no Lineal.* Escola Técnica Superior de Enxeñeiros de Camiños, Canais e Portos. Doctoral Thesis from Universidade da Coruña. In Spanish.

[8] Kuhn H. W., Tucker A. W. [1950] *Nonlinear Programming.* Proc. 2nd Berkeley Sump. Math. Stat. Prob. Berkeley, CA. pp. 481–492.

[9] Vanderplaats G. R. [2001] *Numerical Optimization Techniques for Engineering Design: With Applications.* 3rd edition. Vanderplaats Research & Development, Inc., Colorado Springs, USA.

[10] Nieto F. [2006] *Ph.D. Thesis: Análisis de Sensibilidad y Optimización Aeroelástica de Puentes Colgantes en Entornos de Computación Distribuida.* In Spanish. Construction Technology. La Corunna, University of La Corunna.

[11] Venkataraman S., Haftka R. T. [2002] *Structural Optimization: What has Moore's Law Done for Us?* 43rd AIAA/ASME/ASCE/AHS/ASC Structures, Structural Dynamics and Material Conference, 22–25 April, 2002, Denver, CO, USA.

[12] Grama A., Gupta A., Karypis G., Kumar V. [2003] *Introduction to Parallel Computing.* Pearson Education Ltd, Delhi, India.

[13] MHPCC (2003). *Introduction to Parallel Programming.* MHPC. Centre.

[14] Wilkinson B., Allen M. [2005] *Parallel Programming. Techniques and Applications*

using Networked Workstations and Parallel Computers. Pearson Education, Upper Saddle River, NJ, USA.

[15] Nieto F., Hernández S., Jurado J. A. [2005] *Distributed computing for the evaluation of the aeroelastic response and sensitivity analysis of flutter speed of the Messina Bridge*. Fluid Structure Interaction and Moving Boundary Problems Conference, A Coruña, Spain, WIT Press.

[16] VMA Engineering. [1988] *ADS A Fortran Program for Automated Design Synthesis. Version 3.00*. Vanderplaats, Miura & Associates, Goleta, CA, USA.

Chapter 12

Optimization of suspension bridges with aeroelastic and kinematic constraints

12.1 Introduction

In Chapter 11 the basis of the optimum design method applied to the design of suspension bridges considering aeroelastic constraints was explained. Focus was put not only on the advantages of optimization but also on sensitivity analysis techniques upon trial and error design procedures. Moreover, the general formulation of the optimum design problem was presented and the immense computational power required for solving the computer-aided design problem was pointed out, which entails the implementation of parallel computing strategies in order to tackle with the problem within a reasonable amount of time.

The aim of this chapter is to show the results that can be obtained applying optimization techniques in the design of suspension bridges. The authors have felt entailed to present a real design problem in order to exploit the capabilities of the nonconventional design techniques. Thus, the chosen bridge has been the projected Messina Strait Bridge, which is a state-of-the art bridge as it is the longest as well as the widest projected bridge in the world.

12.2 Messina Strait Bridge general description

The projected Messina Bridge will join the Island of Sicily with the Calabria province in the Italian Peninsula. A key characteristic of this challenging bridge is its deck consisting of three independent boxes joined together by transverse beams disposed each 30 m. The lateral boxes host road traffic, while the central box is devoted to railway traffic. The project of the Messina Bridge has been alive for years, suffering continuous improvements; in fact, the last official design is the one published in 2002, which has been the base for the public international competition held in 2004 for its construction. In Figures 12.2.1–12.2.3 the geometric definition of the bridge is presented, while in Figures 12.2.4–12.2.6 several digital pictures of the bridge are shown.

Suspension bridges, as well as cable-stayed bridges, may be analysed by means of three-dimensional finite element models adopting beam elements[2,3]. Thus, the first step in the optimum design process must be to work out a bridge with a three-dimensional beam elements model, bearing in mind that the great amount of repetitive calculations to be carried out entails the number of degrees of freedom to be moderate[4].

The main geometric and mechanical properties of the bridge that have been used to work out the finite element model of Figure 12.2.7 are summarized in Table 12.2.1. The structural model includes beam elements for representing the three boxes of the bridge

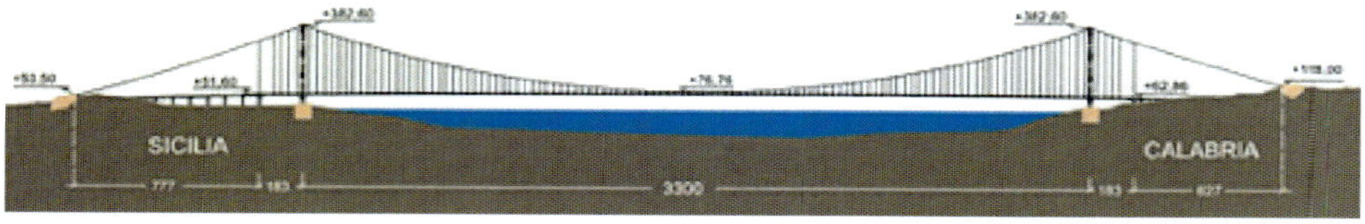

Figure 12.2.1: Layout of the Messina Strait Bridge (from public technical documentation property of Stretto di Messina S.p.A.[11]).

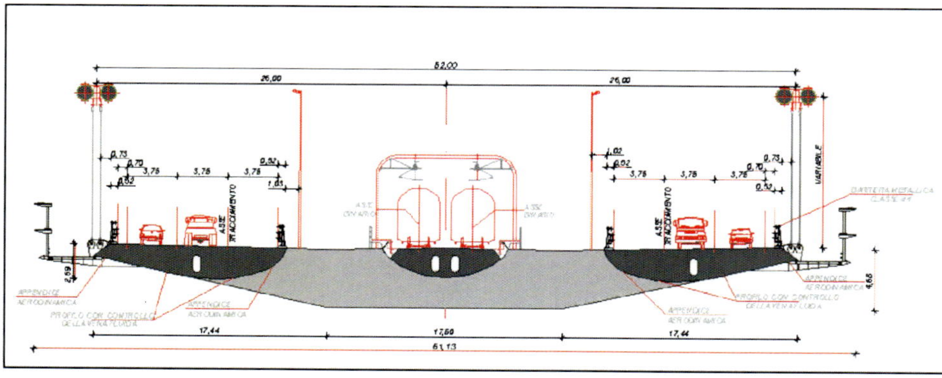

Figure 12.2.2: Messina Bridge deck cross section.

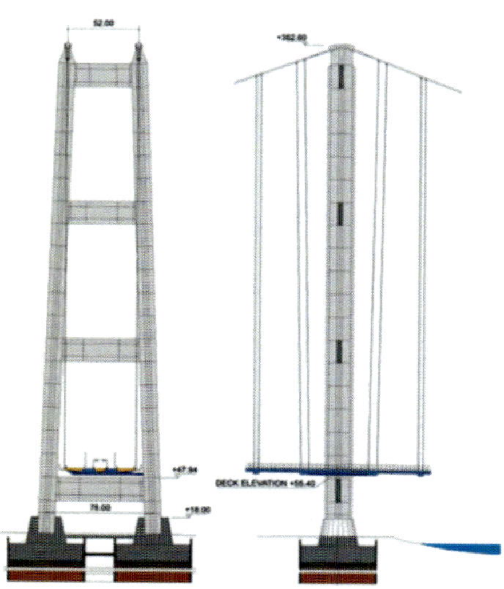

Figure 12.2.3: Geometric definition of the Messina Bridge towers (from public technical documentation property of Stretto di Messina S.p.A.[11]).

deck. Therefore, the influence in the aeroelastic and structural performance of the structure caused by changes in any of the parameters of the central box or the lateral ones can be studied. In this work both the lateral boxes are forced to experiment the same changes along the design process with the aim of keeping the symmetric configuration of the bridge deck. The finite element model presented in this application example has

Figure 12.2.4: Virtual image of the Messina Bridge deck.

Figure 12.2.5: Detail of aerodynamic devices

476 beam elements and 301 nodes. A compromise has been met between the accuracy of the structural model and the size of the optimization problem to be solved as optimum design considering that aeroelastic constraints must be regarded as a *grand challenge problem*.

Figure 12.2.6: Detail of the hangers.

Figure 12.2.7: Finite element model of the Messina Strait Bridge.

Table 12.2.1: Geometric and mass properties of the Messina Strait Bridge.

Total length (m)	3666
Main span length (m)	3300
Secondary spans length (m)	183
Distance tower-anchorage Sicily side (m)	960
Distance tower-anchorage Calabria side (m)	810
Central box cross-section area (m^2)	0.39
Central box cross-section vertical bending inertia I_y (m^4)	0.301
Central box cross-section lateral bending inertia I_z (m^4)	2.12
Central box cross-section torsional inertia J (m^4)	0.738
Central box mass (T/m)	8.79
Central box non-structural mass (T/m)	5.73
Central box polar moment (Tm^2/m)	1112.93
Central box non-structural polar moment (Tm^2/m)	1093.92
Lateral boxes cross-section area (m^2)	0.495
Lateral boxes cross-section vertical bending inertia I_y (m^4)	0.451
Lateral boxes cross-section lateral bending inertia I_z (m^4)	8.404
Lateral boxes cross-section torsional inertia J (m^4)	1.039
Lateral boxes mass (T/m)	7.28
Lateral boxes non-structural mass (T/m)	3.40
Lateral boxes polar moment (Tm^2/m)	590.34
Lateral boxes non-structural polar moment (Tm^2/m)	520.83
Width between cables (m)	52
Total deck width (m)	61.13

The real total deck cross section can be idealized by means of three simplified box symmetric cross sections as shown in Figure 12.2.8.

The length of the box members and the location of thicknesses in the central and lateral boxes are shown in Figures 12.2.9 and 12.2.10. Sub-index c is referred to central box, while sub-index l means lateral boxes. Furthermore, the geometric properties of each box are known.

$$A^c = A^c\left(B_c, b_c, h_c, e_1, e_2, e_3\right) = 0.390 \text{ m}^2 \tag{12.2.1}$$

$$I_y^c = I_y^c\left(B_c, b_c, h_c, e_1, e_2, e_3\right) = 0.301 \text{ m}^4 \tag{12.2.2}$$

$$I_z^c = I_z^c\left(B_c, b_c, h_c, e_1, e_2, e_3\right) = 2.210 \text{ m}^4 \tag{12.2.3}$$

$$J^c = J^c\left(B_c, b_c, h_c, e_1, e_2, e_3\right) = 0.738 \text{ m}^4 \tag{12.2.4}$$

$$A^l = A^l\left(B_l, b_l, h_l, e_4, e_5, e_6\right) = 0.495 \text{ m}^2 \tag{12.2.5}$$

$$I_y^l = I_y^l\left(B_l, b_l, h_l, e_4, e_5, e_6\right) = 0.451 \text{ m}^4 \tag{12.2.6}$$

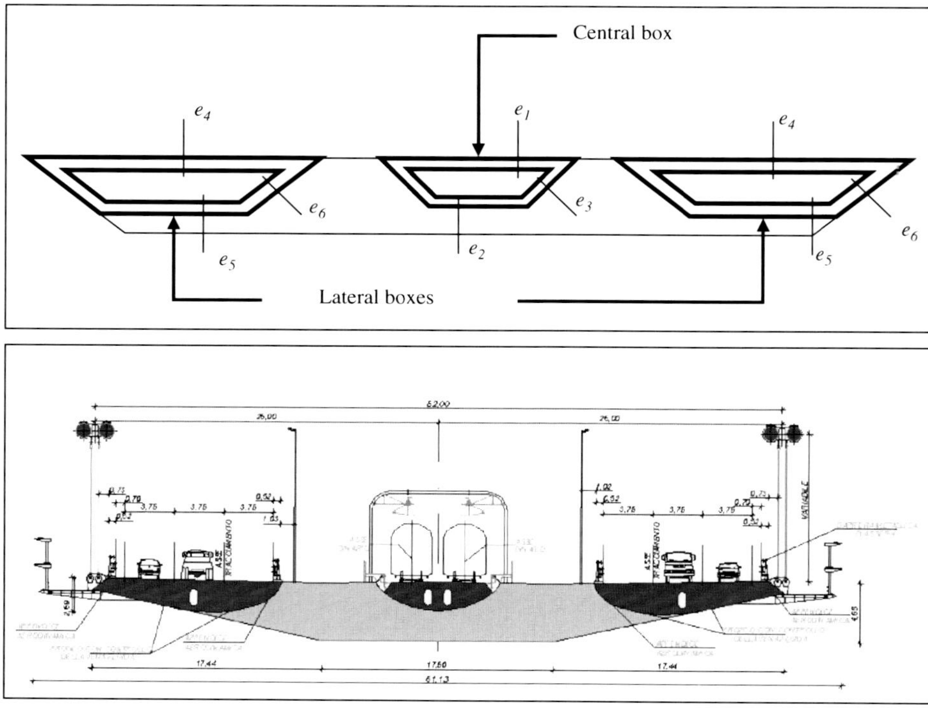

Figure 12.2.8: Messina Bridge simplified deck cross section and real deck cross section.

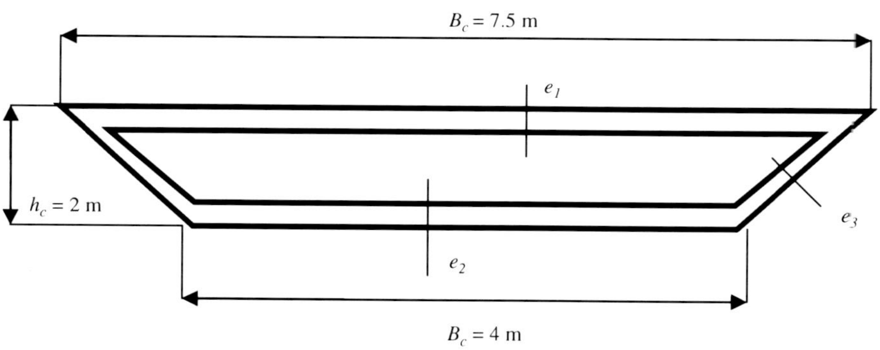

Figure 12.2.9: Central box simplified symmetric cross section.

$$I_z^l = I_z^l\left(B_l, b_l, h_l, e_4, e_5, e_6\right) = 8.404\ \text{m}^4 \tag{12.2.7}$$

$$J^l = J^l\left(B_l, b_l, h_l, e_4, e_5, e_6\right) = 1.039\ \text{m}^4 \tag{12.2.8}$$

Based upon the real boxes geometric and mass properties equivalent thickness for the simplified symmetric boxes, cross sections have been evaluated. The perimeter of the boxes is not allowed to change. Therefore, for each central or lateral box a non-linear

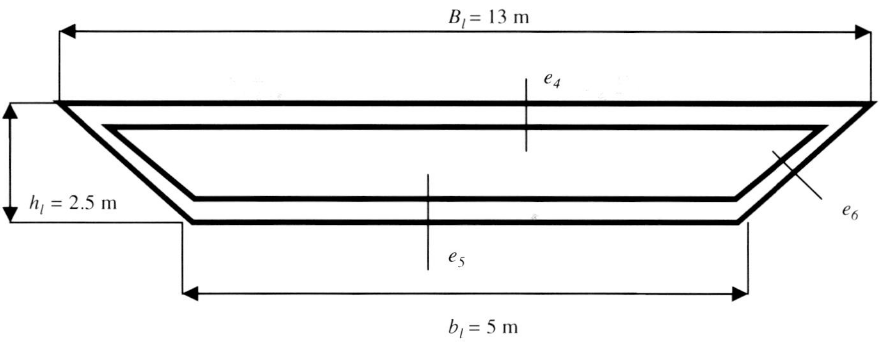

Figure 12.2.10: Lateral boxes simplified symmetric cross section.

Table 12.2.2: Messina Bridge model equivalent thicknesses.

	Central box		Lateral boxes
e_1 (m)	0.021	e_4 (m)	0.011
e_2 (m)	0.028	e_5 (m)	0.013
e_3 (m)	0.029	e_6 (m)	0.032

system containing four equations and three unknowns must be solved in order to obtain the thicknesses of the steel members of each box. It is imposed tha the difference between the area of the considered cross section and the area obtained from the evaluated thicknesses must be minimized. The calculated results are presented in Table 12.2.2.

It is important to bear in mind that the presented simplified box cross sections are used just for evaluating the equivalent thicknesses, while the aerodynamic characteristics, such as the flutter derivatives, are the ones corresponding to the precise box cross sections, tested by means of section model tests considering the scaled geometry of the current design of the bridge deck.

12.3 Messina Strait Bridge optimum design formulation

The optimum design problem of the Messina Bridge can be expressed by means of the following equations:

$$\begin{aligned}\min F = A = A^c + 2A^l = {} & B_c h_c - \left(\frac{B_c - b_c}{2}\right) h_c - \frac{(h_c - e_1 - e_2)^2 (B_c - b_c)}{2h_c} - \\ & -(h_c - e_1 - e_2)\left\{B_c - \frac{1}{h_c}\left[(h_c - e_2)(b_c - h_c) + e_3\sqrt{(B_c - b_c)^2 + 4h_c^2}\right]\right\} + \\ & + 2\left\{\begin{array}{l} B_l h_l - \left(\dfrac{B_l - b_l}{2}\right) h_l - \dfrac{(h_l - e_4 - e_5)^2 (B_l - b_l)}{2h_l} - \\ -(h_l - e_4 - e_5)\left\{B_l - \dfrac{1}{h_l}\left[(h_l - e_5)(b_l - h_l) + e_6\sqrt{(B_l - b_l)^2 + 4h_l^2}\right]\right\} \end{array}\right\}\end{aligned} \tag{12.3.1}$$

subject to

$$0.003 \text{ m} \le e_i \tag{12.3.2}$$

$$g_1 : \frac{U_{f,cr}}{U_f} - 1 \le 0 \quad \text{with } U_{f,cr} = 75 \text{ m/s} \tag{12.3.3}$$

$$g_2 : \frac{w_c}{w_{c,\max}} - 1 \le 0 \quad \text{with } w_{c,\max} = \frac{3l}{1000} = 9.9 \text{ m; } l = 3300 \text{ m} \tag{12.3.4}$$

The aim of this optimum design problem is to obtain the minimum weight deck, that is, the minimum deck cross-section area, compatible with the flutter speed and the cinematic constraints. The adopted initial design has been the year 2002 design proposal. Therefore, any eventually obtained material saving by means of implementation of optimum design techniques would represent a direct economical improvement for the infrastructure owner or the company in charge of the bridge construction. The minimum allowable flutter speed is 75 m/s, as stipulated in the project requirements list. Moreover, the maximum centre span vertical movement has been established for static overloads caused by train and road traffics. The selected design variables have been the thicknesses of the boxes composite members. Thus, three different optimization problems can be solved:

- *Problem C.* The set of design variables is $\mathbf{x} = (e_1, e_2, e_3)$. This means that the optimum design process will be based upon modifications exclusively extended to the central box cross section. Therefore sensitivity analysis of the flutter speed must be evaluated with regard to the geometric properties of the central box cross section before obtaining the derivatives of the flutter speed with regard to the central box thicknesses.
- *Problem L.* The selected design variables in this case are $\mathbf{x} = (e_4, e_5, e_6)$. The central box will remain constant, while changes are going to take place only in the lateral boxes' cross section. Sensitivity analysis of the flutter speed must be evaluated with regard to the geometrical properties of the lateral box section before calculating the required gradients.
- *Problem CL.* A total optimum design process can be carried out when the considered set of design variables is $\mathbf{x} = (e_1, e_2, e_3, e_4, e_5, e_6)$. In this case changes in both central box and lateral boxes will occur. Sensitivity analysis of U_f must be calculated with regard to the geometric properties of both central and lateral boxes as a previous step before evaluation of gradients with regard to the design variables. As a matter of fact computer cost of this optimization will be significantly higher than in the two aforementioned problems C and L.

12.4 Messina Strait Bridge sensitivities results

Along the optimization process recurrent evaluation of flutter speed and its derivatives with regard to thicknesses must be carried out (see Section 11.5.1). In Table 12.4.1 the sensitivity analyses of U_f with regard to the central box thicknesses are presented for the

Table 12.4.1: Analytical sensitivities of the flutter speed with regard to the central box cross-section thicknesses.

	$\partial U_f / \partial e_1$	$\partial U_f / \partial e_2$	$\partial U_f / \partial e_3$
Analytic sensitivity	667.83435	308.41624	411.92169

Table 12.4.2: Analytical sensitivities of the flutter speed with regard to the lateral boxes' cross-section thicknesses.

	$\partial U_f / \partial e_4$	$\partial U_f / \partial e_5$	$\partial U_f / \partial e_6$
Analytic sensitivity	1739.7617	540.47016	380.70779

initial design. It must be borne in mind that previous evaluation of the structure's natural frequencies and mode shapes, flutter speed, derivatives of the natural frequencies and mode shapes with regard to the geometrical properties of the central box cross section as well as derivatives of the flutter speed with regard to those geometric properties are required.

In Table 12.4.2 the analytical sensitivities of the flutter speed with regard to the lateral boxes' cross-section thicknesses are presented.

The high values of Tables 12.4.1 and 12.4.2 make clear the important influence of the thicknesses in the bridge flutter speed. In addition, the positive sign obtained in all cases means that an increment in any of the thicknesses would produce a significant increase in the bridge flutter speed.

12.5 Messina Strait Bridge optimum design results. Problem C

This optimum design problem has been solved by means of the DIOPTICA whose capabilities were explained in Section 11.5.3. In this section the charts and tables containing the main results offered by that piece of software are shown.

The objective function is the addition of the cross-section area of central and lateral boxes. In the initial design the area of the deck cross section is 1.417 m^2; however, once the deck cross section is optimized by means of modifying the thicknesses of the central box, the area reaches 1.290 m^2, which represents a 8.96% of material saving (See Figure 12.5.1). Moreover, the cross-section area of the central box is modified from 0.415 m^2 in the initial design up to 0.288 m^2 in the optimum design, which represents a decrement of 30.6% in the material of the central box. On the other hand, the number of required iterations has been low; in fact, design improves mostly in the first iteration.

In Figure 12.5.2 the evolution of the bridge flutter wind speed along the automated design process is presented. It can be seen how the lower bound of the flutter wind speed is reached from the beginning of the optimum design process.

In Figure 12.5.3 the evolution of both aeroelastic and cinematic constraints is presented. The flutter speed-related constraint has been the one active along the automated design process.

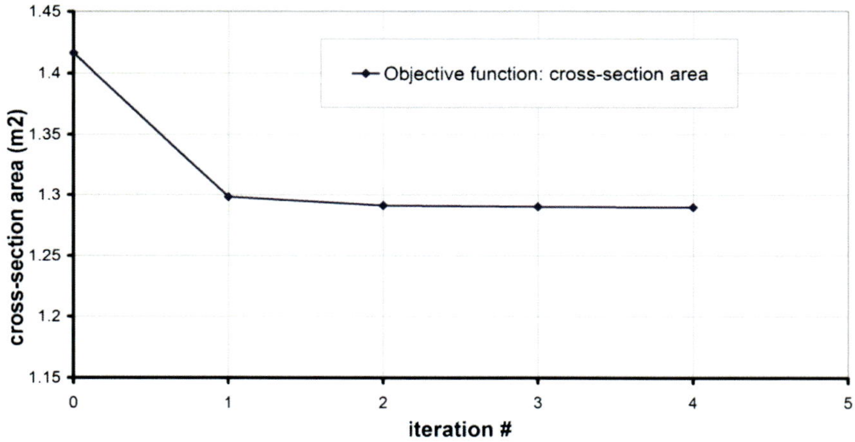

Figure 12.5.1: Problem C. Objective function evolution along the optimization process.

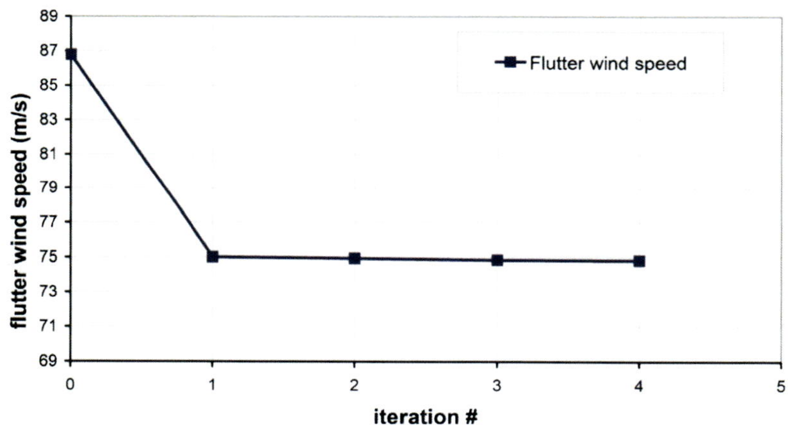

Figure 12.5.2: Problem C. Flutter wind speed evolution along the optimization process.

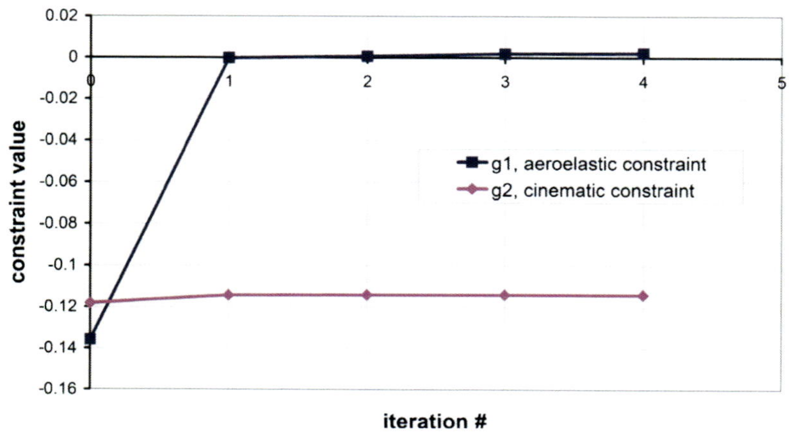

Figure 12.5.3: Problem C. Constraints evolution along the optimization process.

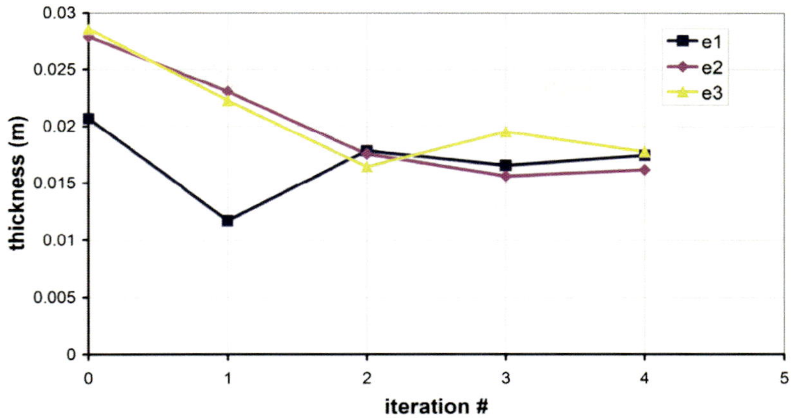

Figure 12.5.4: Problem C. Central box thicknesses evolution along the optimization process.

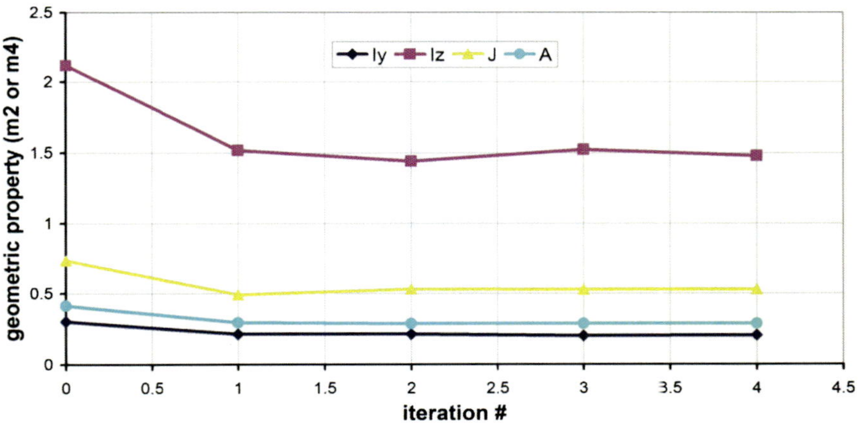

Figure 12.5.5: Problem C. Central box geometric properties evolution along the optimization process.

Table 12.5.1: Problem C. Thicknesses evolution.

	Initial design (m)	Optimum design (m)	% Change
e_1	0.02068867	0.01743807	−15.71
e_2	0.02788813	0.0161712	−42.01
e_3	0.0285524	0.01774415	−37.85

In Figure 12.5.4 the evolution of the equivalent thicknesses of the central box, which is the design variables, along the optimum design process is presented. The thicknesses are lower in the optimum design than in the initial design; however, along the design process each design variable has suffered both increments and decrements. In Table 12.5.1 the values of the thicknesses in the initial design and in the final design are shown as well as the percentage of variation suffered in the design process.

Table 12.5.2: Problem C. Central box cross-section geometric properties evolution.

	Initial design (m)	Optimum design (m)	% Change
I_y (m^4)	0.303263	0.207372	−31.62
I_z (m^4)	2.120037	1.479248	−30.23
J (m^4)	0.736781	0.527903	−28.35
A (m^2)	0.415096	0.28816	−30.58

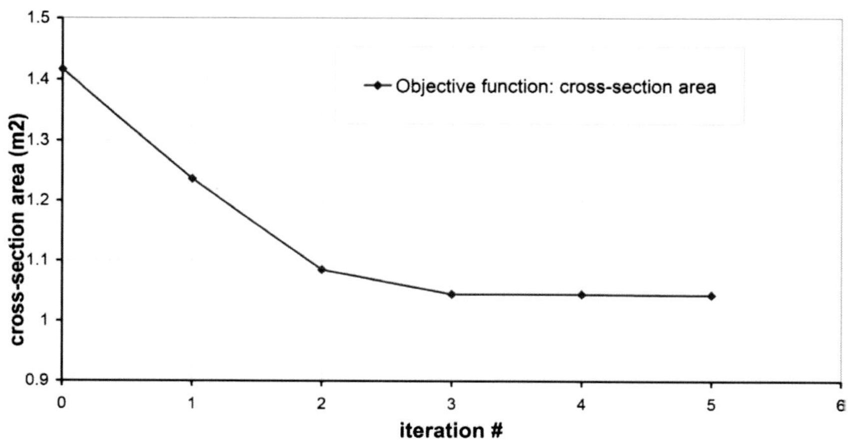

Figure 12.6.1: Problem L. Objective function evolution along the optimization process.

In Figure 12.5.5, the evolution along the optimum design process of the geometric properties of the central box cross section, that is, area, torsional inertia and bending inertias, is presented. In addition, in Table 12.5.2 these geometric properties of the central box in the initial design and in the optimum design are listed along with the percentage of suffered change.

It can be seen that the geometric properties of the central box have suffered decrements close to 30% as a result of the material saving reached by means of the application of optimization techniques.

12.6 Messina Strait Bridge optimum design results. Problem L

In problem L, the chosen design variables have been the thicknesses of the lateral boxes' cross section. Thus the central box cross section in this problem is not allowed to experiment any change. Of course, both lateral boxes suffer the same modifications along the optimization process in order to keep the symmetry of the deck cross section. As in the former example, the results that are going to be presented have been obtained by means of the DIOPTICA code.

In Figure 12.6.1 the evolution of the objective function can be seen. The decrement of the deck cross-section area is important, especially for the iterations #1 and #2. The total number of iterations, 5, has been low.

The deck cross-section area in the optimum design is 1.043 m², while the deck cross-section area in the initial design, which is the 2002 design proposal, was 1.417 m². Therefore the material saving reached in this case is 26.39%, which has a direct impact on the economical cost of the construction of the bridge. If only the lateral boxes are considered, the total cross-section area has change from 1–0.628 m², which represents a decrement of 37.18% in the required amount of steel.

In Figures 12.6.2 and 12.6.3 the evolution of the flutter wind speed for the modified designs along the optimization process as well as the values obtained for the constraints are presented.

In Figure 12.6.4 and Table 12.6.1 the evolution of the equivalent thicknesses of one of the lateral boxes is presented.

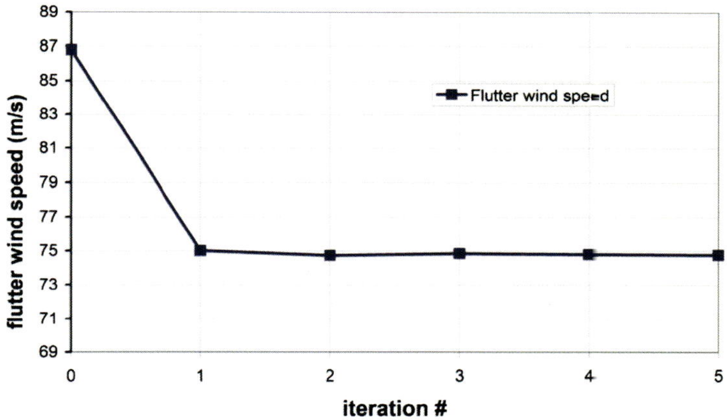

Figure 12.6.2: Problem L. Flutter wind speed evolution along the optimization process.

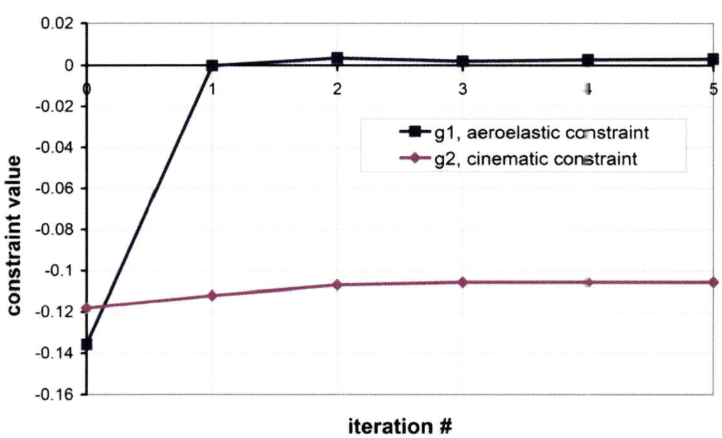

Figure 12.6.3: Problem L. Constraints evolution along the optimization process.

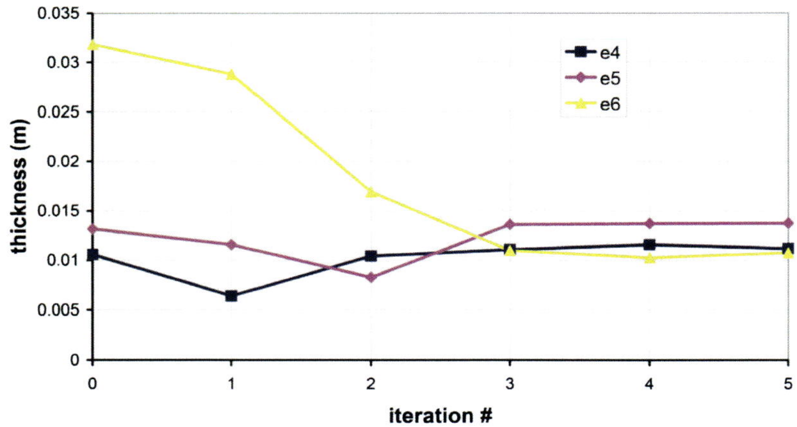

Figure 12.6.4: Problem L. Lateral box thicknesses evolution along the optimization process.

Table 12.6.1: Problem L. Thicknesses evolution.

	Initial design (m)	Optimum design (m)	% Change
e_4	0.0105744	0.0111624	5.56
e_5	0.0131736	0.0137397	4.30
e_6	0.0318247	0.0107284	−66.29

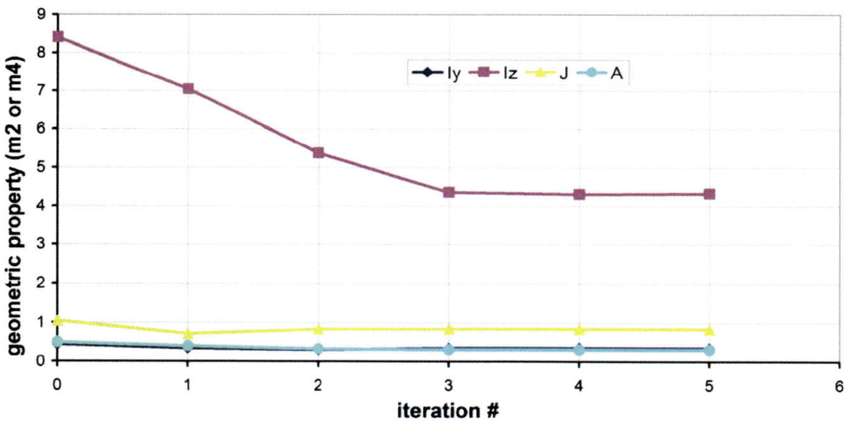

Figure 12.6.5: Problem L. Lateral box geometric properties evolution along the optimization process.

In this problem L, the changes that the design variables have suffered are different from those of the design variables adopted in problem C. In fact, in the optimum, thicknesses in the upper and lower members of the lateral box cross section have increased, while the thickness of the lateral members have experimented an important decrement.

Table 12.6.2: Problem L. Lateral box cross-section geometric properties evolution.

	Initial design	Optimum design	% Change
I_y (m^4)	0.451102	0.353654	−21.60
I_z (m^4)	8.403938	4.333078	−48.44
J (m^4)	1.039133	0.829255	−20.20
A (m^2)	0.500815	0.314115	−37.28

In Figure 12.6.5, the evolution along the optimization process of the geometric properties of the lateral boxes' cross section is presented, while in Table 12.6.2, the values of the geometric properties for the initial and the optimum designs are listed.

In this problem L all the geometric properties have decreased. The percentage of the decrement varies from 21.6% to almost 50%.

12.7 Messina Strait Bridge optimum design results. Problem CL

In this third optimum design example up to six design variables have been considered, which are the thicknesses of the central and lateral boxes cross sections, keeping the symmetric configuration of the bridge deck. Thus a complete optimization of the deck cross section has been performed. This optimization problem is the most complex of the three solved as the number of design variables is bigger and so it is the demanded computer power.

In Figure 12.7.1, the evolution of the objective function is presented. In this case the number of required iterations to reach the optimum design is bigger than in the previous examples.

The total deck cross-section area in the optimum is 0.947 m^2, which represents a saving in the amount of material of 33.17% with regard to the 1.417 m^2 area in the initial design. In addition, it is a wise exercise to analyse the evolution of the cross-section area of the central and lateral boxes along the optimization process. This information is provided in Figure 12.7.2. It can be observed how although the total deck cross-section area decreases the area of the central box cross section increases while the area of the lateral boxes decreases. Thus, by means of this example, it is clear that optimum design techniques allow not only improvements in the amount of material of the structure but also mass redistribution to take place as the material is located there where it is required. In this example, the amount of material located in the central box increases from the initial design, while the material in the lateral boxes decreases.

In Figures 12.7.3 and 12.7.4 the evolution of the flutter wind speed and the values of the constraints along the optimization process are presented.

The evolution of the design variables is particularly interesting as the thicknesses of the central box increase while the ones of the lateral boxes decrease, consistently with the evolution of the cross-section area of each box (see Figure 12.7.5).

In Figures 12.7.6 and 12.7.7, the charts containing the changes in the geometric properties of the central box and one of the lateral boxes are shown. In the same manner, in Table 12.7.2, the values of the geometric properties for the initial and the optimum designs are listed. It can be seen how geometric properties of the central

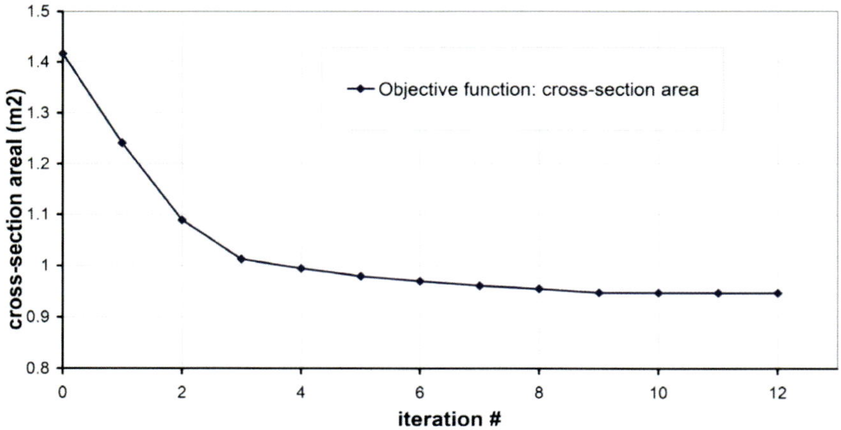

Figure 12.7.1: Problem CL. Objective function evolution along the optimization process.

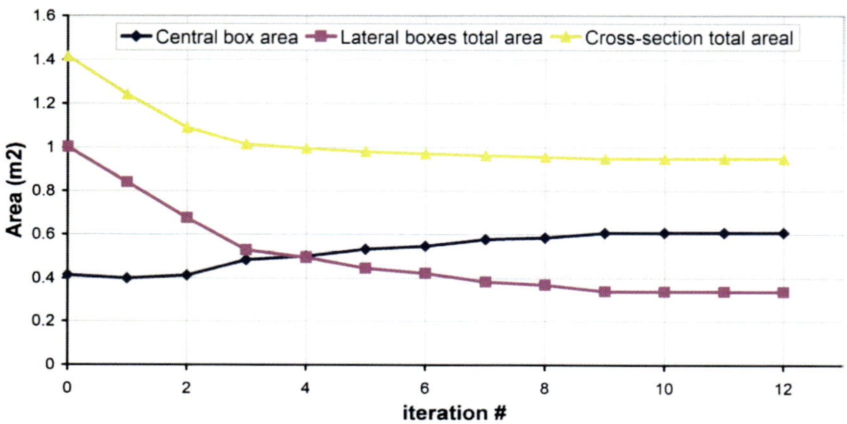

Figure 12.7.2: Problem CL. Cross-section areas evolution along the optimization process.

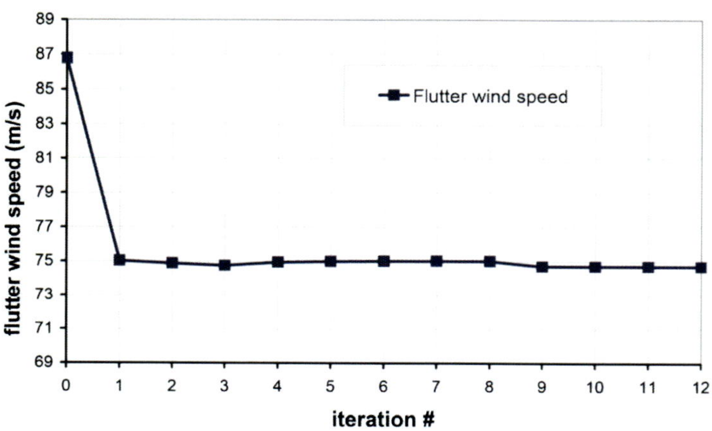

Figure 12.7.3: Problem CL. Flutter wind speed evolution along the optimization process.

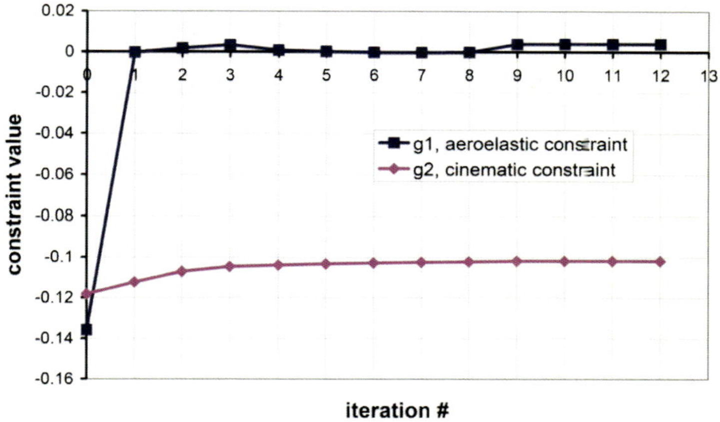

Figure 12.7.4: Problem CL. Constraints evolution along the optimization process.

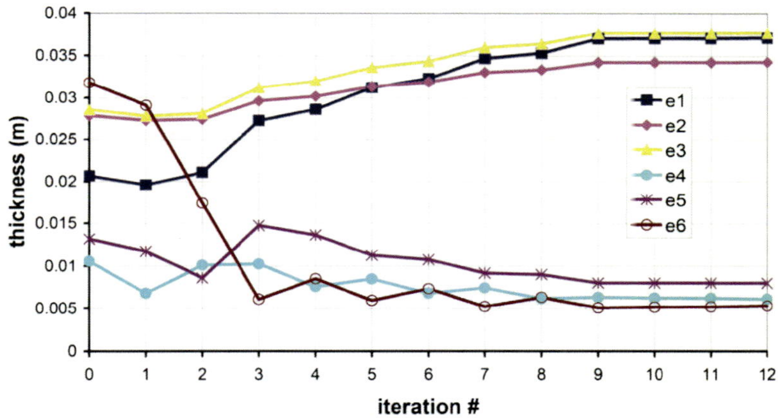

Figure 12.7.5: Problem CL. Lateral and central boxes' thicknesses evolution along the optimization process.

Table 12.7.2: Problem CL. Thicknesses evolution.

	Initial design (m)	Optimum design (m)	% Change
e_1	0.0206887	0.0371048	79.35
e_2	0.0278881	0.0343161	23.05
e_3	0.0285524	0.0376799	31.97
e_4	0.0105744	0.0061111	−42.21
e_5	0.0131736	0.0080028	−39.25
e_6	0.0318247	0.0052853	−83.39

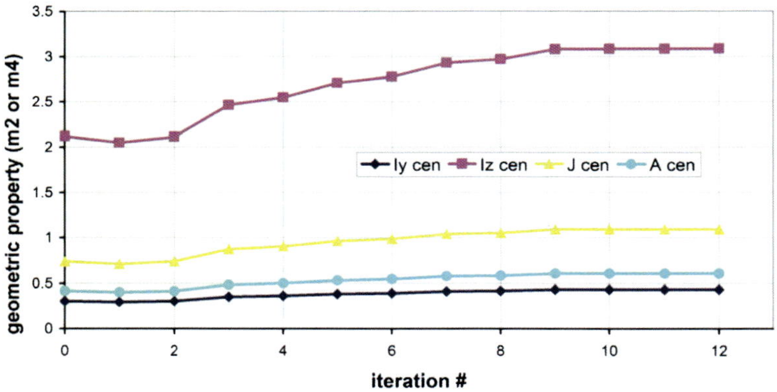

Figure 12.7.6: Problem CL. Central box geometric properties evolution along the optimization process.

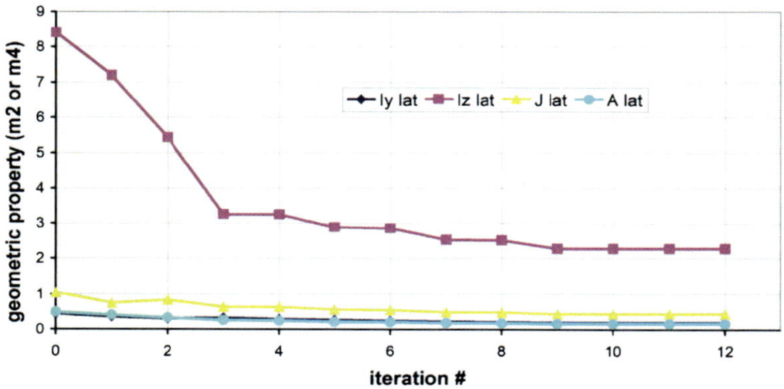

Figure 12.7.7: Problem CL. Lateral box geometric properties evolution along the optimization process.

Table 12.7.3: Problem L. Lateral box cross-section geometric properties evolution.

Central box	Initial design	Optimum design	% Change
I_y^c (m^4)	0.303263	0.430101	41.82
I_z^c (m^4)	2.120037	3.087907	45.65
J^c (m^4)	0.736781	1.089829	47.92
A^c (m^2)	0.415096	0.608495	46.59
Lateral box			
I_y^l (m^4)	0.451102	0.197017	−56.33
I_z^l (m^4)	8.403938	2.268125	−73.01
J^l (m^4)	1.039133	0.442976	−57.37
A^l (m^2)	0.500815	0.169081	−66.24

box increase from the initial design while geometric properties of the lateral boxes decrease.

12.8 Conclusions

Application of optimum design techniques has made possible to reach a bridge design that satisfies both aeroelastic and cinematic constraints while saving up to 33% of the deck material in the 2002 design proposal for the Messina Bridge. In a professional project environment the number of behaviour constraints to be adopted in the design of long-span bridges would be significantly higher, including, for instance, material stress limitations in the cross section. Thus the material savings should be expected not to be so important, although any amount of savings has such an important cost impact in these record-breaking projects that justify the use of optimum design techniques. On the other hand, implementation of structural constraints like the ones related with stresses does not imply additional difficulties arising in the design problem, neither a substantial increment in the required computer power.

Reaching the mathematical optimum in the optimum design problem guarantees that the best possible design for the considered constraints has been obtained. However, when the mathematical optimum could not be obtained, the design improvements that can be reached with two or three iterations may make worthy to apply optimization techniques.

The required computer time for performing the optimum design of a suspension bridge can reach several days, applying parallel programming techniques. Thus sequential computing is unfeasible as computer times can reach several weeks.

The optimum design problem itself can be improved including important structural constraints such as stress limitations, fatigue or many others. In the field of the aeroelastic behaviour, additional phenomena as buffeting or vortex shedding can be added to the list of project requirements. Additionally, optimum design techniques can be applied to cable-stayed bridges or suspension bridges with truss girders. Therefore a lot of research can be done within this realm.

12.9 References

[1] Stretto di Messina S.p.A. [2007] *web site*, http://www.strettodimessina.it/eng/preliminaryproject/88_ENG_HTML.htm

[2] Miyata T., Yamada H., Katsuchi. H. [2003] *Comparative Analysis of Messina Bridge – International Benchmark Study*. Eleventh International Conference on Wind Engineering. 2003. Lubbock, Texas, USA.

[3] Larsen, A. [1993] *Aerodynamic Aspects of the final design of the 1624 m suspension bridge across the Great Belt*. Journal of Wind Engineering and Industrial Aerodynamics, Vol. 48, pp. 261–285.

[4] Venkataraman S., Haftka R.T. [2004] *Structural Optimization Complexity: What has Moore's Law done for us*? Structural and Multidisciplinary Optimization, Vol. 28, pp. 375–387.